Conflict and change in the countryside

GU00382871

For Susan Rebecca

Conflict and change in the countryside

Rural society, economy and planning in the developed world

GUY M. ROBINSON

Belhaven Press
A division of Pinter Publishers,
London and New York

© Guy M. Robinson, 1990

First published in Great Britain in 1990 by
Belhaven Press (a division of Pinter Publishers),
25 Floral Street, London WC2E 9DS and
PO Box 197, Irvington, New York, NY 10533

All rights reserved. No part of this publication may be
reproduced, stored in a retrieval system, or transmitted by any
other means without the prior permission of the copyright holder.
Please direct all enquiries to the publishers.

British Library Cataloguing in Publication Data
A CIP catalogue record for this book is available from the
British Library

ISBN 1 85293 043 8 (Hbk)
ISBN 1 85293 044 6 (Pbk)

Library of Congress Cataloging-in-Publication Data
Robinson, G.M. (Guy M.)
   Conflict and change in the countryside : rural society, economy,
and planning in the developed world / Guy M. Robinson.
      p. cm.
   Includes bibliographical references.
   ISBN 1 85293 044 6 (Pbk)
   ISBN 1 85293 043 8 (Hbk)
   1. Rural conditions. 2. Rural development. I. Title.
HT421.R59   1990   90-321
307.72–dc20   CIP

Filmset by Mayhew Typesetting, Bristol, England
Printed and bound by Biddles Ltd.

Contents

List of plates

List of figures

List of tables

Preface

The title of this book was chosen to reflect two major aspects of the rural scene in the Developed World and demonstrate the extent to which modern rural society must be distanced from traditional views of the rustic idyll. Whilst it is true that many people still retain some mental picture of an overly romanticised unchanging countryside, the reality is of profound change during the last two centuries and especially post-1945. This has permeated all aspects of rural life and has often acted as an adjunct to the types of conflict previously only associated with the growth of cities and the pressures created by urbanisation. Conflict involving different interest groups, ethnic groups, big business, neighbourhood 'communities' and government are familiar facts of everyday life in the major cities. But, increasingly, it has been recognised that similar conflicts, shaped by forces bringing dramatic and far-reaching changes to the whole of society and its institutions, are present beyond the city. Change and conflict have become appropriate metaphors to apply to evolving rural society and economy.

These metaphors can be readily ascribed to modern farming. For example, small farmers have become caught in a financial trap between their suppliers and retailers/wholesalers. Increasingly, it is the large agribusinesses that are overcoming this trap through their extension into the traditional areas of marketing and supply. These large businesses, often supported by favourable legislation, have helped to alter the very nature of the countryside by adopting highly mechanised farming practices. In the destruction of long-standing field boundaries and familiar habitats for flora and fauna lies a major conflict which draws together social, economic and legislative aspects of rural change. This conflict is given expression in terms of battles between different groups in society, for example the 'green' lobby of environmentalists presents arguments diametrically opposed to those of agribusiness. Yet, as with conflict and change in the cities, this is not a simple battle and other groups, such as small tenant farmers, middle-class residents relatively new to the countryside and urban dwellers using the countryside for recreational purposes, must also be fitted into the jigsaw as must restrictions and frameworks imposed by prevailing legislation and the general political context of change.

In seeking to understand this type of conflict and the changing nature of rural areas in general, this book examines the development of the rural areas of the Developed World by means of a three-pronged study of society, economy and planning. Each part contains a strong emphasis upon the two chosen metaphors of the book's title whilst drawing upon literature from a range of disciplines and using examples from throughout Western Europe, North America and Australasia. In particular, the book draws together work by sociologists, economists, geographers and planners to indicate the forces shaping the countryside at the end of the 20th century. Attention is given to the differences between settlement and society in the Old World and the New, but many parallels are also drawn, reflecting the common forces of capitalism, technical change and similar governmental responses to world market forces.

The division of the book into three parts is also deliberate: it is felt that this enables a coherent analysis of the critical elements of rural development. Separation of social, economic and legislative aspects of rural development allows a range of changes and conflicts to be examined from different perspectives and different emphases. It also encompasses a range of methodologies and concepts, though with particular emphases depending upon which aspect of rurality is being considered.

The term 'rural' itself has proved an especially elusive one to define for purposes of academic study despite popular conceptions based on images of rusticity and idyllic village life. In this book the simple conceptualisation of 'rural' shown in Figure P.1 is used. This treats 'rural' as a 'black box' affected by change in certain influential variables. Although portrayed as separate forces, the four variables indicated are interrelated, but represent key factors affecting the separate elements shown within the black box. The elements are conceptualised as 'nesting' within 'rural' in the form of overlapping sets, but are left open-ended to show that they are influenced by the four outlying variables. The three elements are seen as being interrelated, but amenable to separate examination. The simplicity of this arrangement distorts the reality of the numerous interconnections between rural life and urban 'black boxes', and the complexity of relationships between society, economy and legislation. However, the complexities are not overlooked in the subsequent examinations of the individual elements and variables.

The nature of these examinations has varied from discipline to discipline, though with common foci upon particular aspects of rural development, notably the influence of urban and industrial development on rural society, the dominance of agriculture in the rural economy, and the regulation of conflicts between land uses by the formulation of regulatory legislation. It is the intention here to provide a systematic synthesis of the three elements of 'rural' distinguished in Figure P.1, addressing these major foci by way of summarising studies of key issues.

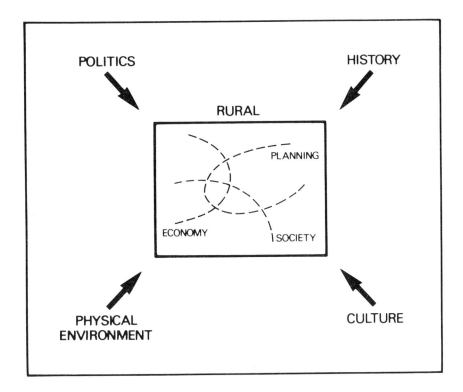

Figure P.1 Three elements of 'Rural'

This is a broad approach which will highlight the contributions of a range of methodologies. It will draw upon specific examples wherever possible, with special emphasis upon Britain, North America and Australasia, reflecting the author's own research experience.

Although some discussion is devoted to defining 'rural' (Chapter 1), it is recognised that this can be a rather arid exercise and one which in the past has tended to ignore common economic, social and political structures in both urban and rural areas. Instead, it is argued that a sufficient reason for a separate consideration of conflicts and changes in the countryside is the clear recognition of a different range of problems in 'non-urban' areas. In simple terms, such 'rural' areas define themselves with respect to the presence of particular types of problems. A selective list of examples could include depopulation and deprivation in areas remote from major metropolitan centres; a reliance upon primary activity; conflicts between preservation of certain landscapes and development of a variety of economic activities; and conflicts between local needs and legislation emanating from urban-based legislators. Key characteristics of 'rural' are taken to be extensive land uses, including large open spaces of underdeveloped land, and small settlements at the base of the settlement hierarchy, but including settlements which are thought of as rural by most of their residents and including settlements recognised as 'market towns' serving a surrounding area dominated by agriculture and/or forestry.

In short, the separation of 'rural' and 'urban' is regarded as important because it is a separation recognised by society, even if their view of the former may be less clearly defined than their view of the latter. Rural areas do have different problems; the nature of conflicts and changes in rural areas are of a different magnitude to those in metropolitan centres; and the inhabitants of rural areas tend to have a different pattern of life experiences from those resident in towns. Thus, despite the current vogue for regarding the whole of society as subject to the influence of over-arching and massive controlling structures associated with capitalism and central government policy, it is still important to recognise significant characteristics of 'rural' and to examine them in a coherent fashion: the obverse of the much vaunted 'urban question' must not be overlooked at a time when conflict and change in the countryside are producing radical alterations to existing landscapes.

Acknowledgements

There are two things that should not be attempted when in the middle of writing a book. One is moving house; the other is getting married. Having done both during the course of writing *Conflict and Change in the Countryside*, I feel duty bound to blame delays in completion of this task on Susan Rebecca who assisted with removals and, more or less willingly, was persuaded to walk down the aisle and make eternal vows to a would-be author who had failed by a mere twelve months to meet his publisher's deadline. Given too that two years of long-distance courtship accompanied the writing of this book, it is also appropriate that the book is dedicated to Susan Rebecca – help and hindrance can be such pleasant companions!

In providing the author with a bride carrying their passports, both Australia and the United States contributed substantially to this book. They did so too in a more academic sense by allowing me free reign in various university libraries. In Australia these included libraries in the Universities of Adelaide, Melbourne and Tasmania, Flinders University, James Cook University, Monash University, the Australian National University and Queensland Institute of Technology. In the United States the institutions were the University of Illinois at Chicago and Northwestern University. During a brief 'holiday' the University of Canterbury, New Zealand, provided similar assistance. Librarians seldom receive acknowledgement for the help they give to their customers, but I am happy to express thanks to all those in these various institutions and in my own university, who made the task of tracking down relevant material so much easier.

The initial work on this book was started whilst I was a Visiting Fellow at the University of Melbourne and Visiting Tutor at University College, Melbourne. I am very grateful to both institutions for the facilities placed at my disposal and especially to the latter for introducing me to my future wife. Michael Webber's consideration in providing me with an office overlooking the University of Melbourne cricket pitch should also not go unmentioned. Work was halted only briefly during a break in New Zealand, thanks to the kindness shown by Barry Johnston in allowing me to have office space in the Department of Geography at the University of Canterbury. The hospitality extended by Edwina Palmer and Eric Pawson was also most appreciated. Whilst in Australia and New Zealand, discussions with Richard Le Heron, Ron Johnston and Rex Honey proved both valuable and entertaining.

In Edinburgh I have been fortunate to receive help and encouragement from a number of my colleagues. Never have so many uttered the fateful words, 'Have you finished your book yet?!' or for so many months! I am particularly grateful for help extended in a variety of ways by David Sugden, Terry Coppock, Mike Summerfield, Peter Furley, Liz Bondi, Ged Martin, Ray Harris and, at the University of Glasgow, Susan Smith. The mysteries of word processing and analysing data on the Departmental computer were explained patiently by Steve Dowers and Bruce Gittings. Jack Hotson produced several maps of agricultural distributions at my request and continued to demonstrate his superiority on the golf course. Anona Lyons and Elizabeth Clark coped admirably with my own inadequate cartography which they then converted magically into the maps and diagrams I had desired. Perhaps most important, though, I have been very fortunate to benefit from a steady flow of ideas from my research students and from undergraduates in my 'Land Use' and 'Rural Development' classes. The students' enthusiasm has been a continuing source of encouragement and a stimulus to delve beyond the obvious and the familiar.

There is no doubt that this work would not have seen the light of day without the assistance of Iain Stevenson. It was he who encouraged me to embark on the project several years before I put pen to paper, and subsequently, as Editorial Director at Belhaven Press, he coaxed and inveigled a completed manuscript from me. This completion was eased by the accompaniment of Glen Campbell, Jussi Bjorling and the Spanish Radio and Television Orchestra amongst others. Last, but naturally not least, my parents continued to provide help and encouragement in ways too numerous to mention but never overlooked.

Edward Arnold Ltd., for Figure 8.1 (from Champion *et al.*, 1987, *Changing places: Britain's demographic, economic and social composition*).

Broads Authority, for Figure 10.10 (from Broads Authority, 1982, *What future for Broadland?*).

Butterworths, for Table 5.8 (from Volkman, 1987, in *Land Use Policy*).

Paul Chapman Publishing Ltd., for Table 8.3 (from Brunt, 1988, *The Republic of Ireland*).

Croom Helm Ltd., for Table 8.2 and Figure 8.2 (from Cloke and Park, 1985, *Rural resource management*); for Figure 3.5 (from Wild, 1983b, in Wild, ed., *Urban and rural change in West Germany*); for Figure 3.7 (from Gilg, 1983, in Pacione, ed., *Progress in rural geography*).

Countryside Service, Hereford and Worcester County, for Figure 10.6 (from *Guide to Clent Hills*).

The Editor, Erdkunde, for Figure 6.7 (from Naylon, 1966, in *Erdkunde*).

Environment Canada, for Table 5.4 (from Lands Directorate, 1985a, *Land use change in Canada, Fact Sheets*, reproduced with the permission of the Minister of Supply and Services Canada).

Geo Brooks, for Figure 3.6 (from Woods, 1985, in White and Van Der Knapp, eds., *Contemporary studies of migration*); for Figure 8.5b (from Breathnach, 1985, in Healey and Ilberry, eds., *The industrialisation of the countryside*); for Figure 11.2 (from Huigen, 1983, in Clark, *et al.*, eds., *The changing countryside*).

Geographical Association, for Figure 7.5 (from Bradshaw, 1984, in *Geography*); for Figure 10.7 (from Robinson, 1986b, in *Geography*).

The Editor, Research Discussion Papers, Department of Geography, University of Edinburgh, for Tables 8.6 and 9.4 and Figures 4.4 and 4.5 (from Nurminen and Robinson, 1985, in *Research Discussion Papers*, Department of Geography, University of Edinburgh).

Information and Documentation Centre for the Geography of the Netherlands, for Figure 12.5 (from *Compact Atlas of the Netherlands*, 1985).

The Editor, Journal of Agricultural Economics, for Table 4.6 (from Gasson, 1967, in *Journal of Agricultural Economics*).

Institute of British Geographers, for Table 4.5 (from Aitchison and Aubrey, 1982, in *Transactions of the Institute of British Geographers*); for Table 8.4 (from Harrison, 1981, in *Area*); for Figure 4.7 (from Warnes and Law, 1976, in *Transactions of the Institute of British Geographers*).

Longman Group UK, for Table 4.1 and Figure 4.6 (from Bohland, 1988, in Knox *et al.*, 1988, *The United States – a contemporary geography*); for Tables 4.7, 4.8 and 4.9 (from Gasson, 1988, *The economics of part-time farming*); for Figure 5.5 (from Mather, 1986, *Land use*); for Figure 6.1 (from Bowler, 1979, *Government and agriculture: a spatial perspective*); for Figure 7.4 (from John, 1984, *Scandinavia, a new geography*).

Macmillan Publishers Ltd., for Figure 12.2 (from Toyne, 1974, *Organisation, location and behaviour: decision-making in economic geography*).

Methuen & Co., for Table 12.1 and Figure 12.1 (from Cloke, 1983, *An introduction to rural settlement planning*), for Table 12.2A (from Cloke, 1979, *Key settlements in rural areas*); for Figure 3.1a (from Lee, 1986, in Langton and Morris, eds., *Atlas of industrializing Britain*); for Figures 11.1 and 11.5 (from Moseley, 1979, *Accessibility: the rural challenge*).

John Murray (Publishers) Ltd., for Table 6.2 and Figure 6.8 (from Winchester and Ilbery, 1988, *Agricultural change: France and the EEC*).

Thomas Nelson and Sons Ltd., for Figure 3.1b (from Lawton, 1964, in Watson and Sissons, eds., *The British Isles – a systematic geography*).

New Zealand Demographic Society, for Tables 3.8 and 3.9 (from Robinson, 1986a, in *New Zealand Population Review*).

New Zealand Geographical Society, for Figure 8.4 (from Robinson, 1988c, in *New Zealand Geographer*).

North British Publishing, for Tables 2.2, 2.3, 5.5, 5.6 and Figures 2.3, 5.6 and 5.7 (from Robinson, 1988b, in Robinson, ed., *A social geography of Canada: Essays in honour of J. Wreford Watson*); for Figures 3.8 and 11.6 (from Carlyle, 1988, in Robinson, ed., *A social geography of Canada: Essays in honour of J. Wreford Watson*); for Figures 5.1, 5.4 and 6.4 (from Robinson, 1988a, *Agricultural change: geographical studies of British agriculture*).

Oxford University Press, for Table 12.2B (from Woodruffe, 1976, *Rural settlement policies and plans*).

Pergamon Press Ltd., for Table 4.4 and Figures 4.2 and 4.3 (from Hugo and Smailes, 1985, in *Journal of Rural Studies*); for Table 5.3 (from Marsden *et al.*, 1987, in *Journal of Rural Studies*); for Figures 1.3 and 10.4 (from Clout, 1972, *Rural geography: an introductory survey*).

Regional Studies Association, for Table 9.3 (from Breathnach, 1984, in *Regional Studies*); for Figures 1.3a and 1.3d (from Cloke and Edwards, 1986, in *Regional*

Studies); for Figure 1.3b (from Cloke, 1977, in *Regional Studies*).

Royal Dutch Geographical Society, for Figure 11.3 (from Johnston, 1966b, in *Tijdschrift voor economische en social geografie*).

Royal Scottish Geographical Society, for Table 9.1 and Figure 9.2 (from McCleery, 1988, in *Scottish Geographical Magazine*).

Royal Town Planning Institute, for Figure 12.3 (from Cloke and Shaw, 1983, in *Town Planning Review*).

Schools Unit, University of Sussex, for Figure 6.3 (from Schools Unit, University of Sussex, 1983, in *Exploring Europe*).

The Editor, Scottish Association of Geography Teachers, for Table 8.7 (from Heeley, 1988, in *Journal of the Scottish Association of Geography Teachers*).

Shillington House, for Figure 2.1 (from Maher, 1982, *Australian cities in transition*).

Swedish Society for Anthropology and Geography, for Figure 3.2 (from Lewis and Maund, 1976, in *Geografiska Annaler*); for Figure 8.7 (from Kariel and Kariel, 1982, in *Geografiska Annaler*).

United States Department of Agriculture (USDA), for Figure 5.2 (from USDA, 1981, *A time to choose*).

Unwin Hyman Ltd., for Table 10.5 (from Adams, 1986, *Nature's place: conservation sites and countryside change*, Allen & Unwin).

Verso, for Table 2.7 (from Wright, 1985, *Classes*).

Westview Press, for Figure 5.3 (from Gregor, 1982, *Industrialization of US agriculture: an interprative atlas*).

PART ONE
RURAL SOCIETY

1

The rural realm

1.1 Rural areas in the Developed World

There is much variation within the Developed World in the proportion of population in individual countries classified as living in rural areas (Table 1.1). From just 5 per cent of population termed rural in Belgium, the other extreme is represented by Portugal where over 70 per cent of the population is rural. There is also much variation in the size of the rural population: from over 60 million in the United States to two other countries with more than 10 million and twelve with fewer than 2.5 million.

Unfortunately the official definitions of 'rural' and 'urban' employed by the 23 countries represented in Table 1.1 vary tremendously (see Table 1.2). For example, at one extreme, Switzerland regards communes of 10,000 inhabitants or less as being rural whereas the critical limit in Norway is under 200 inhabitants. In other cases designations depend upon particular criteria, often archaic, rather than objective definition based on contemporary functions. Yet, despite the fact that the range of definitions contributes to the variability, there is still a strong indication of the extent to which countries in the Developed World have very different proportions of their population living in rural areas. This also tends to reflect variation in the importance of both the rural economy and rural society in national life. For example, Table 1.1 also shows significant differentiations in the national labour forces, contrasting countries such as Eire, Greece and Portugal, where agriculture still accounts for a significant amount of the country's labour force, against most of the other developed countries where it does not. Such variation can be extended to many other aspects of the rural areas of the Developed World, reflecting especially the different economic, social and political backgrounds of the individual countries.

There are certain unifying factors, though, that give credibility to the

Table 1.1 The rural population of the Developed World, 1985

Country	Total popn. (mill.)	Rural popn. (mill.)	% rural	% economically active in agriculture
Australia	15.8	2.3	14.3	6.4
Austria	7.6	3.4	44.9	7.0
Belgium	9.9	0.5	5.4	2.2
Canada	25.4	6.2	24.3	4.3
Denmark	5.1	0.8	16.1	5.9
Eire	3.6	1.6	44.4	15.4
Finland	4.9	2.0	40.2	9.3
France	54.6	14.6	26.7	6.7
Greece	9.9	4.1	41.9	25.9
Iceland	0.2	0.0	10.8	7.8
Italy	57.1	16.0	28.0	9.5
Luxembourg	0.4	0.1	22.2	2.2
Malta	0.4	0.1	12.0	4.5
Netherlands	14.5	1.7	11.5	5.0
New Zealand	3.3	0.5	16.5	11.0
Norway	4.2	1.2	29.3	7.0
Portugal	10.2	7.1	70.3	23.0
Spain	38.6	3.7	9.6	13.7
Sweden	8.4	0.9	11.0	4.7
Switzerland	6.4	2.7	42.9	5.5
United Kingdom	56.1	4.5	8.0	2.5
United States	239.3	62.9	26.3	3.1
West Germany	61.0	8.5	14.0	5.1
	636.9	145.4	22.8	5.9

Sources: United Nations Demographic Yearbooks; F.A.O. Production Yearbooks

task of considering 'rural' within the Developed World as a whole. The dominance of capitalist modes of production is one such factor. This has promoted a common range of industrial and urban influences on the countryside consequent upon similar experiences of development rooted within the economic and social structures associated with capitalism. There are common processes that have given rise to particular changes and conflicts in rural areas albeit with the magnitude, chronology and distribution of these varying considerably from country to country and even between different regions within countries, for example contrast the countryside in rural-urban fringes against that of peripheries remote from urban influence. However, rural areas in all the developed countries have experienced broadly similar effects through the increased separation of workplace from residence, from the labour-shedding nature of modern agriculture, and pressures placed upon wilderness areas and fragile ecosystems. The timing and magnitude of these changes has varied

Table 1.2 Examples of national criteria used in classifying settlements as rural

Australia: Population clusters of less than 1,000 inhabitants, excluding certain areas (e.g. holiday resorts)

Austria: Communes (*Gemeinden*) of less than 5,000 inhabitants

Canada: Places of less than 1,000 inhabitants, having a population density of under 400 per sq km

Denmark: agglomerations of less than 200 inhabitants

Eire: Settlements of under 1,500 inhabitants

France: Communes containing an agglomeration of less than 2,000 inhabitants living in contiguous houses or with not more than 200 m between the houses

Greece: Population of municipalities and communes in which the largest population centre has less than 2,000 inhabitants

Iceland: Localities of less than 200 inhabitants

Luxembourg: Communes having less than 2,000 inhabitants in the administrative centre

Netherlands: Municipalities with a population of less than 2,000 but with more than 20 per cent of their economically active population engaged in agriculture, excluding specific residential municipalities of commuters

New Zealand: Administrative areas with a population of less than 1,000 inhabitants

Norway: Localities of less than 200 inhabitants

Portugal: Agglomerations and other administrative areas with less than 10,000 inhabitants

Scotland: Agglomerations and other administrative areas of less than 1,000 inhabitants

Spain: *Municipios* of less than 2,000 inhabitants

Switzerland: Communes of less than 10,000 inhabitants

Source: United Nations Demographic Yearbooks

considerably, but examples of their occurrence can be found throughout the Developed World.

Before making a systematic examination of rural society, economy and the planning of rural areas, it is important to recognise not only the existence of similarities between rural areas in the Developed World, but also to note important differences or 'discontinuities' that must be considered when attempting to apply general statements. Broad differences can be traced to different historical backgrounds: on the simplest of terms, contrasts between the Old World and the New can be recognised, as can some between North America and Australasia, and others associated with particular 'cultural realms'. For example, colonisation of North America by different groups of people produced regional variation in social and economic development, though Harris (1977) suggests there were actually great similarities in the structure of north-western European societies overseas, based largely upon the application of the desire for the private control of land to situations where land was cheap and markets poor. In such cases, notably Canada and New England,

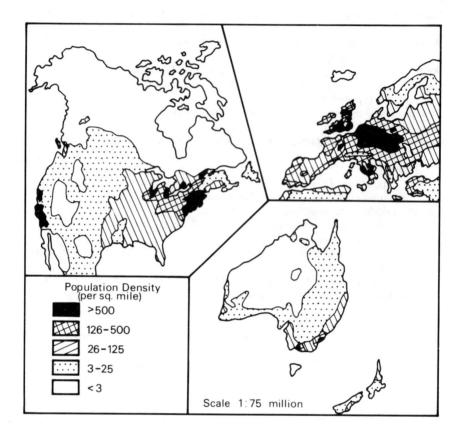

Figure 1.1 Population densities in the Developed World

'remarkably homogeneous and egalitarian rural societies of subsistent farmers emerged quickly' (Harris, 1977: 469). Differences in the origin and background of the settlers were often submerged until access to land became more restricted and new factors of differentiation emerged, promoting critical distinctions within rural societies in North America, but ones also found within Western Europe.

One contrast in terms of the range of different types of rural area found in the Developed World is illustrated by the existence of variable population densities. As shown in Figure 1.1 part of the 'rural' Developed World is characterised by the existence of large sparsely populated regions (see Table 1.3 and Plate 1). These are present in North America and Australia, beyond the areas which experienced rapid colonisation by European settlers in the 18th and 19th centuries, whilst in the nations of the long-settled European heartland only Scandinavia and, marginally, the Scottish Highlands contain extensive low density areas (Trewartha, 1969: 80–90).

Table 1.3 The characteristics of the 'sparselands'

Holmes' (1977) definition of these sparselands includes environmentally inhospitable lands located beyond the ecumene, 'together with the extensive, low-yielding agricultural and improved grazing lands of the ecumene, with densities no greater than 10 persons per square mile (4 per sq km).' A more detailed account lists nine characteristics:

(a) difficult physical environments suppressing economic activity and repelling residential settlement;

(b) economic activity focuses upon exploitation of natural resources yielding high returns but negligible local multiplier effect;

(c) the harsh environment precludes them from participating in many of the trends experienced by other rural districts;

(d) provision of even basic services is difficult and costly and requires governmental assistance;

(e) government influence is very high in terms of providing and maintaining services and also through the frequently high levels of government ownership of land;

(f) populations are often highly mobile, unbalanced in composition, with low dependency ratios and confined to the level needed to sustain limited economic activity;

(g) indigenous, pre-European groups often form an important element in society, and are associated with a range of problems relating to loss of cultural identity, assimilation and economic and social viability;

(h) distance from major metropolitan centres produces feelings of deprivation of important aspects of social, economic and cultural life;

(i) technical advances overcoming some of the most severe effects of isolation have not reduced tolerance levels for the type of isolated living found in sparselands.

Source: Holmes (1981)

Simplistic divisions between Old and New World or populous – versus sparse-lands do little justice to the complex differentiations possible between rural areas in developed countries once economic and social criteria are considered. For example, Australia, Canada, Denmark and New Zealand rely upon staple exports for their economic well-being. Therefore their rural economies have tended to receive a greater share of attention within national economic management than their more industrial counterparts. But, of course, the industrial leaders also have primary product dependencies, and the United States has a significant contribution to its foreign currency earnings from agricultural exports (Hoggart and Buller, 1987: 55).

There are many similar complex differentiations within the rural realm of the Developed World. Yet, the multi-faceted variation is still amenable to systematic analysis and to the drawing of comparisons and formulation of generalities as well as the highlighting of very different experiences. It would be erroneous to overlook the contrasts, especially

Plate 1 Rural? – 1. The Sparsely Populated Regions: Yulara National Park, Central Australia. Spinifex grass and the 'desert oak' dominate the landscape, and for hundreds of thousands of square miles human habitation is limited to a few scattered cattle 'stations'

with respect to the operation of vastly different planning mechanisms, but frequent parallels between countries and individual regions emerge in analyses of rural society and economy, and it is these similarities which are given prime focus here, especially in the first two parts of this book.

To reinforce this demonstration of inter-continent similarities, the results of a cluster analysis (Tryon and Bailey, 1970) of 23 developed countries are shown in Tables 1.4 and 1.5. This analysis, based on 15 selected variables, groups countries in terms of similarities with respect to absolute and relative population characteristics, land use, inputs to agriculture and agriculture's role in the national economy. The list of variables is not intended to be definitive of all possible indicators of rural economy and society, but is simply a general and subjective choice for use as one way of classifying the 23 countries concerned.

Many of the significant correlations between the variables represent very predictable relationships, e.g. the high positive correlation between agriculture's proportional contribution to gross domestic product (AGGDP) and agricultural employment as a proportion of all employment (ECAGRIC), and the inverse relationship between agricultural employment as a proportion of all employment (ECAGRIC) and gross domestic product per capita (GDP) (Table 1.5). However, some are less predictable, for example the positive correlation between percentage area

Table 1.4 Variables used in the classification of developed countries using 'rural indicators'

Variable	Coding
Total population	POPN
Population density	DENSITY
% Rural population	% RUR
% Economically active in agriculture	ECAGRIC
% Arable	ARABLE
% Permanent (tree) crops	CROPS
% Pasture	PASTURE
% Forest	FOREST
% Other land uses	OTHER
Tractors/head of population	TRACTORS
Fertiliser consumption/head of population	FERTS
GDP per capita	GDP
Agriculture as a % of GDP	AG GDP
Food and beverages as % imports	FOODIMP
Food and beverages as % exports	FOODEXP

Table 1.5 Correlations between 'rural indicators'

Positive		Negative	
AD GDP – ECAGRIC	0.86	GDP – FOODIMP	– 0.64
FOODIMP – ARABLE	0.67	PASTURE – OTHER	– 0.62
FERTS – DENSITY	0.62	FOREST – FOODIMP	– 0.60
FOODIMP – DENSITY	0.59	ECAGRIC – GDP	– 0.60
FOODEXP – AG GDP	0.58	DENSITY – TRACTORS	– 0.55
FOODEXP – TRACTORS	0.55		
CROPS – FOODEXP	0.55		
ECAGRIC – %RUR	0.54		
DENSITY – ARABLE	0.50		
CROPS – AG GDP	0.45		

N.B. Only those correlations significant at the 0.01 level are shown

under arable (ARABLE) and imports of food and beverages (FOODIMP), and the negative correlation between tractors per head of population (TRACTORS) and population density (DENSITY). High population densities are associated with high per capita fertiliser consumption, high proportions of arable land, but also fewer tractors per head of population.

The classification itself is produced from the highest correlations between the 23 countries on the basis of their rank order performance with respect to the chosen variables. At the outset all the countries are separate. Once the grouping process is started it could continue step-by-step until all 23 are combined into one giant group. In this case, though,

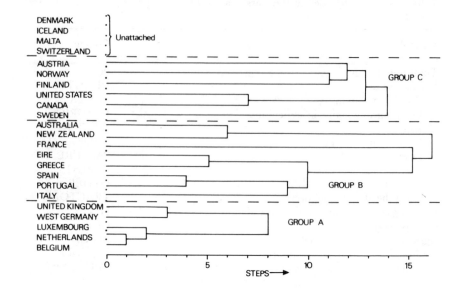

Figure 1.2 Cluster analysis using 'Rural Indicators' as Variables

the process has been continued only for those associations which were significant at the 0.05 level. Hence, the grouping was stopped after 16 steps and with four countries, Denmark, Iceland, Malta and Switzerland unattached to a group. This procedure, known as simple linkage cluster analysis (see Davis, 1973: 465; Mather, 1976: 324–7), produced three groups of countries (Figure 1.2). The high degree of subjectivity involved in the selection of the 15 variables on which this grouping is based precludes detailed examination of this classification, but the three groups do suggest certain broad characteristics of 'rural performance' in the Developed World.

(a) Of the three groups the one that had the closest associations between its member countries was Group A representing *the industrial heartland* of the Developed World: Benelux, West Germany and the United Kingdom. These are countries with small proportions of their population classed as rural, a small agricultural labour force, a small contribution by agriculture to the GDP, high GDP per capita, high population density, and food imports exceeding exports. They are countries typified by great pressure upon the countryside from urban sprawl (Plate 2), and farming activity dominated by high levels of investment and 'industrial' agriculture.

(b) There is a general contrast between Groups A and B. The latter, the largest distinguished by the classification procedure, consists of the *Mediterranean countries plus Eire and*, at a lower level of association, *Australia and New Zealand*. The Mediterranean countries and

Plate 2 Rural? – 2. A 1960s shopping precinct in the 'village' of Pattingham,
five miles (8 km) from the centre of Wolverhampton in the England Midlands,
resembles those found in the centre of many British towns

Eire have a greater proportion of their populations working on the
land than the countries in Group A. Agriculture plays a more significant role in the economy, GDP per capita is lower, food exports are
of greater importance, permanent tree crops play an important role
in the arable sector, and purchased inputs are lower. Australia and
New Zealand share some of these characteristics, but with a smaller
proportion of their labour force employed in agriculture, higher
proportions of food as a component of exports, but lower as
imports, lower population densities, and they have more land
devoted to pasture than the others in the group, excepting Eire. The
Group B countries tend to be those with either remnants of relatively
poor peasant society and farming reflecting limited capitalisation
(e.g. the Gaeltacht and northern Greece) or large areas of extensive
agriculture (e.g. the Darling Downs) or pockets of agriculture and
land use tending towards the conditions found in Group A (e.g. the
Paris Basin, the Po Valley, the Riverina and the Waikato).

(c) The third group, Group C, consists of *Scandinavia* (excluding
Denmark and Iceland), *North America and Austria*. A common
characteristic of these countries is a high proportion of land under
forest (76 per cent in Finland's case). They also have high GDPs per
capita, a low contribution to GDP from agriculture and low population densities. Within the group the United States and Canada stand

out in terms of their greater size of population and higher proportions of exports contributed by food and beverages. The presence of
large areas of land devoted to extensive forms of agriculture, as
indicated by the low input of tractors and fertilisers per capita, also
distinguishes the two North American countries.

The four countries unattached to the three groups stand out as individuals with significantly different 'profiles' to those followed by members
of the groups. The harsh physical environment and small population size
contribute to Iceland's distinctiveness whilst small size, limited
agricultural land use and low GDP per capita characterise Malta's
individuality. Sweden and Switzerland tend towards Group C, though
with certain differences: principally very low proportions of exports
provided by food, a high (Switzerland) or low (Sweden) proportion of
rural population, and a high proportion of land under pasture
(Switzerland).

Whilst it is important to recognise the shortcomings in this type of
highly generalised classification which ignores variations within countries,
the broad indications given of similarities and differences, both inter- and
intra-continental need to be borne in mind when examining rural
problems. The groupings suggest the presence of some common general
dimensions to these problems: in broad terms reflecting the industrialised
core (Group A), a southern and western less industrialised region to
which can be added Australia and New Zealand (Group B), and a
northern and Alpine zone to which can be added North America (Group
C). When particular aspects of rural society and economy are examined
it will become clear how different problems have been of a greater order
of magnitude in the 'core' and peripheries of the classification.

1.2   Defining 'rural'

By continually discussing rural in a taken for granted manner (as in rural
sociology, Rural Sociological Society, etc.), we fail to call into question its
realness and inadvertantly tend to reify it (Falk and Pinhey, 1978: 553).

One of the major problems when referring to the term 'rural society' is
that of a clear definition of 'rural'. Whilst most people have a generalised
conception of what constitutes a 'village', and associate the term 'rural'
with farming or 'unspoilt countryside', a more concrete definition is
much harder to produce. Numerous delimitations of 'rural' appear in
literature within a variety of academic disciplines, although many have
been developed with just one purpose in mind. For many studies a rough
working interpretation of the word may suffice, but Cloke (1980a)
argues that, for rural geography at least, any suitable conceptual
framework must involve understanding what is 'rural', how it differs
from 'urban' and what spatial dynamics are involved in the manifestation
of this difference. Similarly, Larsen (1968: 581) argues that a uniform

definition of 'rural' is vital for the production of valid generalisations from sociological analysis. This view has been reflected since the early 1960s by repeated attempts to formulate the minimum and essential criteria of rural society. However, the reduction in certain rural-urban differences through large-scale migration, the influence of urban-centred mass media, greater interdependence of rural and urban economies, and other forms of increased systematic linkage make such a formulation extremely difficult (see Plate 2).

The very fact that there are growing references to rural geography, rural sociology and rural planning reinforces this desire for a clearer guide to what is rural and what is not, despite the recognition that an over-riding concern with definition may stultify academic endeavours by limiting the scope of enquiry. In general terms 'rural' has been regarded as referring to populations in areas of low density and to small settlements, but, beyond this, division between rural and urban is highly problematic. Three principal reasons for the difficulty in differentiating between what is urban and what is rural are cited by Carter (1981: 16–20). These are the settlement continuum, the changing character of settlements and inadequacy of official designation.

1.2.1 The settlement continuum

The problem of the *settlement continuum* is summarised neatly by a statement in an early edition of the *United Nations Demographic Yearbook*:

There is no point in the continuum from large agglomerations to small clusters or scattered dwellings where urbanity disappears and rurality begins; the division between urban and rural populations is necessarily arbitrary (United Nations, 1955).

This statement, supporting the existence of a continuum of settlements along a line that changes character gradually from rural at one end of the scale to urban at some point, has important implications for the study of both rural and urban societies. It does not help the search for a strict interpretation of 'rural', but the notion of the continuum has been taken up at some length by sociologists, and will be considered later in this chapter and in Chapter 2.

1.2.2 The changing character of settlements

The search for a clear and simple definition of 'rural' is not aided by changes in the nature of settlements which have produced the pheno-menon of urban sprawl, with towns breaking out of their mediaeval walled surrounds. Suburbs have pushed far beyond early town limits and have invaded areas formerly well beyond the immediate urban sphere of influence. Other changes have destroyed close networks of market towns and villages by transferring market functions to larger centres and attaching the urban trappings of modern estates and out-of-town shopping facilities to settlements with only a small population. This has blurred divisions that once appeared much more clear-cut.

1.2.3   Official designation

Official designation has frequently used a critical size of population to indicate when a settlement ceases to be rural and becomes urban (see Table 1.2). This size limit varies enormously and in many cases is supplemented by other criteria. Hence in France the definition of a rural commune is one 'not containing an agglomeration of more than 2,000 inhabitants living in contiguous houses or with not more than 200 metres between houses and communes of which the major part of the population is not part of a multi-communal agglomeration of this nature.' The US Bureau of the Census defines 'rural people' as those individuals living in open country or in towns with a population of less than 2,500, but this means that rurality can vary from the 'extreme rural' of much of the Great Plains to the 'nearly urban' of rural enclaves within the megalopolis of the north-east (Lang, 1986). In Spain rural settlements are those with less than 2,000 inhabitants whereas in New Zealand the upper limit of a rural settlement is a population of 1,000. This last delimitation has led Heenan (1979) to suggest thought be given to breaking down the all-embracing definition of 'rural' using farm and non-farm population as a primary distinction, with further differentiation on the basis of nominated criteria, for example, village and extra-village residence.

The difficulties outlined above have not prevented a wide variety of definitions being employed both officially and for specific purposes. Therefore, to the official designations can be added numerous 'working' definitions of 'rural', e.g.

those parts of the country which show unmistakable signs of being dominated by extensive uses of land either at the present time or in the immediate past (Wibberley, 1972: 259).

in a land use context, rural land encompasses areas which are under agriculture, forest and woodland, as well as wild uncultivated tracts in a natural or semi-natural state (Best and Rogers, 1973: 26).

Both of these make use of the notion of a particular type of land use generally recognisable as a definitive characteristic of rurality. However, as soon as the inhabitants of rural areas are added to the definition the difficulties multiply. There have been several attempts to overcome these problems, with occupational, demographic, ecological, social organisation, and cultural characteristics all being used as defining attributes (e.g. Hoggart and Buller, 1987: 8–28).

In order to simplify the plethora of definitions, they can be grouped into three types: socio-cultural, occupational and ecological.

1.2.4 Socio-cultural definitions

A common way in which 'rural' has been distinguished from 'urban' has been in terms of clear differences in behaviour and attitude between people in small and large settlements. For example, historical accounts of 'rural communities' have stressed the high degree of interaction within such entities, emphasising the way in which everyday life was underpinned by shared values. This accentuation of harmony and consensus contrasted with other 'communities' in which there was conflict and differentiation based on class division and variation in social status.

In the context of the United States, Flinn (1982) recognises three commonly held views of rural areas:

(i) *small town ideology*: democracy stems from the small town which also promotes a 'natural' life-style;
(ii) *agrarianism*: farm life is regarded as the best upbringing for a family, with the family farm ideal for efficient and plentiful food production;
(iii) *ruralism*: the countryside is cherished for its open space, close association with nature and a 'natural' order of life.

These traits suggest a particular view of the countryside, couched in socio-cultural terms. They find an echo in commonly held views of the British countryside, for in Britain the countryside has often been regarded as representing a source of national strength. For the 'Romantic Right' of British politics this rural myth has been embodied in the country house, the church and traditional hierarchical rural society based on the squire, parson and a deferential labour force. For the 'Romantic Left' the myth has been translated into eulogies of rural folk society, the village community, rural crafts and the worthiness of farm labour. The combined consensus has expressed a hostility to materialism, modern urban development and a desire for a perceived Arcadian golden age. Periodically this romanticism has been translated into legislation or populist movements. For example, in the inter-war period a 'back to the land' desire was expressed in the form of the green belt concept, garden cities and the growth of hiking and rambling.

Although these views may not be especially coherent, Palmer *et al.* (1977), working on society's conceptualisation of 'rural', demonstrated that images of the countryside are structured along a number of dimensions which they distinguished in terms of accessibility, activity, degree of crowding, facilities/settlement, scenery, evaluation (e.g. relaxing versus disturbing) and emotion/reflection.

Some of these ideas are elaborated in Section 1.5 in terms of the notion of a differentiation between rural and urban based largely in socio-cultural terms and forming a continuum of change.

1.2.5 Occupational definitions

Whilst socio-cultural definitions often make an implicit assumption about the type of occupation followed by rural inhabitants, this assumption is stated more specifically in occupational definitions. The dominance of agriculture and forestry is the crucial element in such definitions, which stress the distinctive social values shared by agriculturalists (e.g. Glenn and Hill, 1977). Yet, several recent studies of farmers and farm labourers show that they are by no means homogeneous groups, especially farmers who tend to be representative of several social groups and so have significant attitudinal differences (e.g. Bell and Newby, 1974; Buttel, 1982a; Coughenour and Christenson, 1983; Newby, 1980b).

An occupational definition based solely upon the primary sector ignores the important role of certain kinds of industry in 'rural' areas, especially in the sectors of food processing, craft industry and power generation. Marketing and service functions performed within the countryside are also excluded. Thus occupation alone is not an especially decisive definitional variable.

1.2.6 Ecological definitions

Another way of looking at the continuing difference between rural and urban has been the ecological approach in which social forms are seen as arising from and being modified by human cultural adaptation to environmental circumstances. The different 'environmental milieu' of small towns and rural areas is held as responsible for maintaining certain differences between them and large cities. Within this milieu can be included the physical and man-made environment, the population, technology, values and beliefs and social organisation.

The two definitions quoted above in section 1.2.3 could be regarded as ecological definitions in that they express 'rural' in terms of its environmental setting. Concepts such as 'open countryside' and 'small size of settlement' are embraced in definitions of this type. Problems of occupational specificity are avoided, though in 'open countryside' the dominance of primary activity is often implicit. Such definitions tend to be imprecise, though in broad physical terms everyone recognises a difference between rural and urban. As Ford (1973: 3) remarks, 'One doesn't have to be a particularly astute observer to detect that contemporary life in New York City and Los Angeles is still quite different from that in Bug Tussle, Oklahoma or Gravel Switch, Kentucky.' The implication here is that not only is there a huge environmental gulf but that there are also important social and economic differences that may be of more relevance in focusing specifically upon rural problems, e.g. higher rates of males to females, lower incomes, proportionately more families living in poverty, a smaller proportion of women in the labour force, a less well educated population, and an over-representation of the elderly (Willits *et al.*, 1982).

Table 1.6 The variables used in Cloke's 'Index of Rurality'

Variable	Characteristics in 'rural' areas
Population per acre	Low
% change in population 1951–61, 1961–71	Decrease
% total population: over 65 years	High
% total population: male 15–45 years	Low
% total population: female 15–45 years	Low
Occupancy rate: % population at 1.5 per room	Low
Households per dwelling	Low
% households with exclusive use of:	
(a) Hot water (b) Fixed bath (c) Inside W.C.	High
% in socio-economic groups: 13/14 farmers	High
% in socio-economic group: 15 farm workers	High
% residents in employment working outside the rural district	Low
% population resident for < 5 years	Low
% population moved out in last year	Low
% in-/out-migrants	Low
Distance from nearest urban centre of 50,000	High
Distance from nearest urban centre of 100,000	High
Distance from nearest urban centre of 200,000	High

Source: Cloke (1977)

Similar conceptions about the character of rural areas, broadly related to an overall ecological distinction between rural and urban, are utilised by Cloke (1977; 1979: 3–22) who presented an alternative to fixed practical definitions by adopting what he termed an *index of rurality.*

Cloke's index described a continuum with absolute rurality at one end and absolute urbanity at the other. It was derived from a consideration of numerous socio-economic variables processed by means of multivariate analysis. The list of variables employed is shown in Table 1.6 with indicators of the characteristics deemed to be indicative of rurality. This index was applied to England and Wales for data from the 1961 and 1971 population censuses with five categories recognised along the continuum – extreme rurality, intermediate rural, intermediate non-rural, extreme non-rural, and urban (see Figure 1.3a) Comparisons between 1961 and 1971 revealed that lowland areas were being subjected increasingly to urban pressures, causing a decrease in rurality. This led Cloke to condense the number of categories to two – 'pressured' and 'remoter' areas, both including urban and rural characteristics but with the former representing areas where the degree of rurality was declining, and the latter areas where the level of rurality was static or increasing (Figure 1.3b). The areas distinguished in this way are very similar to those depicted in Clout's (1972: 45) map of the commuting hinterlands of England and Wales (Figure 1.3c). Similar attempts to use an index of rurality were also made for England and Wales by the Department of the

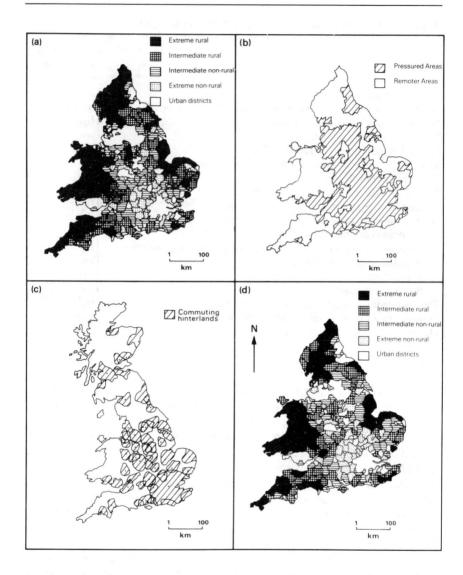

Figure 1.3 (a) 'Rurality' in England and Wales, 1971 (based on Cloke and Edwards, 1986); (b) Pressured and Remoter Areas in England and Wales, 1981 (based on Cloke, 1979); (c) Commuting hinterlands around major British cities, 1961 based on Clout, 1972); (d) 'Rurality' in England and Wales, 1981 (based on Cloke and Edwards, 1986)

Environment (1971) and for Scotland by the Scottish Office (1978: 49), using a similar set of 25 indicators of socio-economic structure to those selected by Cloke.

Despite the problems of changing administrative boundaries, a different census format and of the changing nature of rurality itself, the index was replicated for 1981 (Cloke and Edwards, 1986). Comparisons between the indices derived for 1971 and 1981 showed a movement towards increasing urban influence in the extreme rural areas (Figure 1.3d). This was most apparent in the north of England. Elsewhere, there had been quite widespread movements towards the urban pole of the continuum employed in the analysis. For example, greater urban influence was discerned in East Anglia and in the axial belt between London and the Midlands. Cloke and Edwards (1986: 295–303) recognised the rather coarse nature of this attempt to measure rurality and refined their study of the 1981 data by modifying their set of variables to take account of population mobility and the use of rural areas for tourism and second home ownership. This new index suggested the penetration of urban influences into the South-West and East Anglia might have been overstated previously. It also confirmed the existence of four major blocks of extreme rural areas: the South-West; Mid and North Wales and the Welsh Borders; the agricultural lowlands around the Wash; and the Pennines.

The presence of these definitional difficulties has meant that academic disciplines focusing upon rural areas have also been plagued with problems relating to the scope of their enquiry and the distinction between a 'rural study' as opposed to other systematic studies. This has been the case especially for *rural geography* and, to a lesser extent, for *rural sociology* too. A brief outline of some interpretations of their scope adds further variety to the definitions of 'rural'.

1.3 Rural studies

1.3.1 Rural geography

In recent years rural geography has often been recognised as a separate systematic branch of human geography. Yet, its exponents frequently resort to unsatisfactorily bland descriptions such as 'the study of geographical aspects of human organisation and activity in non-urban areas' (White, 1981), and there has been argument that rural geography does not constitute a systematic study but is more of a consideration of a chaotic range of economic, social, demographic, cultural and resource use questions in the rural environment.

This is a much wider interpretation than would have been made pre-1939 when rural geography, although well to the fore within the discipline, was associated with the study of rural settlements and field systems, using a largely historical approach. This was stimulated by the work of Vallaux (1908), examining the origin, structure and patterns of rural settlement, and was continued primarily by the French geographers, Vidal de la Blache (1918) and Demangeon (1935). This strand of rural geography might be more clearly recognised today under the heading *rural settlement*, and is still associated primarily with the historical development of settlement and settlement patterns in rural areas (e.g. Roberts, 1977). Evidently, the changing content of rural studies within disciplines can add to the definitional confusion.

The increased scope and clearer recognition of a sub-set of human geography called 'rural geography' has developed from the 1960s, with several attempts to define the scope of the subject and, in so doing, clarify the qualifying adjective. Five such examples are shown in Table 1.7, the variation between them partly reflecting the different purposes behind their definitions – for example, Sylvester (1969) seeking to portray the elements incorporated by the term 'rural landscape' and Clout (1972) focusing on the demands upon rural land.

The combinations of topics represent a wide coverage of subject matter that collectively suggest something of a 'rag-bag' of economic, social, political, environmental, resource and ecological phenomena. Although in the past two decades rural geographers have focused more closely upon specific problems, e.g. depopulation, declining services in peripheral areas, the influence of urban sprawl on the countryside and conflicts between recreation and conservation, they have mirrored human geographers as a whole by 'branching into anarchy' in terms of their kaleidoscope of foci, methodologies and general ethos (Johnston, 1979: 189).

Within this 'anarchy' two divergent views of 'rural' have emerged. One might be termed 'the traditional' which reflects Clout's focus on land use and has links with earlier work on rural landscape and settlement. This recognises characteristic types of rural land use as discrete spatial units for investigation, and possessing traits such as marginality and the presence of special types of community. This approach, as represented in both Gilg's (1985) and Pacione's (1984) texts, has tended to eschew theory and closer links with developments within other social science disciplines. However, it has had a quite strong 'applied' component, strengthening the link between rural geography and planning.

In contrast, a strong line of argument in the 1980s has combined a broader inter-disciplinary approach with a philosophy that has regarded 'rural' as something of a false theoretical category. Although this strongly reflects views from sociology that all locations, be they urban or rural, are subject to the same basic underlying structures and processes, a geographical context has been provided by a specific focus upon the locality and the particular political economy in operation (e.g. Cloke, 1989b; Marsden, 1988). This approach mirrors human geography in its faltering attempts to develop a philosophical base that will allow it to embrace a suite of methodological possibilities incorporating humanistic and empirical elements as well as the more restrictive, and generally more

Table 1.7 Definitions of rural geography

Sylvester (1969) Components of the rural landscape	Clout (1972) Demands on rural land	Pacione (1984) Synthesis of major developments in rural geography
Economic system (production & trade) Social system (people) Physical geography (landscape & climate) Strategic system Ecclesiastical system Civil administrative system	Food production Manufacturing industry Housing Communications Water gathering Recreation Nature conservation Military requirements Mineral extraction	Evolution of the settlement pattern Spatial organisation of settlement Settlement planning and change Agriculture Structural change in agriculture Agriculture & urban development Population dynamics Rural communities Metropolitan villages Seasonal suburbanisation Quality of life Housing Employment and rural development Service provision Transport and accessibility Resource exploitation & management Conservation Leisure and recreation Power and decision-making

Bowler (1975)/Robinson (1987)
Categories of research in rural geography

Agriculture
– Enterprise and farm systems; structural change; land use patterns; land use competition; marketing and distribution; social geography.
Forestry, Fishing and Mineral Extraction
– Forest-based recreation; social/employment impact of forests; economics of forestry; commercial fishing; mineral extraction; derelict land.
Recreation and Tourism
– Second homes; resorts and tourism; recreation; impact studies/land use effects; National and Country Parks; demand assessments; sports.
Rural Settlement
– Economic structure/function; historical development; housing.
Rural Population
– Distribution, growth/depopulation; employment structure; social geography; the elderly.
Rural Transport
– Modal studies; impact studies; transport policy.
Rural Development, Planning and Conservation
– Rural development policies; resource evaluation; institutions; planning control of land use; settlement planning; planning of services; conservation issues; ecology and landscape changes.

Gilg (1985)
Review of topics & themes in rural geography

Agricultural geography
Forestry, mining & land use competition
Rural settlement & housing
Rural population & employment
Rural transport, service provision & deprivation
Rural recreation & tourism
Land use and landscape
Rural planning & land management

utopian, concepts, within structuralism. It is possible to view the apparent chaos of the breadth of scope and plethora of methodologies in rural geography with some alarm, and yet the growth of the subject and its close links with a range of disciplines seems a healthy situation for those wishing to encourage greater academic interest in rural affairs (see also Pacione, 1983).

Cloke (1988a) refers to the continuing 'fog of ignorance' surrounding geographical studies of rural areas. He champions the adoption of a stronger theoretical base via the embracing of critical social theory. However, as much of this theory puts forward arguments that rural areas and rural development have no causal significance then this may be a self-defeating 'saviour'. Indeed, work by Hoggart and Buller (1987), which seeks to draw on these theories, rather side-steps the issue of how 'rural' fits into the theoretical picture. They treat rural as a sub-set of locality types, but avoid what Cloke terms an 'untheorised descriptive research agenda.'

The impact of critical social theories upon studies of rural society is considered in more detail in Chapter 2. The current state-of-play in their use by rural geographers seems to reflect a growing division within rural geography, although the actual subject matter for investigation remains relatively stable. According to White (1981: 296–7) the particular aspects of rural life that have attracted the most attention from geographers have been:

(a) the causes and consequences of depopulation;
(b) the growth of the influence of urban-related populations in rural areas;
(c) the pattern of recreation and tourism in the rural environment;
(d) structural changes in agriculture and their social and demographic implications;
(e) rural planning.

Not surprisingly it is these elements that lie at the heart of the topics covered in this text.

1.3.2 Rural sociology

Systematic study of rural society as a specialised area of sociology developed pre-1940 primarily in the United States, but gained institutional support post-1950 in many countries. Its basis has been the social organisation of rural society, its sub-systems and its relationships with the rest of society and especially those elements living in urban areas. Ecological, cultural and behavioural emphases all featured in the initial development of this specialisation whilst links with other disciplines were formed by foci upon demographic analysis, settlement patterns and land tenure (e.g. Constandse et al., 1964). In the 1960s emphasis upon the adoption and diffusion of technological innovations broadened the scope of study. At this time Larsen (1968) listed four broad categories for the sociological investigation of rural society:

(a) studies using rurality as the independent variable;
(b) comparative studies of rural societies;
(c) studies for which rural society is the setting within which selected phenomena are analysed;
(d) studies of social change within each of the other three categories.

Fairweather and Gillies (1982), analysing the content of articles in *Rural Sociology* and *Sociologia Ruralis* in the 1970s, found a focus upon social psychological subject matter, often with the use of questionnaire surveys, statistical data analysis and an ahistorical basis. Both journals were representing what Newby (1980a) terms 'conventional rural sociology' in that they were largely empiricist, inductive and positivist. However, the vogue in the late 1970s and 1980s has been for anti-positivist approaches in which formulation of theory and theoretical problems have been more to the fore. In this it bears remarkable similarities to the developments occurring within rural geography and urban sociology (e.g. Pickvance, 1976).

Especially given the growth of Marxist and realist approaches within sociology, it shares with geography the problems of working in a context recognising the differentiation between urban and rural as a significant factor affecting social development. A common theme within sociological literature has been the way in which the social relationships of individuals are much less governed by place, proximity and tradition than was formerly the case. For example, mechanisation in agriculture has reduced communal activity whilst the growth of new functions within the countryside has reduced social integration and homogeneity. In effect, 'community' can now be conceived of distinct from a specific geographical context. Yet, at the same time, it has become fashionable to recognise the existence of local social systems and social processes, if only as local deviations of processes operating on a larger scale (Dickens *et al.*, 1985: 18).

In North America emphases within rural sociology have tended to retain a greater focus upon empirically verifiable developments within society (e.g. Dillman and Hobbs, 1982). However, new and imaginative schema have also been devised for future studies. For example, Carlson *et al.* (1981) extend their work on American rural society to include the dynamic environmental setting within which rural society is placed. They set five goals for their work:

(a) to describe the contemporary status of society;
(b) to develop an environmental perspective from which to examine rural social issues;
(c) to examine the impacts of rural change;
(d) to suggest procedures with which to resolve conflicts and deal with impacts;

(e) to consider a future for rural areas in which public policies and
procedures can effectively respond to contemporary and emerging
social and environmental opportunities.

These goals are only achievable by extending beyond sociological source
material.

If there are core areas around which the new approaches within rural
sociology have developed, it would seem to be a focus upon landholding
and the social structure, commodity production and social relationships
(Buttel and Newby, 1980; Newby, 1980a). Within each of these there
has been a growing concern to move away from sentimental or idyllic
evocations of rural life towards a more structured and critical approach
to rural problems resting upon an understanding of the social processes
within advanced capitalism (e.g. Newby, 1980b; 1983; 1987b). These
processes have often been examined with respect to the belief in the
important role played by the very nature of agricultural activity in determining property relationships within the countryside. Thus land as a
factor of production has played a crucial role in shaping rural society,
and it is theory relating to landholding and social structure that has
pushed some rural sociologists away from the formerly close affinity with
the rural economists (e.g. Buttel, 1980; 1982a; Gasson, 1971) for whom
sociological variables have often been merely residual factors. Yet, 'we
can only come to terms with the changes which are currently occurring
in English rural society if we first analyse the economic pressures to
which it is subject' (Newby, 1980b: 24).

1.3.3 Rural economics

Traditionally, much of the work done by economists pertaining to rural
issues has focused specifically upon the problems of agriculture.
Agricultural economics have a particular niche within economics because
of the special character of the 'agricultural problem': the low level of
income within agriculture compared with other sectors of the economy,
plus problems of overproduction, and government intervention to help
farmers by supporting agricultural prices. Work by economists has
concentrated upon the short-term and cyclical fluctuations in prices and
incomes, and upon the wide range of agricultural stabilisation
programmes. In a recent examination of the scope of agricultural
economics, Veldman (1984) viewed agricultural economics as the study
of agricultural prosperity, covering the returns to the agricultural production systems in relation to farm size, the degree of mechanisation and
rationalisation, and land management.

Agricultural economics has been a highly applied field of study, with
relatively little development of theory and a limited contribution to the
development of economic thought (Houck, 1986), with the possible
exceptions of work related to fluctuation and uncertainty in prices,

production and markets. However, there have been major analytical contributions in the areas of the measurement of productivity change (Hayami and Ruttan, 1971), the adoption of innovation (Summers, 1983), the understanding of the rationality and efficiency of economic units, and the relationships between population change and agricultural technology (Boserup, 1981; Johnson, 1986).

A significant trend has been to extend economic studies to encompass all aspects of the food chain, giving attention to marketing and consumption as well as production (e.g. Kohl and Uhl, 1985; Burns et al., 1983). The role of agriculture within the overall rural economy has also been considered, but with a recognition that an understanding of agriculture requires a consideration of non-economic systems, e.g. socio-cultural, political and technological systems (Louwes, 1977; Hill and Ray, 1987: 402-25). This also recognises the weaknesses of a mono-disciplinary approach and champions the need for explanations combining economics with work from sociology, political science and the physical sciences (Newby, 1982).

A major area of research within rural economics has been that of the economics of rural resource development, using economics to assist decisions concerning the allocation of public resources in rural areas (e.g. Whitby et al., 1974). This type of work follows that of the geographers and sociologists in emphasising a concern for issues affecting the inhabitants of rural areas, and has often cut across traditional divisions between disciplines (e.g. Cockin et al., 1987; Hodge and Whitby, 1981). It has extended economists' focus beyond agriculture and has led to greater consideration of the rural employment market (e.g. Barrows and Bromley, 1975), economic multipliers (e.g. Craig et al., 1982; Garrison, 1968) and factors promoting rural change (e.g. Macmillan and Graham, 1978).

One set of arguments in the analysis of competing interests for available resources has suggested minimal regulatory intervention in the land market (e.g. Frankena and Scheffman, 1980; Stiglitz, 1979). This view holds that market adjustments will occur to ensure sufficient resources are allocated to the appropriate types of production, e.g. food and timber. Other arguments suggest the pursuit of economic efficiency may produce shortages for future generations (e.g. Briggs and Yurman, 1980; Plaut, 1980).

The extent to which separate sub-sets of the social sciences can be combined into a coherent and recognisable body of 'rural studies' is problematic. Publications such as the *Journal of Rural Studies* (commencing in 1985) and the *International Countryside Planning Yearbook* (formerly the *Countryside Planning Yearbook*, commencing in 1980) suggest that at least the sub-sets are not being viewed in isolation and that there is a strong inter-disciplinary component to studies of rural problems. This has incorporated work by geographers, sociologists and economists, but also ecologists and professional planners (Cloke, 1985). The growth in awareness by researchers from a range of disciplines that they are working on common problems, albeit from different viewpoints, is producing an acceleration in knowledge and understanding of these problems.

Much work from a variety of disciplines has addressed particular issues with a planning or applied dimension. Thus work in *rural planning* has extended beyond that from the planning profession. Geographers, sociologists and economists alike have had an input to planning for rural areas and to monitoring the results of particular pieces of legislation. This planning-related work has extended the scope of work from individual disciplines and given an added dimension to 'rural studies', as considered in detail in Part 3. First, though, attention is devoted to the main bases of rural society and rural economy, drawing principally upon literature from the three disciplines considered briefly above. For example, both rural geography and rural sociology have made use of the concept of the *rural-urban continuum*, not only as a means of recognising something distinctively rural, but as a vehicle for investigating the evolution of rural society. This use is now examined and the concept of the continuum is developed with respect to notions of specific 'processes' acting upon and altering rural society.

2

The urbanisation of rural communities

2.1 The process of urbanisation

The concept of the urbanisation of rural or peasant society in the 19th and 20th centuries has formed a central role in investigations of rural change in the Developed World. Yet, surprisingly, many studies have not produced an explicit definition of '*urbanisation*' and the specific changes that are subsumed within this broad term.

Johnston (1983: 363–4) recognises three different ways of referring to urbanisation:

(a) As a *demographic phenomenon*, in which an increasing proportion of the population is concentrated in urban areas. This concentration is produced both by migration from rural areas and differentials in fertility and mortality between rural and urban areas. Table 2.1 shows how this process has proceeded apace throughout the Developed World since 1950.

(b) As a *social and economic phenomenon* inherent in capitalist industrialisation. Urban areas are the vehicles for trade and commerce as they can facilitate efficient concentrations of linked production, distribution and exchange processes. This interpretation links urbanisation to industrialisation and the evolution of complex urban systems.

(c) As a *behavioural phenomenon*, in which urban areas, especially large cities, act as centres of social change, transmitting this change throughout the urban system and possibly beyond.

These notions of urbanisation have been applied within a variety of contexts, with differing emphases placed upon the processes involved. For example, describing the 'suburbanisation process' for Australian cities in which the countryside around major cities has been increasingly altered by urban influences, Maher (1982) identifies the key influences as being technological advances, demographic change, behavioural factors, the role of public policy and changing relationships between these influences and economic and social structures:

Table 2.1 The march of urbanisation in the Developed World, 1950–85

	% Urban		Rural Population (millions)	
	1950	1985	1950	1985
Australia	75	86	3.5	2.3
Austria	49	55	3.5	3.4
Belgium	63	95	3.2	0.5
Canada	61	76	5.3	6.2
Denmark	68	86	1.4	0.8
Eire	41	56	1.8	1.6
Finland	32	60	2.7	2.0
France	56	73	18.4	14.6
Greece	37	58	4.8	4.1
Iceland	74	89	0.0	0.0
Italy	54	72	21.5	16.0
Luxembourg	59	78	0.1	0.1
Malta	61	88	0.1	0.1
Netherlands	74	88	2.6	1.7
New Zealand	73	83	0.5	0.5
Norway	32	71	2.2	1.2
Portugal	19	30	6.8	7.1
Spain	52	90	13.4	3.7
Sweden	66	89	2.4	0.9
Switzerland	44	57	2.6	2.7
United Kingdom	84	92	8.1	4.5
United States	64	74	54.8	62.9
West Germany	72	86	14.0	8.5

Sources: United Nations Demographic Yearbooks; F.A.O. Production Yearbooks

(a) *Technological advances* have been offered consistently as the single most important factor in producing centrifugal tendencies in urban growth. The basic stages in this growth are detailed by Clark (1958) who describes transport as the 'maker and breaker of cities'.

Clark placed great emphasis upon the way in which changing transport technology permitted urban centres to become more and more diffuse. From the highly concentrated nature of urbanism in cities where people walked to work or relied upon horse-drawn carriages, the railway in the 19th century brought a dispersal of economic activity to areas previously denied it through lack of access to sea, river or canal transport. It also extended the urban form through the growth of the suburban rail network in the larger cities (e.g. Gilmore, 1953; Warner, 1963). Subsequently, the development of the motor car and its increased ownership by a large proportion of the population has facilitated a tremendous growth in personal mobility. This has given greater stimulus for the separation of workplace and residence, contributing to a marked centrifugal movement of people from the inner cities to new suburbs and locations beyond the continuously built-up area. This tendency has been further accentuated by the changes in telecommunications which have reduced the need for central locations of industry and services (Taafe *et al.*, Tobin, 1976).

(b) The key *demographic changes* have been those affecting the rate of household formation and therefore the demand for housing. The baby boom of the late 1940s and 1950s in most developed countries has added to the numbers wanting housing, whilst the propensity of individuals to form single-member households has risen and so created extra demand more recently. Melbourne, Australia, has been a typical example. In 1947 48 per cent of the city's population lived in municipalities that were beyond an approximate 4.5 miles (7 km) radius from the central business district. By 1976 the figure was 79 per cent (Beed, 1981: 131) (see Figure 2.1a and b). Not only had people relocated from the inner suburbs, but there had been a greater autonomous growth of population in the outer suburbs and beyond. The post-war baby boom was a major factor in the initial flow of migrants to these areas. In Melbourne's case, post-war immigration also played an important role as assimilation of these immigrants tended to foster movement away from the inner city through the operation of invasion and succession processes (Stimson, 1982).

(c) Translating *increased demand for housing* into a demand for housing beyond the confines of the urban area implies either the development of a particular desire for residence in such an area or an inability to live elsewhere. For example, the latter might be the case for people on a local authority housing register who are allocated accommodation in a 'commuter village'. However, the majority of residents in such villages are usually owner-occupiers or renting property, so a definite *element of choice* is involved. Various push and pull factors can be in operation, contributing to movement out beyond a particular urban area. The chief 'pushes' have been the negative aspects of life in cities: high density living and poor environment would be two common factors, the latter including a variety of urban ills, e.g. pollution, presence of heavy industry, traffic congestion and strains on services and utilities. 'Pulls' have been exerted by the general appeal of a different life-style offered by 'the countryside': open space, clean air, quiet and, increasingly, an overcoming of problems of access to work from outlying districts as both retailing and industry have moved to 'out-of-town' locations. Whilst, initially, only the wealthy could afford such a move, the growing affluence of society as a whole has permitted a greater proportion to aspire realistically to a suburban or commuter village life-style. Indeed, living in such areas has acquired a high status perpetuating their favourable image.

(d) The aspirations and expectations involved in conscious decisions to move to rural and semi-rural locations are intertwined with

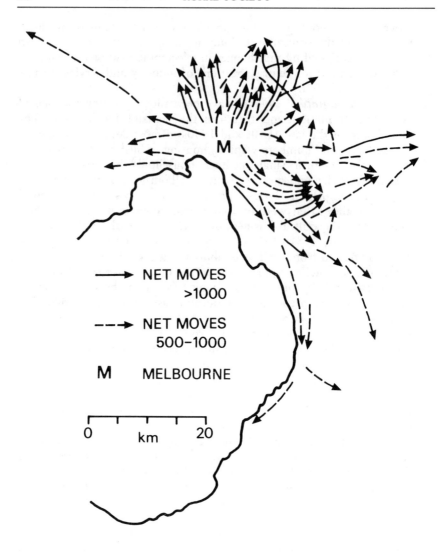

Figure 2.1 Out-migration promoting urban sprawl: Melbourne, 1971–6 (based on Maher, 1982)

prevailing social and economic conditions. In turn, these conditions are related closely to a range of *government policies*. Thus, for example, Popenoe (1980) has suggested that different spatial patterns of suburbanisation in Britain, Sweden and the United States can be traced directly to variations in government influence. More specifically, government decisions on housing policies, public utilities and transport are significant in influencing urban form. For example, housing policies have frequently been a positive encouragement for suburban and ex-urban forms of development. Such encouragement has operated in a variety of ways, through policies adopted by finance institutions, banks and building societies to tax relief and limited controls over land development on the urban periphery. The role of these policies and the various bodies that operate them has been emphasised in a variety of studies concentrating upon economic and social structures underlying spatial patterns (e.g. Harvey, 1975). Such work has reduced the stress upon the role of transport as an independent variable in the urbanisation of the countryside, viewing urbanisation as a manifestation of the process of capital accumulation arising from socio-economic transformations (e.g. Walker, 1979; 1981). This approach does not view the spread of the private ownership of automobiles as an independent influence causing suburbanisation.

Investigations of factors such as the four listed above have contributed to a better understanding of the urbanisation of rural communities. Two areas of study have received special attention: relatively recent changes associated with post-1945 urban growth, and a longer-term social change within the countryside, in which a stable peasantry or folk society has been significantly modified by successive changes deriving from the onset of large-scale industrialisation which developed initially in the 18th and 19th centuries. These two aspects of urbanisation of the countryside will be considered in turn as part of the two aspects of social change in rural areas which, traditionally, have been regarded as complementary but contrasting aspects of urbanisation. These are the 'suburbanisation' of communities around towns and cities, and the effects of rural depopulation caused largely by an exodus to growing urban areas and the transformation of rural peasant society. The latter is discussed in chapter 3 whilst the suburbanisation process is dealt with below. Initially, to highlight the type of changes occurring post-1945, some examples are drawn from North America, and southern Ontario in particular.

2.2   Social change in the commuter village

Figure 2.2 illustrates some of the basic population trends within rural America this century. Although the large-scale addition of new settlements ended around 1910, the number of people in non-metropolitan places with under 2,500 inhabitants did not fall and showed an upturn in population growth in the 1970s at a rate greater than that for the population as a whole. The primary cause for the upturn was in-migration (McCarthy and Morrison, 1977; 1979), reflecting shifts in the location of economic activities from cities plus people's desire and ability to live outside urban areas (Bradshaw and Blakeley, 1979; Zuiches, 1981).

This in-migration brought significant social changes within the host

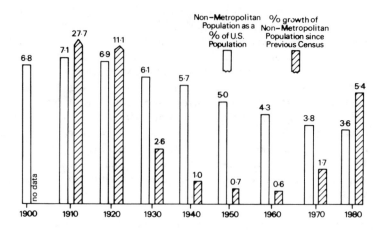

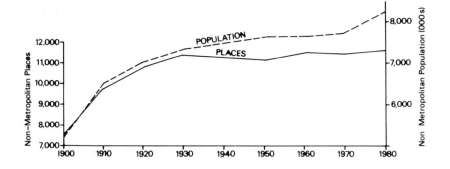

Figure 2.2 Major population trends in rural America in the 20th century (*Source:* Johansen and Fuguitt, 1984)

communities. For example, Wilkinson (1978) referred to five major changes in community interaction in rural USA consequent upon urbanisation:

(a) The scale of local interaction necessary to retain a relatively complete community has increased, i.e. small places cannot provide sufficient activities and services to meet local demands.

(b) To meet demands in rural communities there has been an increased dependence upon outside sources of goods and other resources.

(c) The nature of decision-making in public affairs has been changed by greater influence of the agents of the larger society, i.e. the agenda of public action is often set by the employees of state and federal agencies.

(d) Structural mechanisms producing well-coordinated community action have not developed in rural areas.

(e) Collective action by local citizens has become highly specialised and fragmented. It has not been task oriented, but has become excessively defensive, e.g. the growth of resistance to plans for development.

These changes have been associated with a growing distance between place of work and place of residence. For example, for New York City the maximum extent of the metropolitan commuting field rose to between 50–60 miles (80–100 km) in the 1960s and subsequently more than doubled. Within this field the influence of the metropolitan centre upon incomes depended primarily upon the extent of net in-migration, job substitution and increased female participation rates. However, Mitchelson and Fisher (1987) show that growth of income in the commuter belt in the 1960s and 1970s was affected increasingly by non-metropolitan employment centres. At the same time, metropolitan influence was extended in distance terms, especially for white-collar workers and associated high female participation rates.

2.2.1 The example of Southern Ontario

The nature of the demographic and social changes associated with the increased extent of metropolitan influences has been illustrated in several studies set in southern Ontario where Toronto has exerted a growing control over life within at least a 50-mile (80 km) radius (see Figure 2.3). One of the best examples of the nature of social changes in commuter villages is provided in Walker's (1976; 1977) studies of Bond Head, a community of 683 people (in 1971), located 40 miles (c. 65 km) north of Toronto. Like many small settlements in the area, Bond Head comprises two kinds of communities: the old village with its close ties with communities in the surrounding countryside, and a new housing estate, constructed in 1968, containing 110 houses occupied mainly by people from the professional-managerial group, Toronto-born and commuting to Toronto to work. In 1977 this new 'suburban' population outnumbered the old 'rural' villagers by approximately 2:1. However, even in the old village there had been penetration by former urban residents in the form of retired people from Toronto or urban workers who had purchased houses within the nucleus of the old village.

Walker's examination of the social networks of the inhabitants of Bond Head demonstrated that there were two distinguishable communities there: a rurally oriented group, consisting largely of retired people, and a group of 'suburbanites' living in the new housing estate. Yet, in effect, the latter consisted of several different social sub-groups with minimal inter-group cohesion. Typically, these 'had a small group, short distance, friendship and middle-class orientation more characteristic of an urban community' (Walker, 1977: 347). This meant that the overall pattern of community and territory on the housing estate closely resembled that found by Clark (1968) for suburban Toronto. Amongst the suburbanites,

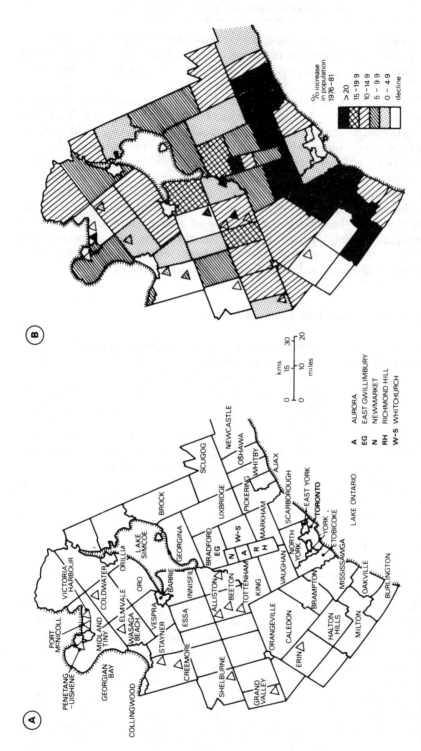

Figure 2.3 Population change in Toronto and its hinterland, 1976–81

half the households derived their incomes from blue-collar labour, although they had largely shed working-class patterns of interaction. So they were part of a private and familistic life-style in which social ties often mirrored the link with Toronto, retained by virtue of commuting to work there. In contrast, the villagers had relatives and friends locally and had reinforced their sense of a local community 'by creating an institutionally complete world' of church, school and a few shops. Change within Bond Head was continuing in the form of the gradual disappearance of the old village.

Walker's (1974; 1975; Walker and Beesley, 1984) further studies of three townships in Simcoe County, 50 miles (80 km) north of Toronto, suggest that the farming community and the former urban residents are pursuing different objectives. The urban-centred focus of the ex-urbanites even extends to the political sphere as it is the farmers who still control local politics in the three townships. This may be a special case where conflicts between the established population and newcomers have not occurred, but there is much potential for such problems to arise. For example, the presence of an influx of young families, putting pressure on existing educational provision, can raise local tax rates as can the need to provide other services and utilities for the newcomers.

In their study of Elora, a settlement of 1,500 people located on the fringes of Toronto, Guelph and Waterloo, Sinclair and Westhues (1975) illustrate the type of conflicts that can occur. In Elora, conflicts arose between the newcomers, who opposed a new apartment development, and the established population who generally looked favourably upon the scheme. It seems the newcomers were seeking what Pahl (1965a) termed 'the village in the mind', i.e. a particular type of desirable residential environment in 'rural' surrounds. Whilst that specific environment may have been altered significantly by their very presence, they still wanted to protect it from any further changes and hence their opposition to the development scheme.

Changes in population occurring in Toronto and its hinterland from 1976 to 1981 illustrate the pattern found for many other metropolitan centres – a decline in the inner city and inner suburbs whilst the outer suburbs and adjacent smaller urban areas grew quite rapidly (Robinson, 1988b) (see Figure 2.3, Tables 2.2 and 2.3). In Toronto's case it is the communities of the adjacent Regional Municipalities that have grown the fastest (York, 23.6 per cent; Peel, 30.5 per cent; Durham, 14.6 per cent; Halton, 11.1 per cent, compared with 0.6 per cent for the Toronto Metropolitan Area itself). But within the Regional Municipalities the greatest proportional increases have been in those parts which form a part of the Toronto Census Metropolitan Area (CMA), e.g. Pickering (42.2 per cent), Ajax (24.2 per cent) and Brampton (38.0 per cent). This is the commuter belt, receiving the latest wave of migrants to leave the city's suburbs. Once separate centres, such as Oshawa, Pickering and Whitby to the east, and Mississauga, Oakville and Brampton to the west, have become part of the general urban sprawl around Lake Ontario. To

Table 2.2 Population Change in the Toronto Census Metropolitan Area (CMA), 1976–81

	Change by locality 1976–81							
	CMA		City		Suburbs		Fringe	
	n	%	n	%	n	%	n	%
Population 000s	2998.9	7.0	599.2	−5.4	1538.2	3.2	861.5	26.9
Households 000s	1040.3	14.4	241.3	4.7	535.1	6.2	263.9	34.3
Size of household	284	−6.3	2.41	−9.7	2.84	−6.9	3.23	−5.3
% non-family households	75.7	−4.2	46.1	12.7	24.1	20.5	14.4	15.2
Family size	3.2	−3.4	3.06	−3.8	3.14	−3.6	3.37	−4.3

Source: Simmons (1984: 26)

the north the receipt of out-migrants has been pronounced in small towns, such as Markham, Newmarket and Vaughan, again all within 30 miles (c. 50 km) of downtown Toronto.

Moving further north beyond the increasingly intensive agricultural production of this fringe belt the pattern of growth is more irregular. Growth has tended to be concentrated more in the small towns and villages, such as Bradford (45.1 per cent), Beeton (24 per cent) and Alliston (13.4 per cent), rather than the smaller farming communities. This partly reflects the recent location of manufacturing activity on 'greenfield' sites near these small towns, whilst the influence of Highway 400 from York to Barrie and Orillia has also been important in influencing growth as far as the north-western shores of Lake Simcoe. Further north, holiday developments and retirement homes have swelled numbers in some communities near Georgian Bay, e.g. in Port McNicholl.

Dahms (1977; 1980) showed that for Wellington County, on the western periphery of Toronto's urban field, part of the spatial organisation of small communities still exists largely independent of the core city. Several small settlements in the area had not only gained population in the 1960s and 1970s, but they had also acquired new functions in the form of specialised retailing and services plus associated employment. Access to such functions proved as convenient as access to work or shopping for residents of several such communities, and hence these smaller centres tended to subsume some of the roles of the higher order centres. In particular, he noted the establishment of a new range of businesses, often of a specialised nature or associated with regional and national franchises. The employment afforded by such establishments meant that small towns of about 1,000 people could often provide close to 300 jobs and encourage daily in-commuting to work (Hodge and Qadeer, 1983: 91–2).

In summarising the social changes occurring on the fringes of Canadian cities, Gertler and Crowley (1982: 282) recognised five different groups of people who now inhabit these fringes. In addition to the advancing

Table 2.3 In-migration to Toronto and vicinity, 1976–81 (for settlements of over 5,000 inhabitants)

	A	B	C	D
I *Increase in population > 20 per cent*				
Ajax	68.3	84.2	9.5	55.8
Whitby	66.9	88.4	4.9	52.3
Milton	63.1	84.9	6.1	49.0
Vaughan	54.6	87.6	7.6	29.5
Brampton	53.0	88.7	11.3	45.3
Pickering	51.5	86.1	5.7	59.4
Mississauga	42.2	67.0	16.4	65.3
Markham	39.2	74.6	13.3	63.6
Newmarket	35.4	89.5	4.0	80.6
Bradford	31.0	85.4	10.7	73.0
II *Increase in population of 15–19.9 per cent*				
Innisfil	105.9	88.9	3.3	54.8
East Gwillimbury	105.7	89.4	5.2	52.5
III *Increase in population of 10–14.9 per cent*				
Tecumseth	215.8	91.9	4.2	16.1
Vespra	101.3	88.2	5.3	45.8
Oro	88.6	93.7	2.6	63.6
Scugog	65.5	90.9	4.8	67.4
Burlington	53.8	73.8	8.2	79.0
Caledon	50.0	86.4	6.3	73.3
Oshawa	34.3	82.9	9.9	84.0
Aurora	30.2	85.4	8.7	86.6
Barrie	20.3	85.4	5.6	92.4
Scarborough	0.4	58.3	27.4	99.7
Orangeville	*†	88.8	4.7	107.1
IV *Increase in population of 5–9.9 per cent*				
Tiny	158.5	93.9	1.9	49.7
Brock	112.8	92.4	1.8	80.0
Whitchurch-Stouffville	86.2	89.1	4.2	83.7
Orillia Township	56.5	95.6	2.2	33.2
Oakville	50.4	60.3	14.4	81.6
Georgina	32.3	92.1	2.5	90.0
Richmond Hill	27.6	81.8	7.9	90.5
Collingwood	20.5	91.0	3.9	91.7
King	6.0	84.7	6.8	98.1
V *Increase in population of 0–4.9 per cent*				
North York	157.6	50.2	34.3	97.3
Etobicoke	94.1	61.1	24.3	95.8
Uxbridge	17.4	90.9	4.8	98.6
Midland	*	88.1	5.9	127.9
Newcastle	*	87.8	5.7	106.5
Halton Hills	*	74.8	9.0	126.7

continued overleaf

Table 2.3 contd

	A	B	C	D
VI *Decrease in population*				
Tay		95.3	0	79.9
Penetanguishene		94.8	1.7	87.7
Essa	*	87.7	1.9	108.5
Orillia City	*	84.4	8.3	151.8
York		53.7	32.8	99.0
East York	53.1	34.5	55.3	
Toronto	*	43.8	36.9	265.1

* = Net out-migration
† = Boundary changes affect data
A = Net internal in-migration as a percentage of population increase
B = Proportion of in-migration involving migrants from same province
C = Proportion of in-migration involving migrants from outside Canada
D = Out-migration as a proportion of in-migration (excludes movements to
    outside Canada)

Source: Census of Canada 1981 (Catalogue 95-945, Vol. 3, Census Sub-
divisions of 5,000 population and over: selected social and economic
characteristics)

tide of *ex-urbanites* who still work in the city and retain their social ties
with the city, they distinguished *part-time dwellers* such as urban-based
owners of cottages, second homes and hobby farms; *part-time farmers*
who live on a farm, operate it but have a job in the city; *full-time
farmers* or those continuing in other forms of full-time primary activity
in the urban fringe; and other *ruralities* who are the remnants of traditional rural society, such as farm labourers and those involved in the
provision of services in a network of hamlets and villages. Thus,
urbanisation was seen as bringing new groups of people into 'rural' locations, transforming not only the appearance of the location, e.g. through
new housing estates, retirement homes and smallholdings, but also
reshaping the community via the formation of multiple social groupings.

The growing numerical superiority of many of the ex-urbanites and
people retaining close links with the large cities has fundamentally altered
the character of communities in the rural-urban fringes. Indeed, given the
dominance and degree of spread of the urban influence over so much of
the countryside of the Developed World, the extent to which clearly
recognisable 'rural' communities have persisted must be called into question. To understand fully the extent of social change in the rural-urban
fringe and beyond, and the degree to which urbanisation has transformed
rural communities in general, it is still useful to return to a traditional
concept, the *rural-urban continuum* – using this as a 'point of departure'
for further study.

2.3 The rural-urban continuum

When examining rural society, its composition and its changing nature, one of the most common starting points has been the *rural-urban continuum*, a concept expressed in a variety of ways in sociological and geographical literature. Essentially, it represents a polar typological approach, and so is a simplified model of the social and cultural system under examination. In its original form the typology merely distinguished the extremes, but more recent interpretations have emphasised the transformation which occurs from one pole to the other. The concept of the continuum has been used by sociologists to distinguish a rural culture and society from an urban one, but often by breaking it down into a simple dichotomy.

One of the earliest accounts of the clear distinction between the two poles appears in Ferdinand Tonnies' work, *Gemeinschaft und gessellschaft*, written in 1887 (Jobes, 1987; Loomis, 1957). His two poles were termed the *Gemeinschaft* (association; community; held in common) and the *Gessellschaft* (society; private). In his description, the rural pole was typified by dominant social relationships based on kinship, locality and neighbourliness, fellowship, a sharing of respon-sibilities and a furthering of mutual good through familiarity and understanding. This was a society in which the exercise and consensus of natural wills and sentiments was expressed in common evaluations, assessments and decisions. There were common goods and common evils, common friends and common enemies. Within this rural society, individuals' highest priorities were attached to family life and the life of the community itself. The community was dominated by the family in the form of extended kinship groups and the interaction of such groups through village organisations, customs and mores, and religious affilia-tion and worship. In such a society, everyone's role was fully integrated in the local system, with the status of each individual being clearly ascribed. It was a very homogeneous and cohesive society in which the characteristic form of wealth was land.

Similar descriptions have accompanied more recent formulations of the polar typology. Examples include work by Max Weber (1922) and Robert Redfield (1947). The former contrasted the traditional with the rational whilst the latter distinguished between folk society and urban society. Redfield (1941) argued that the two opposites were in fact polarisations of a successive pattern of change through which the folk society became urbanised. Support for such a view is provided from work on American cities by Queen and Carpenter (1953), though Milner (1952) pointed out that Redfield's argument was not amenable to quan-titative evaluation.

The opposite set of characteristics were set out in Louis Wirth's seminal paper, 'Urbanism as a way of life', in which he stated,

The bonds of kinship, of neighbourliness, and the sentiments arising out of living together for generations under a common folk tradition are likely to be absent, or at best, relatively weak in an aggregate, the members of which have such diverse origins and backgrounds. Under such circumstances competition and formal control mechanisms furnish the substitutes for the bonds of solidarity that are relied upon to hold a folk society together Wirth (1938: 11).

Wirth felt that urban relationships were impersonal, superficial and transitory. These social relations alienated individuals from their folk or rural backgrounds, destroying the sense of belonging to an integrated community and creating a state of 'anomie' or a sense of being lost in 'the lonely crowd' (Fischer, 1973). Other urban characteristics were the growth of competition, self-aggrandisement and mutual exploitation which, together with the diversification associated with high density living, fostered contrasts of wealth and poverty. Without formalised controls he felt such societies would be unable to maintain any form of social order. A significant contrast with rural societies was the way in which the urban dweller could achieve a particular social status as a result of evolving relationships with others in urban society.

Wirth's contrasts between urban and rural society have given rise to the use of the term *rural-urban dichotomy* to distinguish clearly between two ideal constructs of the social situation. However, the ideas embodied in Wirth's depiction of urban life have stimulated a wealth of discussion and critical review. One of the most coherent critiques has come from Gans (1962; 1968: 98–9) who made three fundamental criticisms:

(a) The conclusions are derived from a study of the inner city and may not apply to the rest of the urban area.
(b) The social relationships are attributed to number, density and heterogeneity, but these may or may not produce the social consequences he proposes.
(c) Even if a causal relationship is demonstrated between the three variables and social consequences, other urban residents are isolated from these consequences by social structures and cultural patterns brought to the city from elsewhere or else developed within the city itself.

Wirth's views seem to imply that cities exist as independent variables outside 'ameliorating' influences of particular social, cultural and historical orders. Yet, work on the pre-industrial city and cities outside the Developed World suggest this is not the case, as the transformation of social values by urban living is not universal nor has urban life always had the disruptive effects described by Wirth (Alam and Pokshishevsky, 1976; Cox, 1965; Sjoberg, 1960). Both criticism and development of the continuum/dichotomy have prompted investigation of particular areas and groups of people, within both urban and rural situations, that have not conformed to the characteristics prescribed in the theoretical typologies.

An important weakness of the polar typology is its implicit assumption of the stability and universality of the two poles. However, if closer examination is made of the rural pole then its dynamic nature over time becomes apparent. Even for the period when Tonnies was describing his archetypal concept of rurality, the influences of urbanisation and industrialisation had effected major changes throughout the countryside, making significant alterations to rural society. So there may well be an over-romanticised view of rural society being presented in his work.

This tendency to look backwards towards an idyllic countryside, often associated with the author's childhood or youth, is repeated again and again in a wide range of literature. For example, in the 1930s Leavis and Thompson (1932) were claiming that 'the organic community of old England has disappeared and the change is very recent indeed.' Yet, just two decades earlier, George Sturt (1912) pronounced that 'rural England is dying out now,' a death attributed to the effects of parliamentary enclosure and suburbanisation. Similarly, in Hardy's novels of the 1870s and 1880s, often set in the 1830s, and in George Eliot's *Mill on the Floss* and *Felix Holt*, published in the 1860s, the 'demise' of rural communities is portrayed. The process can be continued by considering Thomas Bewick's *Memoir*, looking at changes in rural communities since his own village boyhood of the 1770s. All these authors are chronicling the rural changes wrought by a series of changes that significantly altered the nature of village life in Britain and which were generally repeated in many other parts of the Developed World.

For Britain, the elements of change were the enclosure movement, extending back before the parliamentary enclosures of the late 18th and early 19th centuries; the use of steam power which promoted industrial development in the countryside; the coming of the railways; large-scale factory development which was part of the early stages of large-scale depopulation of the countryside; and the coming of the motor car (see Lee, 1959; Thompson, 1959; Whitlock, 1988). These factors can be compared with those cited previously in Section 2.1.

2.4 Metropolitan villages and urban villages

2.4.1 *Evidence from studies in London's commuter belt*

In the 1960s the concept of a rural-urban continuum was considered at length by Pahl (1965a; 1965b; 1967; 1968) and Frankenberg (1966a) amongst others. More recently, other approaches to the study of rural society have relegated its importance (e.g. Newby, 1980b). Its simplicity and misleading nature is asserted by Mitchell (1973) whilst Pahl has sought to redefine the nature of the continuum, introducing greater complexity. He has been critical of its simplicity which has tended to give rise to broad and inaccurate generalisations, especially as the description of the rural pole has too often been held to be universally valid. He has argued that, by implying the transformation from rural to urban is linear

Plate 3 Urbanisation of the Countryside – 1. Another commuter village? A new housing development in Pencaitland, 15 miles (24 km) south-east of Edinburgh, a village designated for 'some limited expansion'

and continuous, this has led to the variety of changes between the two poles being ignored.

Much evidence has been found to show that the social organisation of the inhabitants of some central areas of certain cities was very different from the typical urban way of life as distinguished by Wirth, with its emphasis upon social relationships that were impersonal, superficial, transitory and segmented. Pahl (1968) used the term *urban villages* to describe the characteristics of London's East End in Victorian times. He also referred to similar examples in Boston, Delhi and Mexico City when portraying city dwellers who live in the city but are not of the city.

He cited a similar discrepancy for people living in rural areas but who are really part of the urban system. The term used in this case was *metropolitan villages*, which have a rural setting, surrounded by agricultural land, but which are inhabited by numbers of middle-class, city-born commuters and hence the other commonly used term, *commuter villages* (Plate 3). The presence of this element brings about an important change in the social composition of what previously had been settlements dominated by the families of farmers and farm labourers.

Nevertheless, the presence of urban villagers and commuter villagers does not entirely negate Wirth's ideas of urbanism as a way of life. The development of commuter villages can be catered for in his view of the way in which the city would extend its dominance over its hinterland

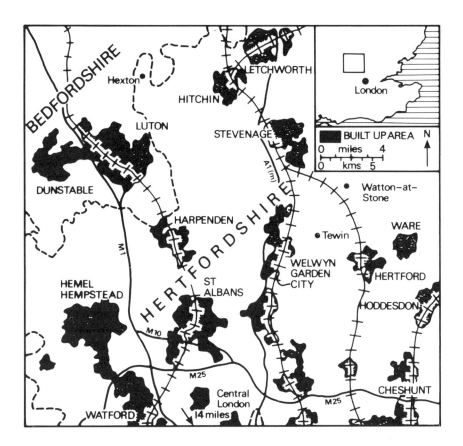

Figure 2.4 Pahl's study area in Hertfordshire

whilst the urban villages can be regarded, perhaps, as one of the 'mosaic of social worlds' developing in the city and in which groups of similar class, race and status will congregate (Wirth, 1938). What these variations do, though, is to make Wirth's concepts less useful. His 'urbanism' merely becomes an ideal type from which urban villages can be seen to diverge, and it must also be seen as internally evaluative, emphasising certain aspects of urbanism whilst ignoring others (Saunders, 1986: 106–7).

Pahl's principal study of metropolitan villages was for three settlements in Herefordshire, 25 miles (40 km) north of central London (Figure 2.4). He identified numerous distinctive groups of people living in what he termed the 'dispersed city'. Middle-class inhabitants with urban life-styles and points of view were dominating these three villages, both in terms of numbers and in their control of social life. In all, six different social groups were recognised (Table 2.4), the largest being the *salariat*. These were the middle-class commuters, often professional people who were of

Table 2.4 Inhabitants of the metropolitan village (after Pahl)

1. Large property owners
2. *Salariat* (business/professional people) – living in a 'village in the mind'
3. Retired urban workers (with capital)
4. Urban workers (limited capital; cheap housing)
5. Rural working-class commuters
6. Traditional ruralites

Source: Pahl (1965a)

the city but not resident in it. They had the desire to live away from some of the undesirable aspects of urban life and the money to be able to realise this wish. Their freedom of choice was expressed in the desire to live in pleasant surroundings in the countryside. They sought a rural home and different life-style from that of the town-dwellers, but, in effect, were only living in a *village in the mind*. Their way of life differed very little from that of people living in a suburb, but their location gave them greater ease of access to the countryside, and the settlement itself was a more distant and possibly smaller entity than a suburb.

Another middle-class group consisted of retired former urban dwellers with sufficient capital to afford to live beyond the confines of the built-up zone. These people can be contrasted with urban workers with limited capital and income, forced to live away from the city to take advantage of cheap housing. This housing could take the form of new housing estates with semi-detached dwellings or government-owned (council) property. The chief social division in the communities was between this group and the middle-class newcomers.

The life of the settlements tended to be dominated by the middle-class who ran the local clubs and societies such as the Women's Institute, as well as being prominent in the life of the church. They were more mobile and more easily able to retain links with the urban area through maintaining a widely dispersed social network. In contrast, the working-class group relied more heavily upon public transport, especially those wives without paid employment. They had a social network focused more closely upon the 'village', although strong links were often retained with family in London. This group tended to be active only in the local soccer and gardening/horticultural clubs.

Three smaller groups represented the remnants of a former pattern of society. Pahl termed these groups the large property owners, traditional ruralites and rural working-class commuters. The property owners represented either the remnants of the former top echelons of rural society in the form of the landed gentry or newcomers with sufficient capital to purchase large estates. In some cases this group still retained its importance within village affairs, acting as leaders of local interest groups. Traditional ruralites were represented by the few surviving farm labourers and local tradesmen. Some of this group had ceased to work in the village or its vicinity and had become commuters with a place in the community similar to that of the lower-class urban workers. The main differences between life in these metropolitan villages and in the city arose because of the smaller size and greater social heterogeneity of the villages, and the development of what Pahl referred to as a deviant form of urbanism. This he associated with the salariat or *cosmopolites* who developed a distinctive pattern of social relationships. These relationships primarily reflected a mixing only with people of their own class. This meant they regarded the village as a friendly place and possessing many of the characteristics they were seeking in their 'village in the mind'. They sought to prevent changes to this situation by stopping further influxes of newcomers which would alter the village's existing character, even though their own presence was relatively recent and had drastically changed the nature of the local community.

The idea of the 'village-in-the-mind' has been revealed to encompass a range of different attractions, generally reflecting a belief in the countryside as possessing a superior natural environment and better housing, and yet still permitting relatively easy access to employment and services (e.g. Blackwood and Carpenter, 1978; Hareven, 1977; Stevens, 1980).

2.4.2 Evidence from other commuter belts

In effect, Pahl was discovering both gemeinschaft and gessellschaft relationships in the same group of people in one single village. So he argued it was difficult to recognise a rural-urban continuum with respect to geographic, demographic or economic indices. One suggested alternative was to focus upon changing social relationships, which he attempted to do when asking the question, 'is the rural-urban continuum a typology or a process?' To answer this he examined empirical material gathered outside Europe and North America which suggested there were fundamental discontinuities between urban and rural life. For example, he referred to studies for India (Bailey, 1957; Majumdar, 1958), Turkey (Striling, 1965), Sudan (Barclay, 1964) and the Central African Copper Belt (Watson, 1958). For the latter, it was apparent in the late 1950s that urban values were beginning to permeate the countryside, though it was still possible to measure differences between urban and rural in the hinterland of the main towns.

In a similar context, a more wide-ranging change was revealed by Epstein's (1962) study of Dalena, a village in southern India. Here the social system was being modified by the development of commuting from the village to a nearby urban factory. An example of the modification was the replacement of the joint or extended family as the mainspring of community life by the nuclear family. However, Dalena still maintained its social identity despite this alteration of relationships within the village and the changed economic role of the factory workers. This was just one of many complex illustrations used by Pahl to show that villages in the Third World also change once 'the economic frontier has hit them'. But, he argued that economic development does not always lead to social change and can often produce unexpected results. So there are examples of reactions against urban work and life, strengthening the traditional values of village life. With urbanisation there are observable changes away from what might be regarded as the 'traditional rural pole', but the actual type of change can vary considerably.

A conclusion from this work was that the isolation of rural communities as separate systems for the purposes of academic study had become an increasingly unreal exercise. Whilst it was possible to note changes away from a rural pole, generalisations were extremely dangerous given the variety of change that could occur. There did not appear to be universal evidence of a rural-urban continuum, and, as a classificatory device, it was of little value. Instead, a more appropriate distinction may be between *local and national values* in the world. Local values are characteristic of small-scale society whilst national values are those of the large-scale society. The former society nests within the latter. People who live in the countryside but commute to work in towns are members of the national system not the local. They impose national values upon village organisations. The chief conflicts between the two sets of values tend to occur in the urban villages and the metropolitan villages.

Pahl felt that if use was to be made of the rural-urban continuum, this concept must be viewed as a process rather than a typology. With change through urbanisation occurring throughout the world, it was unrealistic to think of distinct peasant, folk, tribal or rural societies in isolation. If there was a true continuum from rural to urban it could be identified in temporal terms not spatial terms. So he argued that rural sociologists must concentrate on the social processes which lead to differentiation between and within societies. The implication of this is that individual geographic, economic and demographic indices are not so important as analysis of process.

In social terms, Pahl's (1965a; 1965b) work on commuter villages on the outer fringes of London portrayed the demise of existing geographical and social hierarchies through selective in-migration. The arrival of mobile, middle-class commuters from a different social and economic world to the established population brought a new class structure reflected in housing segregation. Typically, large blocks of one class or price of housing would be established, paralleling suburban development. This was especially so in settlements also in receipt of working-class in-migrants whose mobility was generally more constrained and whose presence in commuter villages tended to reflect the availability of local authority housing stock.

Within the British context many of these points have been supported by studies in a range of 'urbanising' communities (e.g. Radford, 1970; Ambrose, 1974; Connell, 1978; Lewis, 1967; Pacione, 1980; Robin, 1980). Pacione's study, in particular, confirms the extent to which social segregation can occur between different social groups as recognised by the divide between private and local authority households. His study of Milton of Campsie, to the north-east of Glasgow, highlighted the same type of divisions in the life-style of the middle and lower classes found in Pahl's studies in the mid-1960s. Major differences in mobility were revealed, with the middle class being more open in their social activities, both socially and spatially.

Attitudes of the two groups to their 'village' also differed. The private householders emphasised notions of 'the peace and quiet of rural life' whilst those living in local authority housing added the value of a 'close-knit community', reflecting the way in which they looked upon the village as a place of residence, though lacking social and recreational opportunities. For the middle class, the attractions of 'living in the countryside' outweighed the disadvantages of costs and time involved in commuting to work in the city. In fact, studies of commuting have suggested that the cost of travel is rarely regarded as important (e.g. O'Farrell and Markham, 1975). For example, Richardson (1971) suggested that commuting costs are subordinate to choice of house, area and environmental preferences as the principal determinants of residential location. This contradicts earlier work suggesting costs of travel may be 'traded off' against housing costs (e.g. Alonso, 1960; Hoover, 1968; Wingo, 1961). Richardson's ideas seem to be especially applicable in North America and Australasia where petrol costs have been cheaper than in Europe, and where greater availability of land and fewer planning controls on urban sprawl have encouraged greater separation of residence from work-place.

Renewed interest in the social composition of settlements in the rural-urban fringe has occurred in the 1980s, indicating the continuation of trends first recognised two or three decades earlier. For example, Harper (1987; Harper and Donnelly, 1987), on the basis of work in two English study areas, identifies three broad groupings of settlement at the rural-urban interface: the small agricultural settlement, the urbanised commuter village and settlements in the process of consolidation between the two. This division, based largely on population size, social characteristics and infrastructure, was carried further in terms of distinguishing the diverse groups of people living in these settlements. Nine broad groupings emerged from this analysis, with reference to socio-economic characteristics and ways of life, and they form a useful comparison with those groups distinguished by Pahl in his English commuter villages in the mid-1960s (see Table 2.5).

Whilst there are undoubted similarities between the effects of rural suburbanisation experienced in Britain and in other parts of the Developed World, there are certain differences between countries, as revealed by numerous studies for parts of Western Europe (e.g. Andrian, 1981; Berger et al., 1980; Fricke, 1971; Geoffroy, 1982; Kunst, 1985; Mougenot, 1982; Taffin, 1985; Wild, 1983a). The work of Herden (1983) and Fricke (1976) in the Rhine–Neckar portion of the Rhine Rift Valley has shown the growing importance of single-family housing as a

Table 2.5 Inhabitants of the metropolitan village (after Harper)

1. Tenants – living in tied or privately rented property
2. Local authority renters
3. Locals – resident all their lives (e.g. farmers)
4. Principal relocating group – young families, social class I to III, resident < 15 years
5. Spiralists – migrate at frequent intervals for purposes of employment
6. Semi-relocators – migrate to purchase cheap property, but retain close ties with place of origin
7. Wealthy late age – social class I or II, able to purchase large country properties on or near retirement
8. Mobile retired – social class II or III, migrated on retirement
9. Long-term residents – arrived in late 1950s/early 1960s

Source: Harper (1987)

factor in rural suburbanisation and also differences between communities at varying distances from major cities. Wild and Jones (1988) extend this work in their cross-border comparison of the hinterlands of Freiburg and Colmar in the Rhine Rift Valley (Plate 4). They showed that there were important differences between French and West German villages in terms of community level experiences of rural suburbanisation. These differences related to variations in village configuration and density, degree of modification and intensification of the village cores, diversity and quality of new residential construction, and the integration of new and old quarters. Primarily, it is differences in national planning legislation and policies that have caused cross-border differences, but there is much room for similar comparisons to reveal inter-regional and international variation in the suburbanisation process.

2.5 Using the continuum

Demonstrations of the complexity of the rural-urban continuum provide grounds for thinking of the continuum not as uni-dimensional, but in terms of several dimensions. It could be argued that a series of continua exist, depending upon which sociological, demographic, cultural, political and economic variables are being considered. Certain elements will not be present in any given situation and the variables will not necessarily change along the continuum in a regular and predictable fashion, giving rise to a variety of spatial variations in the patterns.

This type of complexity was conceptualised by Burie (1967) in terms of physical, cultural and structural dimensions. This idea is portrayed in Figure 2.5 in which any group of communities can be placed with regard to one another in a three-dimensional construct. Another possible approach using the continuum is that suggested by Lockwood (1964) who felt that change along the continuum could be viewed in terms of

Plate 4 Urbanisation of the Countryside – 2. Many Alsation villages, such as Ricquewihr, retain their tightly clustered form and picturesque mediaeval character. Their economy is still dominated by the production of wine, though in some cases commuter 'appendages' are being built for workers in Strasbourg, Colmar or Mulhouse

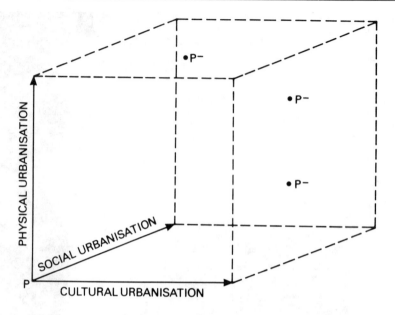

Figure 2.5 The process of urbanisation for an imaginary community (P) as represented in three-dimensional space. Over time P can change to any of the P⁻ points. (based on Burie, 1967)

how 'core' or 'dominant' institutions changed, and measuring this against the fixed positions of the two poles. In effect, the complexity of such an operation has prevented its use. Similarly for Burie's work, few empirical studies using this concept have been forthcoming. Instead, to realise the complexity of the multi-dimensional continuum via its direct as a practical concept, the example of Frankenberg's (1966a; 1966b) work can be given.

Frankenberg used the continuum as a framework in which to explore and interpret the nature and variations of a defined social system. This took the poles of the continuum as ideals against which actual communities should be compared and ordered. He felt that it was this comparative role that was the continuum's chief value, i.e. the continuous scale from urban to rural provided the most satisfying device for classifying communities according to their urban or rural tendencies. In practice, this comparison was effected using economic organisation and technology as key variables, resurrecting the familiar dichotomies from classical sociology. The characteristics recognised as forming the urban and rural poles in Frankenberg's study are indicated in Table 2.6.

Frankenberg made use of a series of studies of rural communities in Britain carried out mainly in the 1940s and 1950s. Therefore, it could be argued that subsequent changes in these communities have eroded some of the characteristics he regarded as typical of 'rural'. His examples portraying this pole refer to a series of well-known *community studies*

Table 2.6 Characteristics of 'rural' and 'urban' societies

Rural	Urban (less rural)
Community	Association
Social fields involving few	Social fields involving many
Multiple role relationships	Overlapping role relationships
(different roles to the same person)	(different roles to different persons)
Simple economy	Diverse economy
Little division of labour	Extreme differentiation/specialisation
(uniformity of individuals)	(diversity/complementarity)
Ascribed status	Achieved status
– total status	– partial status
education according to status	status derived from education
Role embracement	Role commitment
Close-knit networks	Loose-knit networks
Locals	Cosmopolites
Economic class is only one division	Economic class is main division
Conjunction	Segregation
Integration re-work	Alienation re-work

Source: Frankenberg (1966a)

(see Douran-Droughin et al., 1981; 1982; 1985; Jones, 1973). The earliest example employed was from field survey in the 1930s by Arensberg and Kimball (1940), who studied Ryamona and Lough in County Clare, Eire. This was an area where the economy was almost entirely agricultural, the farmers used virtually no hired labour and only simple cultivation techniques. This was a community well removed from most urban influences as was Llanfihangel, mid-Wales, for which Frankenberg used Rees's (1950) work carried out in the 1940s. Whilst Llanfihangel was 'more integrated into the national economy' (Frankenberg, 1966a: 13), it was still a community dominated by family farming and with a social structure little changed from the 19th century. In contrast, Gosforth, West Cumberland, had wage-earners in industry, an atomic power station, businessmen and landowners as well as hill farmers operating in a similar fashion to the Welsh example. Williams' (1956) study of this community concentrated upon its traditional elements, but the intrusion of urban influence was more pervasive than in the previous two examples and enabled Frankenberg to use it to illustrate certain elements of change along the continuum (see also Williams, 1963).

Further integration into the urban system was introduced by reference to Frankenberg's (1957) own study of Glynceiriog, north-east Wales, and the study of a mining village in Yorkshire (Dennis et al., 1957). Other studies referred to urban communities, although Frankenberg felt Banbury, Oxfordshire, included all social classes and status groups to be found in Britain in the late 1950s. As such it represented the mid-point of the continuum, the meeting point of urban and rural, containing both

traditional rural characteristics such as agricultural employment and a
new industrial branch plant (Stacey, 1960).

He included elements of the urban village in selected urban studies,
referring to studies of Bethnal Green in London's East End. Bethnal
Green was predominantly a poor, working-class, metropolitan borough
with strong, local kinship ties, loyalties to the neighbourhood and cohesion arising out of the shared poverty and lack of social and geographical
mobility of its residents. Housing and industry were closely intermingled,
accentuating the self-contained 'village' characteristics of the community.
This contrasted sharply with life on urban housing estates. Several
examples were used to portray the classically anomic societies of council
housing estates, e.g. Watling, north-west London (Durant, 1939), and
estates on the outskirts of Liverpool (Lupton and Mitchell, 1954) and
Sheffield (Hodges and Smith, 1954). As in Bethnal Green, these estates
were dominated by working-class manual workers, but these workers
had been allocated their homes rather than selecting them themselves.
Without special commitment to a place, and with commuting to work
over a longer distance the norm, few local ties were established.

Separated from work and from play, from beer and from books, paved with
good intentions, the social life of housing estates seems as far as it is possible to
get from the utopian dreams of a village fellowship made urbane by town
civilization (Frankenberg, 1966a: 201).

It appears from the selected 'community studies' that social class and
life-style characteristics are inadequate on their own to explain ways of
life and patterns of behaviour. In addition, whilst many parts of Britain
seemed rural in visual terms, i.e. when considering dominant land use,
they contained very few social characteristics that pointed to the
existence of a community in the sense of gemeinschaft.

Frankenberg attempted to construct a morphological continuum along
which the communities could be ranged, but with the implication that
one particular community in the continuum would not necessarily
become successively like the more urban communities represented. He
stated, 'I do not visualise the continuum as a neat Euclidean straight line
. . . but rather as a rather messy squiggle, blurred here and sharp there.'
To indicate the nature of this multi-dimensional continuum, he
considered certain key general concepts such as community, role,
network, social conflict, class and status-group (Table 2.6). These
enabled him to distinguish between the rural and the urban or less rural,
and so fit the various chosen examples into position between these two
ideal extremes. The exercise seems to have worked quite well, but there
was no representative of the commuter village amongst his chosen
examples to add additional complications to this many faceted
continuum.

2.6 The changing concept of rurality

2.6.1 Re-evaluating community studies

There are several examples of significant re-evaluations of the community studies of the 1950s and 1960s (e.g. Harper, 1989). These reappraisals have arisen partly because of the problem of the formulation of different conceptions of *community*. Three stand out:

(a) community as locality, i.e. merely a geographical expression;
(b) community as a local social system (e.g. Stacey, 1969);
(c) community as communion, i.e. a sense of common identity which may or may not have a specifically local basis (e.g. Bell and Newby, 1971; 1976).

'Community', like 'rural', has been given numerous definitions. Hillery's (1955) study of its use revealed 94 specific definitions, though he concluded that there was basic agreement that a community 'consists of persons in social interaction within a geographical area and having one or more common ties.' This can be compared with Thorns' (1976: 15) argument that 'community' has 'one unifying theme, that of a cohesive group of people, held together by different things which they share, for example, territory, ideas, work, skills.' It must also be noted that the sociological connotations associated with the term community are often replaced by a looser usage of the word in making it synonymous with 'settlement' or 'population' (Stacey, 1969; Bell and Newby, 1971). Generally, though, a community can be regarded as a local social system, with variation that in some cases has been labelled as 'rural' and in others 'urban'. However, the extent to which a highly integrated rural social system or community has now been altered radically by increased mobility and the influence of national rather than local values means that the concept of a community has also changed. It may now be more often applied to shared interests rather than to a recognisable and cohesive social system (Scherer, 1972).

A weakness of the community studies of the 1940s and 1950s was to describe communities as objects of study so as to provide a catalogue of 'folk life'. According to Newby (1986), it is more fruitful to regard community studies as a method of obtaining data on those processes of social change for which the locality is the appropriate level of analysis. In other words, it is the snapshot or 'little picture' of the study of an individual community versus the 'big picture', with the former characterised by being 'impressionistic, overly descriptive and alarmingly non-cumulative' (Newby, 1986: 211).

Instead, Newby's work, grounded in studies of agrarian class rejections, has demonstrated a different approach to rural society. This has focused upon property rather than occupation, and has stressed the importance of farming's development as a business rather than 'a way of life.' This approach has become essentially a sociology of agriculture and has rather lacked any explicit focus upon a specific locality (Bradley and Lowe, 1984). Meanwhile, Urry (1984) has argued that economic restructuring in recent years in the manufacturing and service sectors has influenced local systems of social stratification in rural areas over and beyond those based solely on differences between a new middle-class and an old established, agrarian, rural class structure. This is a more holistic view of the community or locality (e.g. Urry, 1986), though it remains to be seen whether 'locality studies' manage to avoid the limitations of the earlier community studies (Day et al., 1989; O'Neill, 1989).

Too many community studies viewed rural life in over-romanticised and nostalgic terms with an inherent anti-urban bias. Yet, the rosy view of the traditional rural community has been dispelled by studies such as that by Williams (1973), revealing poverty and exploitation for the mass of the population, the successors of the rural peasantry. Other studies, such as Newby's (1977) for East Anglia, reveal a continuation of rural poverty and class division in rural areas today. Such divisions are often referred to in terms of differences between long-term rural residents and newcomers from urban areas, but this is an over-simplification ignoring differentiation within both groups (e.g. Berger, 1960).

A common view in social studies in the 1980s has been that the social relationships of individuals are much less governed by place, proximity and tradition than was formerly the case. For example, mechanisation in agriculture has reduced communal activity whilst the growth of new functions within the countryside has reduced social integration and homogeneity. In effect, community can now be conceived of distinct from a specific geographical context. Yet, it has also become fashionable to recognise the existence of local social systems and social processes, if only as local deviations of processes operating on a larger scale (Dickens et al., 1985: 18).

Cloke and Thrift (1987) have argued that some of the social changes described with respect to metropolitan villages in studies in the 1960s and 1970s were over-generalised. In particular, they view the emphasis upon 'newcomers versus locals' as hiding a range of class 'fractions' not fully appreciated despite some of the breakdowns of the two groups into smaller sub-groups. They follow Wright's (1985) work based on considerations of capital, organisation ('what people do') and skills or credentials (Table 2.7). This gave a 12-fold division in which a *service class* within the middle class was highlighted as the major motive force for change in the countryside. This group, consisting of people working in managerial capacities, has exerted a major economic influence and, via their social and cultural hegemony, also obtained political power to alter social and economic structures.

2.6.2 Rural as an 'incidental' variable

A fundamental aspect of the concept of a rural-urban continuum/

Table 2.7 Typology of class locations in capitalist society

Organisation assets	Assets in the means of production		Skill/credential assets	
	Owners of means of production	Non-owners (wage labourers)		
Owns sufficient capital to hire workers and not work	Bourgeoisie	Expert managers	Semi-credentialled managers	Uncredentialled managers
Owns sufficient capital to hire workers but must work	Small employers	Expert supervisors	Semi-credentialled supervisors	Uncredentialled supervisors
Owns sufficient capital to work for self but not to hire workers	Petty bourgeoisie	Experts non-managers	Semi-credentialled workers	Proletarians

Source: Wright (1985: 88)

dichotomy is that it has been regarded as an appropriate analytic framework for examining pre-industrial societies. In these societies, economic activity was associated extremely closely with the two socio-cultural systems at either pole of the dichotomy. Given a clear spatial distinction between town and country in this pre-industrial period, two separate 'ways of life' existed, though with a certain amount of interdependence between them. As demonstrated in this chapter, this separation may then be regarded as being broken down in a variety of ways during the course of large-scale industrialisation. This process, affecting both poles of the dichotomy, changed both the urban and rural ways of life, restructuring social and economic arrangements and blurring the distinction between urban and rural.

The exact nature of this blurring, and of how a workable distinction between urban and rural may be developed, has certainly proved problematic. However, it is a distinction which seems to have been of less concern to those interested in urban studies, perhaps because they have been able to tackle a more lengthy and pressing agenda of issues clearly related to the growth of cities and to urbanisation. A common attitude amongst urban geographers is expressed by Dunleavy (1981: 2–3):

The contemporary problem in defining urban studies is not to distinguish 'urban' from 'rural' areas of the country, for no such geographical distinction is possible. Nor can this problem be reinterpreted in terms of an 'urban/rural continuum'. Instead the key problem is to separate out a small set of 'urban' concerns from the great mass of social activities which are neither 'urban' nor 'rural'.

This relegates the importance of a distinction between rural and urban to a minor role and gives much greater weight to specific processes within the industrialisation and urbanisation of society, e.g. production, exchange, consumption, socio-cultural life, state intervention and external affairs. It also eliminates the notion of distinctive urban and rural socio-cultural systems, arguing that these have become part of a set of general social processes contributing to a single overall system.

Under this argument, the subject matter of rural studies would be those aspects of the system clearly recognisable as having closer links with the rural than the urban: agriculture and forestry, for example, or planning related to preservation of wilderness and provision of services in small, isolated communities. The argument implies that, in the destruction of the rural-urban dichotomy, the multi-faceted continuum developed is either too complex or irrelevant to the problem of defining rural studies.

Even if it is accepted that the rural-urban continuum is essentially spurious, it must be acknowledged that its two poles, the urban and the rural, are distinctive. They are also concepts that still occupy a central place in western culture. For example, Williams (1973: 297) illustrated the sharp distinction between the commonly held images of urban and rural:

The pull of the country is towards old ways, human ways, natural ways. The pull of the idea of the city is towards progress, modernisation, development.

He argued that, for the majority of people in the Developed World, their everyday lives are built around capitalist materialism with 'modes of detached, separated, external perception and action: modes of using and consuming rather than accepting and enjoying people and things' (Williams, 1973: 298). Hence he regarded rurality as a 'humanity' all but submerged by a capitalist society, but with this rural idyll retaining a powerful hold over workers in the capitalist system. It is a moot point to consider whether such rural idylls are not also present under 'non-capitalist' systems and whether his emphasis upon the mode of production is more of a political statement than a demonstrable causal agent. However, there has been a large body of literature in the 1970s and 1980s, broadly related to or drawing on Marxist theory to analyse society, voicing critiques of capitalism and its role in the urbanisation of society.

Within most of the work by Marxists, though, and also in the work of Durkheim, Engels and Weber, a recurrent theme has been that 'the city in contemporary capitalism [is] not a theoretically specific object of analysis', and so, by implication, neither is the countryside (Saunders,

1986: 16). Despite their different viewpoints and methods of analysis, their common concern was with the development of capitalism and not specifically with urban or rural change which tended to be regarded as secondary influences on the development of fundamental social processes generated within capitalist society.

Whilst for Weber neither the city nor the village represented the basis for human association, Durkheim also relegated the importance of divisions between rural and urban, arguing that localism had been undermined by the occupational and social division of labour (Durkheim, 1933: 27–8).

Similarly, Marx took the view that no aspect of society could be analysed independently of the totality of social relations, but a 'reality' which he held to be vital to social organisation was the division between urban and rural because it was an essential expression of the social division of labour. This urban-rural division had to be investigated with respect to the underlying mode of production which it sustained and which is sustained by it. Moreover, the argument that 'modern history is the urbanisation of the countryside' (Marx, 1964: 77) hints at the centrality of the division of urban and rural in the evolution from a feudal, peasant society to an urban, industrial, capitalist one. Initially, this division can be seen in terms of industrial 'bourgeoisie' versus feudal landowners. However, as agriculture itself became a part of capitalist social relations, the nature of the division was altered. Marxists would argue therefore that whilst a city can be differentiated from the countryside, this division is unimportant in terms of social relationships. Both urban and rural workers are part of the capitalist structure, and it was to the urban workers not the rural that Marx and Engels looked to lead their vaunted struggle for socialism. Only from the late 1960s has Marxist theory been extended and developed to reconsider the 'capitalist city' and its importance in the evolution of capitalism. In this neo-Marxian work, further statements have been made concerning rural-urban relations.

A more recent extension of the view of the unimportance of a rural-urban division to an understanding of society is in the work of Lefebvre (1976; 1977). He argued that modern capitalist production has an urban base (the urban revolution) supporting an urban society in which 'reproduction, the relations of production, not just the means of production, is located not simply in society as a whole but in space as a whole' (Lefebvre, 1976: 83). Whether that is urban space or rural space is largely immaterial. A similar conclusion obtained from a different neo-Marxist perspective appears in Castells' work in the 1970s:

As soon as the urban [or rural?] context is broken down even into such crude categories as social class, age or 'interests', processes which seemed to be peculiar to particular urban [rural?] areas turn out to be determined by other factors (Castells, 1976: 40).

Thus Castells' claim that urbanism constitutes neither a 'real' nor a 'scientific' object of sociological study precludes a role for the countryside too as such an object. Other neo-Marxists, however, reject this view and at least accept the city as a worthy subject for study (e.g. Dear and Scott, 1981; Harvey, 1973; 1982; Roweis and Scott, 1978; Smith, 1984).

The Marxist or structuralist view also sees the separation of work-place and residence as a result of changes in the productive process which led to greater social differentiation of the city. The more affluent moved away from poor inner city areas which had been produced because of insufficient wages paid to manual labourers. So, in this interpretation, the decision to move away from the inner city is related specifically to the social process creating social stratification and great differences in the distribution of wealth in society. This process has continued throughout the 20th century, though with a higher proportion of the population gaining the means to participate in the move to the suburbs and beyond.

The structuralists couch their arguments in terms of controls forming and determining the level of constraint on individual action (e.g. Thorns, 1980; Bassett and Short, 1980). For example, for the United States, Checkoway (1980) showed how in the 1950s large building contractors capitalised on economies of scale to develop large suburban sites whilst government supported this initiative by guaranteeing mortgages and enhancing the profitability of such schemes. With the construction industry and government giving such an impetus to suburban develop-ment, a major push was given to out-migration from the city over and above any general desire for such movement by the populace.

Checkoway (1980) stresses that suburbanisation and distancing of residence from work-place must be seen as a product of decisions and institutional interactions. Indeed, the significance of the role of institu-tions or 'gatekeepers' (e.g. Ambrose and Colenutt, 1975; Palmer, 1955; Pahl, 1975) has increasingly been recognised. For example, in the immediate post-World War Two housing boom in the United States, decisions by builders were vital in furthering suburban development. Checkoway shows how larger builders tended to have a 'suburban orien-tation'. Large builders favoured suburban development as mass produc-tion required 'large, less expensive tracts of land typically found near the city limits or in suburban areas beyond' (Checkoway, 1980: 25). Restric-tions tended to be less in suburban areas, transportation was often good, prospects of space for expansion were attracting retail, manufacturing, wholesale, office and service establishments, and there was support for such developments from federal government programmes. Together these factors helped produce a growth rate for suburbs in America ten times greater than that of central cities in 1950.

Orange County, California is a good example of this growth: a population increase of 65 per cent in the 1940s, but double this rate in the new suburbs outside Los Angeles, e.g. Monterey Park (+ 140 per cent), Arcadia (+ 154 per cent), Manhattan Beach (+ 175 per cent) and Hawthorne Covina (+ 350 per cent). On Long Island this process was represented by Levitt and Sons building 17,000 identical houses for over

70,000 inhabitants (Gans, 1967). other examples of the importance of large builders to this 'drive' towards suburbanisation include Blietz (Chicago), Taylor (Washington DC) and Gellert and Storuson (San Francisco).

Maisel's (1953: 95) description of the builders is apt:

These are the new giants in an industry once populated by pygmies. Here at the very peak of the housebuilding pyramid, are the leaders of construction who are not content merely to build houses. They construct communities.

But this construction was also driven by federal support which gave grants and loans for slum clearance and urban redevelopment, e.g. the 1949 Housing Act. Mortgage insurance authorisations were increased, specifically for 'suburban and outlying areas' in the 1954 Housing Act, and home ownership for a greater proportion of the population was a definite aim. Other federal programmes reinforced this, e.g. the highway programme (Gelfand, 1975: 222–35; Muller, 1976) and tax policies promoting suburban construction (Arnold, 1971).

Whilst the distinction between urban and rural has often been relegated by both Marxists and non-Marxists to a secondary or incidental role in recent studies of social differentiation, location, seen as a highly significant variable in explaining how social phenomena develop, has been especially prominent amongst those working within a realist epistemology (e.g. Sayer, 1984a; 1984b; Urry, 1981; 1985).

Realism is defined by Gregory (1982; 1985) as a philosophy based on identifying causal mechanisms (e.g. how does something happen?) as opposed to empirical regularities (e.g. how widespread is something?). Realists argue that empiricism and positivism erase this distinction whilst realism seeks to reinstate it. It regards the world as differentiated and stratified but consisting of events as well as mechanisms and structures (Sayer, 1985).

The realists' argument is that society only acts in a particular manner when in particular spatially-conjectural relations with other things. Space has no general effects, but a particular spatial context, rural or urban for instance, affects how and whether certain social processes develop. This would seem obvious to most geographers, brought up to acknowledge the distinctiveness of place but familiar with the generalities of central place theory or the core-periphery model which suggests how locational patterns will tend to occur until distortions are promoted by specific spatial contexts. This familiarity has led the neo-Marxist geographers, Harvey and Massey, to champion the importance of space within social theory, but whereas Harvey (1982) seems unable to go beyond a generalised attack on capitalism, Massey (1984: 58) argues that 'the reproduction of social and economic relations and of the social structure takes place over space, and that conditions its nature.' She then attempts to demonstrate how location shapes or affects social developments, and class relations in particular. For example, one area may support a radical branch of agricultural workers' trades unionism while another does not.

Hence, distinctions between places appear as important considerations if we wish to understand social relationships.

This argument lies at the heart of the rationale for the locality studies referred to earlier (Section 2.6.1), but it is not fully accepted in the work of Giddens (1979; 1981; 1984) on the spatial aspect of social organisation. He suggests that our ability to overcome the 'tyranny of distance' means differences between places, especially urban and rural, are far less important today than they were before the development of capitalism in Western Europe from the 18th century onwards:

The old city-countryside relation is replaced by a sprawling expansion of a manufactured or 'created environment' (Giddens, 1984: 184).

As this created environment is everywhere, be it city blocks or enclosed fields, social life has transcended both the temporal and spatial dimensions. This view implies that the nature of a particular locale is relatively unimportant in determining how and what we do there. So this 'treat(s) space as a "backdrop" against which social processes develop' (Saunders, 1986: 285).

* * *

The preceding discussion demonstrates that there are a variety of views of how distinctions between rural and urban might be recognised, and of how important such distinctions might be to the understanding of social evolution and differentiation. One extreme is to regard urban and rural as insignificant categories alongside the over-arching and controlling structures of the capitalist system. Other views place greater stress upon the uniqueness of place or locality as a factor in maintaining social variability and differential development of society. The latter provide greater scope for maintenance of specifically 'rural' or 'urban' studies, though there still seems to be uncertainty about the nature of causal links between the nature of 'rural' society and the character of 'rural' places. Further studies of how society views itself, of its own recognition of labels such as 'urban' and 'rural', may well provide more fruitful avenues of investigation than some of the recently overtly politicised enquiries more concerned with self-justifying '-isms' equally as self-deceiving as approaches they are rejecting.

3

Rural depopulation

3.1 Rural depopulation as a component of urbanisation

Some of the most profound changes in the countryside have been produced by the migration of the rural population to urban areas. This
process has occurred over a period of more than a century in many rural
areas. It has tended to be most pronounced in those areas physically
remote from cities, though not always so. For example, in northern
Norway the population in the three northernmost counties, those most
remote from metropolitan influence, has continued to grow and the
number of communes losing population fell during the 1970s (Almedal,
1983).

In Britain many rural parishes had fewer people in 1901 than in 1801
(Lawton, 1986). One of the principal reasons for this depopulation, both
this century and last, was the decline in agricultural employment accompanied by the concentration of industrial employment in urban areas
(Figure 3.1). Even post-1945 the decline in agricultural employment has
continued to foster *rural out-migration* (Table 3.1). This phenomenon
and others associated with rural depopulation are examined in this
chapter, which, in many ways, complements the discussion of urbanisation of the countryside in the previous chapter.

The widespread loss of population from rural areas throughout the
Developed World is suggested in broad terms in Table 2.1, though the
full extent of depopulation is masked by definitional problems of 'urban'
and 'rural'. Perhaps of more significance as a guide to the extent of
depopulation is data for the decline in the proportion of the labour force
engaged in agriculture (Table 3.1). When the demise of the farm labour
force is considered, it is clear that post-1960 dramatic falls have occurred
throughout the Mediterranean, the Alps (Plate 5), certain parts of Scandinavia and Eire. In the core areas of Western Europe, this movement
from the land came much earlier, e.g. in the United Kingdom it is
estimated that the maximum agricultural labour force was reached in the
1860s (Orwin and Whetham, 1971: 203–39). Yet, as suggested in Table
3.1, there has still been a continued loss of rural population from

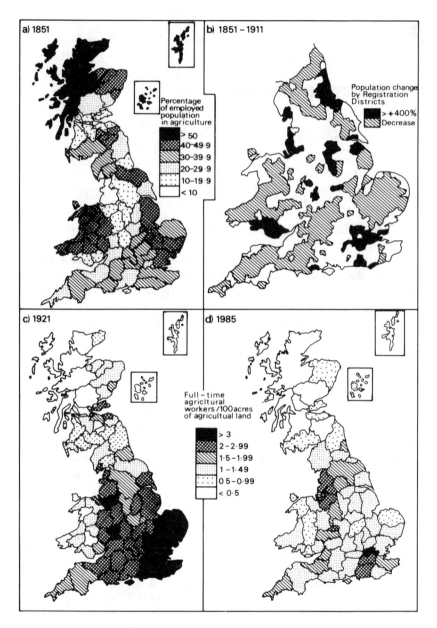

Figure 3.1 Employment in agriculture in Great Britain, (a) Percentage of employed population in agriculture, 1851 (based on Lee, 1986: 31); (b) Population change by Registration Districts, 1851–1911 (England and Wales only) (based on Lawton, 1964); (c) Full-time agricultural workers per 100 acres of agricultural land, 1921 (*Data source:* Agricultural Census); (d) Full-time agricultural workers per 100 acres of agricultural land, 1985 (*Data source:* Agricultural Census)

Table 3.1 The decline in the proportion of the labour force employed in agriculture, 1950–86

	% Working population engaged in agriculture			% Change	% Change
	1950	1970	1986	1950–70	1970–86
Greece	52.6	47.2	29.9	– 5.4	– 17.3
Spain	49.7	29.6	18.0	– 20.1	– 11.6
Eire	48.7	27.5	17.1	– 11.2	– 10.4
Finland	46.0	19.6	9.1	– 26.4	– 10.5
Portugal	44.3	33.0	23.5	– 11.3	– 9.5
Italy	42.8	19.6	12.4	– 23.2	– 7.2
Iceland	40.0	17.2	7.6	– 22.8	– 9.6
Austria	32.3	14.8	6.5	– 17.5	– 8.3
France	28.7	14.0	7.9	– 14.7	– 6.1
Norway	25.9	11.8	6.1	– 14.1	– 5.7
Denmark	24.5	11.4	7.6	– 13.1	– 3.8
West Germany	24.0	9.0	5.6	– 13.0	– 3.4
Sweden	20.3	8.3	4.3	– 12.0	– 4.0
Luxembourg	19.1	11.1	4.7	– 8.0	– 6.4
New Zealand	16.2	11.9	9.7	– 4.3	– 2.2
Netherlands	14.3	7.2	5.0	– 7.1	– 2.2
Canada	13.5	7.8	3.8	– 5.7	– 4.0
Belgium	12.2	4.8	3.0	– 7.4	– 1.8
United States	12.2	4.3	2.6	– 7.9	– 1.7
Malta	12.2	6.9	4.2	– 5.3	– 2.7
Australia	11.8	8.1	5.5	– 3.7	– 2.6
Switzerland	11.0	7.8	4.4	– 3.2	– 3.4
United Kingdom	6.1	2.9	2.7	– 3.2	– 0.2

Sources: Agricultural Situation the Community (EEC, 1986); Winchester and Ilbery (1988: 12); *FAO Production Yearbooks*

countries in these core areas post-1950. In recent years only four countries have recorded increases in their rural population: Portugal, Canada, Switzerland and the United States. For the latter three this has largely reflected urban sprawl, producing commuter settlements in areas still designated as 'rural', though there has also been an element of counter-urbanisation (see Chapter 4) and, in Canada, some extension of the agricultural frontier, e.g. the Peace River district.

Whilst the urbanisation of the countryside has often been regarded as a phenomenon involving short-distance out-migration of urban residents into 'rural' locations, there is another side to the urbanisation process that has also had dramatic effects on rural communities. Prior to any widespread formation of commuter villages, the dominant focus of migration was directed from the countryside to growing urban and industrial areas. Hence, the initial phases of urbanisation were fuelled by rural out-migration which led to widespread rural depopulation. This can

Plate 5 Rural Depopulation – 1. The Salzkammergut, Austria: Out-migration has affected many of those Alpine valleys lacking tourist development, producing a marked contrast with valleys where tourism has been developed

be contrasted with later phases when there were various reversals of the migration focus leading, amongst other things, to the development of commuter villages.

At the heart of widespread depopulation has been out-migration from the countryside to growing towns and industrial centres. Within this outflow, it may be possible to distinguish between *occupational and non-occupational migration*. The latter refers to a protracted movement away from an area by young people who are unable to find employment there. This is a general movement and not restricted to members of specific occupations. It reflects great pressure of population on local resources, implying limited economic diversification in the area. One example of this is the exodus from Ireland during the Great Famine of the 1840s. In contrast, occupational migration only affects specific rural groups, for example prospective farmers forced to migrate to acquire a holding of their own or landless labourers having to move from one agricultural area to another to acquire a job. Much out-migration from rural areas has affected agricultural labour through the labour shedding nature of farming, but other groups have been affected too as rural industries have declined. Hence this type of distinction on an occupational basis is not always significant (Johnston, 1966a).

The notion of rural depopulation as an integral part of urbanisation has been presented in the form of a spatial framework by Lewis and Maund (1976; Lewis, 1979: 40–4). They viewed urbanisation of the countryside as a process with three essential components which have both temporal and spatial aspects:

(a) *Depopulation* – loss of population related to the increased mechanisation of agricultural production and increased specialisation associated with industrialisation. These changes have been associated with a marked division of labour, socio-economic differentiation, greater wealth (especially in the tertiary sector), decreased hours of work and upward social mobility associated with the transformation of the education system. These familiar developments can be recognised as having had significant social and spatial consequences for communities in the countryside, especially through their influence upon the migration of people from rural areas; with the pull factor of available jobs in towns and the push factor of a declining number of jobs in agriculture. So, one element in the urbanisation process has been rural depopulation, associated initially with the out-migration of the young and socially ambitious to produce an aging community. Such communities have often become dominated by agriculture, with a preponderance of aging or retired farm labourers, and a decline in the size of the agricultural workforce.

(b) '*Population*' – commuting to towns by people who have moved to rural dormitory settlements which, in effect, have acted as extensions of the suburbs. This development around many cities in the western world may be seen as reversing the older phenomenon of rural depopulation. As discussed in Chapter 2, it is associated primarily with the young middle class who quickly dominate their new chosen community, not only in terms of controlling social institutions but also by producing children and so changing the demographic structure. It has also been recognised that in the British case a working-class element can be involved, by way of movement to council estates within the commuter zone.

(c) *Repopulation* – the movement of retired people into rural areas and the phenomenon of second home ownership. Communities receiving these newcomers have developed an aging middle-class population structure.

Pahl's concepts of national and local values may be applied to these processes, with population and repopulation imposing national,

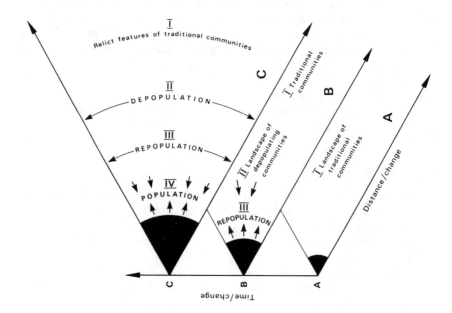

Figure 3.2 A time-space order of urbanisation (*Source:* Lewis and Maund, 1976)

essentially urban, values upon local ones to increase segregation within the countryside by socio-economic status. Lewis and Maund suggested this could be viewed in spatial terms as shown in Figure 3.2. In the initial stages of their model, through limited communication and transport linkages, depopulation only occurs relatively close to urban areas, as a small rural to urban migration flow develops. There may be some repopulation too at this stage. Over time the area experiencing depopulation grows to produce the contemporary landscape, with areas dominated by one of the three processes and with relict features of traditional communities surviving only in very remote areas. They also attempted to demonstrate that this arrangement would be much more complex in reality because of the intermeshing of the spheres of influence of different towns (Figure 3.3). This latter idea is supported by Clout's (1972: 44) reference to major differences between villages within the commuter hinterland, related to ease of communication with major towns, planning regulations, spatial variations in property ownership influencing release of land for development, and provision of local authority housing. The temporal aspect of the Lewis and Maund model can be compared with Zelinsky's (1971) summary of changes in dominant migration patterns, as summarised in Figure 3.4.

Both Lewis and Maund's framework and the traditional rural-urban continuum assume there is a fixed, clearly recognisable rural pole at one extreme. Obviously, this is not a realistic assumption as the character of

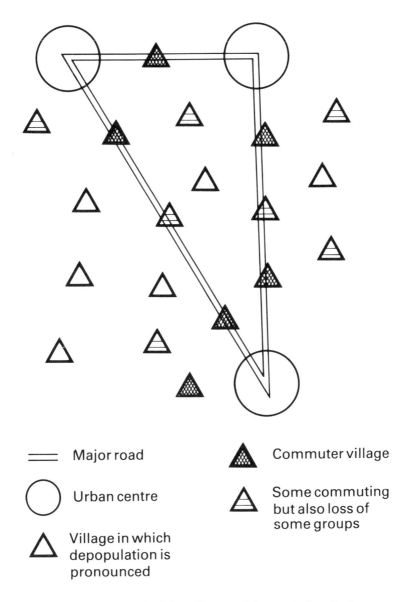

═══ Major road	◬	Commuter village
◯ Urban centre	△	Some commuting but also loss of some groups
△ Village in which depopulation is pronounced		

Figure 3.3 Schematic portrayal of the influence of the proximity of urban centres upon rural settlements

settlements is never strictly static either in time or space. In particular, the model gives a message of a 'depopulating periphery' which has not always been apparent in reality. For example, Drudy (1978), in a study of Norfolk, England, demonstrated that rural depopulation has not been confined to marginal regions. In northern Norfolk there were numerous

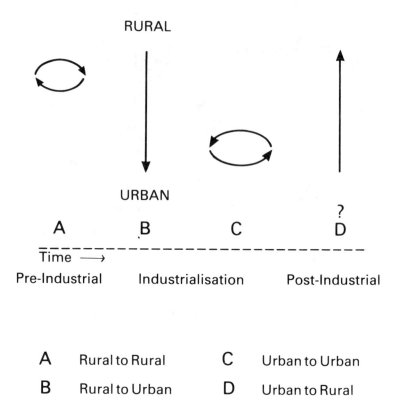

A   Rural to Rural   C   Urban to Urban

B   Rural to Urban   D   Urban to Rural

Figure 3.4 The temporal evolution of dominant migration patterns (based on Zelinsky, 1971)

indicators of prosperity amongst the farming communities which had shared in the 'boom' in arable production of the post-war years. Yet, of 70 parishes studied not directly influenced by Armed Services bases, only five showed an increase in population from 1951 to 1971. Amongst the contributory factors, Drudy cited agricultural adjustments which had caused a high rate of redundancy and the lack of employment alternatives to agriculture. A result of population loss was the reduction in services in the area, notably the rail network. These reductions in turn prompted further out-migration by rendering the district less attractive as a place to live.

Comparison of north Norfolk with the marginal Galway Gaeltacht in western Eire revealed a similar feed-back loop or cumulative causation linking declining services to loss of population. In both cases the lack of non-agricultural employment, especially in skilled or professional jobs, was a major influence on out-migration of school leavers (Drudy and Drudy, 1979).

The traditional view of a close association between remote rural areas and high levels of out-migration has also been questioned by Grafton (1982). His analysis of British census data for the period 1966–71 suggested that remote areas were certainly characterised by net migration losses, but these arose from a small volume of in-migration rather than a high level of out-migration. Levels of out-migration found in less remote rural areas were offset by complementary inflows and it was this lack of inflow, both in terms of total numbers and the age structure of the in-migrant population, that was affecting the dynamism of the peripheries.

Jobes (1987) argues that remnants of gemeinschaft social structure have been preserved in remote ranching communities in the western United States. He refers to these communities as 'vestigial places' where the population has been small, stable and isolated. Yet, few of these settlements remain unaffected by large-scale development, and he notes the destructive impacts of mining, power plants and public works projects. These have overwhelmed traditional values and activities, thereby involving disintegration of existing social structures, e.g. as has occurred in Decker and Colstrip, Montana (Gold, 1985).

Thus although Lewis and Maund's model represents a useful process-oriented approach rather than the somewhat stultified traditional focus upon place, work and folk, its over-simplification limits its practicability. However, it does highlight a continued need for work on the relationship between related processes operating upon rural communities. Meanwhile, more definition can be given to the spatial and temporal characteristics of rural depopulation by considering those processes giving rise to depopulation – initially by examining long-term loss of population from the countryside and its role in the decline of the peasantry in the Developed World.

3.2 Urbanisation, depopulation and the peasantry

3.2.1 The 'pre-industrial peasantry' in Western Europe

In social and economic terms, the initial stages of the urbanisation of the countryside in Western Europe involved the transformation of the *peasantry* into a new type of society – either producing its dislocation via forced or self-motivated migration to urban centres or significantly altering its characteristics *in situ* in the countryside.

It has been argued in Chapter 1 that the nature of rural society in the 18th and 19th centuries cannot be regarded as static, especially in Britain where the progress of enclosure and, later, changes in the Poor Law, brought some rural dislocation and the dominance of the tripartite system of land cultivation: landlord, farmer and labourer (Mills, 1972: 70–5). However, in Western Europe the greater persistence of a distinctive peasantry mitigated against change to some extent. This, coupled with the slower onset of large-scale and widespread industrialisation, promoted a slower rate of rural social change. Indeed, there is more evidence to suggest that urbanisation of this peasantry marked a more significant leap from a semi-feudal set of social and economic relationships to a set based upon the dominance of industrial capitalism (Lopreato, 1967).

Unfortunately, the nature of this change is complicated by the fact that the usage of the term 'peasantry' has been imprecise in academic literature, giving rise to many misconceptions and confusion (Ennew *et al.*, 1977; Shanin, 1972). Marx (1976) treats the peasants as a distinctive class, for example, whereas Redfield (1956; 1973) refers to them as a culture, and Chayanov (1966) as part of a particular form of economic system. However, a common denominator is the dominance of the family unit, the core of what might be termed 'peasant attributes' (Harriss, 1982).

The family in peasant society in 19th century Europe was the unit of production and basis of multi-dimensional social organisation. All labour on the farm was provided from within the family group, working on a small unit of land which may have been tenanted, owner-occupied or share-cropped. The family's prime objective was to provide for its own subsistence, occasionally and irregularly selling surpluses when they arose. The presence of a surplus and a cash return meant that the peasantry were not just subsistence producers, although the majority of their labour may have been utilised to satisfy their own subsistence needs. Shanin (1971; 1983) extended these characteristics to include the existence of a specific traditional culture related to the way of life of small peasant communities, and the dominance of the peasantry by a landlord or 'gentry' class. This dominance was symbolised by the existence of a political stability and control of local markets by a ruling class (Redfield, 1973).

Within the peasantry the dominance of the family operated in a particular fashion. Not only did it control production and exchange, but it also dominated ceremonial activity and relations with the wider social order. The family set the norms of behaviour to the old, the young and the opposite sex. It defined kinship groups and generally produced a strongly cohesive society, though Shanin (1987) also referred to the co-existence of 'analytically marginal groups' which shared most of the peasantry's characteristics but not all, the commonest exception being the lack of possession of land or access to it. Smaller variations have also been recognised within the peasantry itself (e.g. Kroeber, 1948). Attempts to summarise the basic features of an idealised 'peasantry' are illustrated in Table 3.2.

It is evident that an often unstated characteristic of the peasantry has been their persistence through time in a great variety of locations, so that it is still possible to state that 'the peasantry overwhelms the earth with their numbers' (Garst, 1974). However, a basic characteristic of the urbanisation of the countryside in Western Europe has been the extent

Table 3.2 Characteristics of the 'peasantry'

1. Small agricultural producers, who, with the help of simple equipment and the labour of their families, produce mostly for their own consumption, direct or indirect, and for the fulfilment of obligations to holders of political and economic power
2. Peasant family farm as the basic multi-dimensional unit of social organisation
3. Land husbandry as the main means of livelihood
4. Specific cultural patterns linked to the way of life of a small rural community
5. The 'underdog' position – the domination of peasantry by outsiders

Source: Shanin (1987)

to which the long-established rural peasantry has been modified by economic and social change (Reissman, 1964). Consequently, today it is only in the countryside of the Developing World that the population is overwhelmingly from a clearly recognisable peasant class.

In Western Europe the typical peasant community of the 19th century consisted of a nucleated village with homesteads clustered in the midst of economically exploited fields providing subsistence, but increasingly evolving into production for the market without necessarily completely disintegrating the society. Such communities were governed by the political dimensions imposed by the local norms of social control, through long established patterns of individual and collective behaviour. But there was also some link with an all-embracing political control, usually expressed through a small elite exercising dominance in the form of master and serf, or more democratic control, perhaps with a council of elders, or some form of religious control. Indeed, it has been the over-throw of this subordinance to an elite that has so often been associated with bloody revolution in various parts of the world.

3.2.2 The destruction of the peasantry

Changes in the old-established peasant society have occurred in several ways to produce societies less exclusively agricultural, more productive and richer (Galeski, 1968). Harriss (1982: 37) distinguishes three:

(a) the development of capitalist farming;
(b) state initiatives towards collectivisation and the formation of co-operatives or state farms;
(c) the growth of capital-intensive, small-scale farming.

Not all theorists recognise the last change, which is often referred to solely with respect to Japan and Taiwan (Griffin, 1973; Johnston and Kilby, 1975). However, in Western Europe it was (a) that was the domi-nant force as part of the general capitalist penetration of the whole economy. In agriculture this meant a shedding of labour, and so a 'drift from the land' became part of the destruction of the peasantry.

Many of the ideas about the specific nature of change away from a feudal peasant society relate to the English historical experience in which capitalism tended to involve the concentration of land into relatively large-scale holdings. This involved the formation of a distinctive rural triumvirate of landlord, tenant or yeoman farmer, and landless labourer (Mills, 1980). However, even within England, there were distinctive variations in this triumvirate, e.g. between areas where the common-field system had been most persistent and those where hamlets and a more enclosed landscape had been more dominant (Baker and Harley, 1973). Despite such variation, Mills and Short (1983) refer to a 'peasant system' as forming a significant element of 19th century rural life in England, representing not just the landless labourers, but also small entrepreneurs (both within and outside agriculture). Whilst this system may have represented something significantly different from the peasantry in continental Europe and, according to some, did not merit such terminology (e.g. MacFarlane, 1978), there is no doubt that it was radically altered in the 19th century by urbanisation and industrialisation.

As the development of capitalism progresses in agriculture, the feudal and communal modes decompose, releasing the peasantries which become incorporated in the capitalist mode as a highly unstable class of direct producers subject to differentiation based on their access to the means of production and the subsequent sale or purchase of wage-labour (Deere and De Janvry, 1979: 610).

In the English context these changes, in which rural depopulation played such a key part, can be examined through consideration of changes to clearly recognised village and community types. The two basic types, described, for example, in Holderness's (1972) work, refer to the extent to which rural society was or was not dominated by the gentry and large landowners. Some villages were dominated by a single landowner or small group of like-minded landowners whilst others had no such dominant ownership structure. In the former case, the *closed village*, the landowner or a small group of owners could exert a powerful control over the labouring classes. In contrast with multiple owners, in an *open village*, control was less easy.

Critically, this control, where possible, could be exercised with respect to *poor rates*. From 1601 these were expended 'for the apprenticeship of children whose parents could not maintain them, work was to be provided for the unemployed, and assistance given to those unable to work through illness or old age' (Adams, 1976: 192). In the closed villages, landowners used their control of cottage accommodation to limit the size of the labouring population, thereby protecting their tenant farmers from high poor rates. This then enabled the tenant farmers to pay higher rents to the landowners. Furthermore, social control of the tenantry could be more easily maintained in the closed villages by the

Table 3.3 Characteristics of 'open' and 'closed' villages

Open	Closed
Numerous landowners	Concentration of landownership
Small farms	Large farms
Late enclosure	Early enclosure
Farmers have dual occupations	Limited labour supply
Some labourers are part-time farmers	Labour supply augmented by open villages
Crafts and trades practised	Few trades and crafts
Supply services to other villages	Rely on services from other villages
Lack of control over settlement	Control over cottage accommodation
Self-governing village organisations	Paternalism from 'squire'
Non-conformist religion	'Squire' is patron of church; few non-conformists
Radicalism in politics	Politically conservative ('Squire' is MP/peer)
Development of manufacturing industry	Lack of manufacturing industry
Large population	Small population
High poor rates	Low poor rates

upper echelons of society – the gentry, clergy and yeoman farmers (Mills, 1972: 64).

In contrast, in open villages the multiplicity of ownership was unable to exert such a control upon poor rates. This often encouraged settlement by labourers and the poorer elements of society as well as other groups benefiting from the relative lack of close social control, e.g. people belonging to non-conformist religious groups, tradesmen and craftsmen. The types of differences between the two types of villages are summarised in Table 3.3.

One of the main effects of this development of very different types of villages was seen in the demographic history of the villages, especially in the 19th century. At a time when the population of much of rural lowland England was growing, it was the open villages that grew most rapidly whilst the closed villages, with their restricted entry to newcomers, stagnated or grew only slowly (see Table 3.4). When rural depopulation and reduced rural population growth became the dominant trend in the last quarter of the 19th century, it tended to be the open villages, with their larger numbers, that fuelled the out-migration to the cities. Thus, overall, between 1851 and 1901 rural population in England and Wales fell by 0.6 million, from 46 to 21 per cent of the total (Lawton, 1986: 10). Yet, if sufficient numbers were retained in an 'open' community, they could maintain a community spirit and degree of social cohesion that helped to reduce out-migration.

The legacy of open and closed villages is indicated in Mitchell's (1950; 1951) study of population change in south-west England. He stated that different social outlooks affected the perceived importance of the disadvantages of rural living. Classifying these outlooks on the basis of

Table 3.4 Population changes in selected 'open' and 'closed' villages in the English Midlands, 1801-1981

Parish	1801	1851	1871	1901	1931	1961	1981
'Open'							
Aldington/Badsey (Wo)	367	521	634	931	1339	1835	2310
Barlaston (St)	349	617	821	744	1202	2459	3249
Breinton (H)	238	366	445	417	511	840	709
Bidford-on-Avon (Wa)	928	1537	1727	1369	1842	2436	3172
Burghill (H)	639	946	1036	1303	1553	1552	1400
Claverley (Sa)	1328	1613	1733	1358	1215	1304	1306
Offenham (Wo)	264	400	500	617	917	892	1342
Swynnerton (St)	648	946	778	811	896	3022	4679
Worfield (H)	1354	1735	1676	1448	1416	3572	2068
'Closed'							
Abbot's Morton (Wo)	191	235	213	146	131	149	156
Balterley (St)	237	299	273	253	217	183	207
Bickmarsh/Dorsington (Wo)	100	54	187	183	149	179	69
Callow (H)	109	129	116	78	70	63	52
Felton (H)	107	112	127	69	90	113	105
Hinton-on-the-Green (Wo)	196	192	183	209	213	268	214
Kenchester (H)	85	98	103	121	97	89	92
Patshull (St)	169	112	208	222	204	154	153
Throckmorton (Wo)	150	153	170	109	118	152	90
Upton Cressett (Sa)	53	58	63	52	32	39	26

H – Herefordshire; Sa – Shropshire; St – Staffordshire; Wa – Warwickshire; Wo – Worcestershire

Source: Population census

whether societies were integrated or disintegrating and whether they were 'closed' or 'open' (Table 3.5), he found that the most dynamic type, the socially-integrated, open society was not common. When present, these communities were typically a large village with a stable or growing population which maintained a high level of civic pride, with an active parish council and good ranges of services and local entertainment.

Mitchell found that socially-integrated, closed societies were common. These tended to have people with a very traditional outlook on life, often with the presence of close social links revolving around the activities of the local church. However, the closed nature of the society could contribute to depopulation if local leadership was insufficient to generate a strong feeling of 'community'. In the open, socially-disintegrating societies there was a high turnover of population and an unstable social organisation. Frequently, this was associated with changes in the economic structure of the community, for example the opening of a new factory. This type of study, though subjective, does stress the value of 'community' in determining the amount of depopulation at a local level.

Table 3.5 Depopulation and rural social structure

	Recognised categories of rural settlement	
A	B
1. Socially integrated	Closed society
2. Socially disintegrating	Open society

Source: Mitchell (1950)

Table 3.6 Model of the development of 'open' and 'closed' villages

Date	Village X ('open')	Village Y ('closed')
1086	Not recorded in Domesday Book	Recorded in Domesday Book as a prosperous place – serfs, plough-teams, meadow, pasture; owned by a magnate; tenant-in-chief resident
C13th	Created during mediaeval expansion of land settlement; Colonised by free-holders and tenants; no church	Has church, manor house, manor farm
C16th	Steady growth; increased number of cottages; crafts flourish numerous landowners	Limited growth; dominance of one or two landholders
C19th	Growing population of day labourers and paupers; Small industry present; village sprawls, with an assortment of cottages; religious dissent; provides services for neighbouring villages	Building restricted; old cottages pulled down to limit poor relief; orderly, well designed village
late C20th	Opportunities for in-filling and rounding-off; Expansion may be permitted; May retain service provision	Too coherent and traditional to receive planning approval

Source: Emery (1974: 230–1)

It shows too how demographic trends dating from capitalist penetration of the countryside in the 18th and 19th centuries continued to operate subsequently to produce significant differences even between neighbouring parishes. This continuity, stretching back even to feudal times, is summarised in Table 3.6, based on Emery's (1974: 230–1) graphic account of village evolution in Oxfordshire.

3.2.3 Worker-peasants

Whilst the pace of both agricultural and industrial development in Britain brought sweeping changes in rural society in the 18th and 19th centuries, the peasantry of much of Western Europe generally remained less disturbed and retained a cohesiveness that helped maintain a recognisable 'peasant class' even into the 20th century. Although the penetration of capitalism broke down the self-sufficiency of the peasantry, some forms of peasant society were retained. Indeed, in continental Europe the transition from agrarian to industrial society has been marked by the development of a particular group of farmers, seen by Franklin (1969) as being in a transition between the capitalist, industrial sector and the remnants of the peasant, agrarian sector. This group have been termed the *worker-peasants, five o' clock farmers* or *part-time farmers* and have a long history in some of the coal mining districts of Europe, though some have developed in response to new industrial development post-1945.

The formation of a worker-peasantry has been identified as part of a distinct 'proto-industrial' phase in European economic development. For example, Mendels (1981) formulated a model based on the development of part-time industrial work in 18th-century Europe in conjunction with subsistence farming on fragmented holdings, but with a rapidly growing population and a pool of surplus labour. In this situation merchants sought to increase manufacturing production by drawing in more workers using traditional techniques of production and spreading production from urban areas into the countryside. Gullickson (1986), in a study for the Pays de Caux in Normandy, extends this general situation by emphasising the importance of the under-employment of women and the seasonal character of work in the farming system in the late 18th century. Thus women and men, at certain times of the year, were able to participate in the local cloth trade based on Rouen whilst also being employed on the land.

For West Germany the stability of part-time farming has been described as a 'standing wave' in which the number of part-time farmers has been continually replenished by full-time farmers who change to part-time farming. Important characteristics have been high labour input and gross income lower than that for most of the non-agricultural population. But part-time farming has guaranteed employment plus a measure of subsistence. It has also offered a high degree of autonomy in the work process and, perhaps, offered an important degree of independence and social cohesion (Mrohs, 1983).

In the 1960s it was estimated that one-quarter of West German farms were run by worker-peasants, with a large concentration in the Saarland. Other areas in Western Europe associated with this phenomenon have included the French Massif Central, within commuting distance of the steelworks at Les Ancizes and Issoire, the rubber factories of Clermont Ferrand, and industries in the towns of Alsace-Lorraine and several of the Alpine Valleys (Mignon, 1971).

Worker-peasants have a long history in Germany where, in areas of inheritance amongst all male siblings, smallholdings were rarely large enough to support a family, and income supplement was sought in the form of manual labour in industry. The building trade often provided suitable casual employment, but mining too, with shift work, could fill the required purpose. Post-1945, farming has been combined with factory and office work, especially employment in car factories such as the Mercedes plant at Gosheim, Baden-Wurttemberg (Franklin, 1964). The dual character of the worker-peasants' economic activity has been useful in acting as a cushion against depression, for example during the economic downturn of the 1920s. It has added to the demand for services in the countryside whilst also helping to retain sizeable communities instead of ones greatly reduced by rural depopulation. The composition of these communities can be distorted though, Mignon (1971), for example, referring to the large numbers of male worker-peasants in the Massif Central who remain unmarried, possibly related to the large amount of time spent in working activity. Indeed, this emphasis upon time devoted to work may be a factor in the decline of the worker-peasant. Today the average age of worker-peasants tends to be above the average for full-time farmers whereas the reverse used to be true. In some cases, though, the farming element of the arrangement has tended to become little more than *hobby farming*, making very little additional contribution to the income derived elsewhere.

The presence of worker-peasants has favoured the retention of fragmented smallholdings, leading to inefficiency. The limited amount of labour input possible from a worker-peasant has meant low productivity, and often the presence of uncultivated land. This 'social fallow' (*sozialbrache*) has occupied up to half the holdings in some parts of West Germany despite the fact that worker-peasants' farms are frequently over-mechanised as the additional income obtained from the farm is spent on purchasing machinery, often as a status symbol (Hartke, 1956; Krocher, 1953; Labasse, 1961). For example, the introduction of tractors to the traditional, relatively isolated, peasant village of Talheim in Schwabia in the 1950s was accompanied by an over-mechanisation relative to the productive capabilities of existing land resources. The use of tractors was seen by villagers as essential to retaining their small 4 and 5 ha (10 and 12 acre) holdings, despite the fact that most were run only on a part-time basis (Schwarzweller, 1971). The development of an additional source of income was relatively recent in this example and hence land was still regarded as the fundamental basis for economic security. Therefore, the use of tractors implied 'good management' of this vital resource to the farmers, even though it may have been inefficient on a strictly cost-account basis.

In West Germany, despite long and still continued heavy government involvement in farm reorganisation and agrarian reform, just over 50 per cent of farming households are still less than 10 ha (25 acres). Yet, it is amongst this group of farmers, the *small peasants* or *kleinbauer*, that the development of social fallow has been most common (Wild, 1983b). This phenomenon has been closely correlated with the movement of small farmers out of agriculture. Often, a transition has been effected from full-time farming to complete reliance upon non-farm employment. During this transition, when the farm operator has become, effectively, a *part-time farmer* (*arbeiterbauer*), the amount of social fallow has increased. With the lure of greater incomes obtainable from non-farm employment, full-time farmers, and those who have for a long time operated as part-time farmers, have forsaken agriculture. This process has been especially rapid post-1945, with the West German agricultural labour force falling by 2.2 million between 1950 and 1965 (Wild, 1983b: 207).

The distribution of social fallow has largely coincided with that of the small farmers in those areas where the continuity of peasant traditions and the worker-peasantry has been strongest, e.g. Saarland, the Rhine-Neckar area and the adjoining part of Rhineland-Palatinate, the lower Main Valley, the Siegerland and the western part of the Westerwald (Figure 3.5). However, as the growth of social fallow continued in the early 1970s, new areas were affected. This more diffuse pattern is attributed by Wild to three developments:

(a) the extension of commuter zones, permitting greater separation between farm and non-farm employment;
(b) the growth of *fernpendelnwanderungen* or long-distance and cross-country movements of unskilled rural workers taking up periodic labour contracts;
(c) the growth of farm-based tourism in resort areas such as the Bavarian Alps, the Black Forest and the Moselle Valley.

It would seem that these three factors have promoted the spread of part-time farming beyond the confines of the traditional worker-peasantry. But, from the mid-1970s the formation of social fallow has developed only slowly as government subsidy has raised farm incomes at a time when opportunities for industrial employment have been reduced, and when old social fallow is increasingly being 'recycled' for farm enlargement/consolidation or for non-agricultural purposes, e.g. recreation schemes.

Franklin's (1969) survey of the worker-peasantry, made in the late 1960s, reported that the generation who had followed this life-style from the 1940s were gradually reaching retirement age and were not being replaced. Amongst the reasons cited for this decline were the inferior status of the worker-peasants with respect to full-time farmers and full-time office or factory workers, the difficulty of maintaining a factory-based job and still having sufficient time to operate a smallholding in a way capable of yielding a satisfactory financial return, and unwillingness to undertake the high amount of work associated with the combination of two jobs. Franklin concluded that the worker-peasantry was not part of a stable community. He felt that there were technological forces at

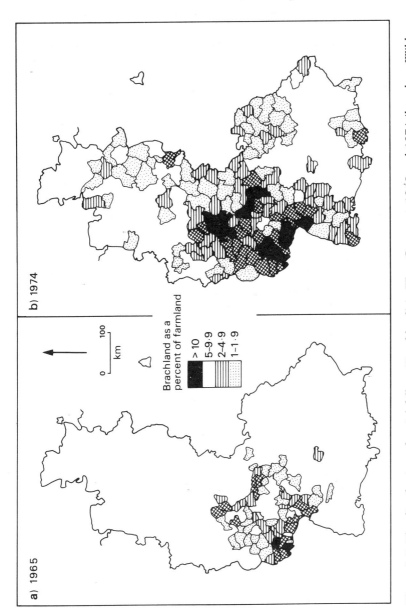

Figure 3.5 The distribution of social fallow (*Brachland*) in West Germany, 1965 and 1974 (based on Wild, 1983b)

work within the capitalist system which would ensure the capitalist element would prevail within the industrial sector of the binary economy. So, increasingly, one of the two sides of the worker-peasant's duality would be rejected. This would tend to be the farming side because of the higher incomes available from industrial employment and also because the worker-peasants tended not to possess a long-term tie to the land. The actual decision to leave farming was dependent upon the income differential between the two types of work combined and industrial work alone, the ease of disposal of the farm property and the stage in the life cycle.

Bringing Franklin's study up to date, in Thieme's (1983) review of West German agriculture in the late 1970s, he states that little remains of the traditional augmentation of farming incomes via work in mining or cottage manufacturing. However, he does recognise some continuation of the worker-peasantry through participation in local manufacturing industries and tourism. This continuity is patchy, though, with upland areas, such as the Black Forest, the Harz, the Fichtelgebirge and the High Eifel, probably having greatest 'continuity of peasant involvement in binary economies' (Thieme, 1983: 243). It would seem, therefore, that the worker-peasant has not been eliminated entirely, though whether such persons now belong to a 'peasantry' as defined in a variety of ways by numerous scholars, is questionable. Both sides of the binary economy have altered dramatically over the past few decades whilst the rural community too has been 'urbanised' in very many different ways, so as to destroy the fundamental attributes of any post-1945 vestiges of a peasant society (e.g. Thieme and Paul, 1980).

3.3 Rural depopulation and out-migration

3.3.1 Push and pull factors

In studies of rural depopulation, two of the main foci have been upon the factors promoting out-migration and the characteristics of the migrants themselves (White and Woods, 1980a; 1980b). Examination of these can effectively build a picture of the social changes occurring in depopulating communities and the concomitant economic responses to these changes.

In general terms, the characteristics of migrants and their reasons for migration are well known (Phillips and Williams, 1984: 74–96; Van Der Knapp and White, 1985), having been the subject of 'laws' formulated by Ravenstein (1885) in the 1880s and much subsequent work on *migration theory* (e.g. Grigg, 1977; Woods, 1982) (see Table 3.7). The most important variables in considerations of differential migration propensities have been age, socio-economic status, past migration experience, familial and social contacts and the differential 'attractiveness' of one area over another. These variables, as part of the process of migration,

Table 3.7 Ravenstein on migration

1. Majority of migrants go only a short distance
2. Migrants going long distances generally go by preference to one of the great centres of commerce or industry
3. Females are more migratory than males within their own country, but males more frequently venture beyond
4. The major causes of migration are economic

Source: Grigg (1977)

are often considered in terms of *push and pull factors*. The former are factors contributing to people's desire to leave a particular area and the latter are the attractive forces operating in the area of receipt for migrants. Major push factors have been the labour-shedding nature of farming over the past one and a half centuries, the lack of alternative opportunities because of the changing economic structure of rural communities, rural deprivation and the pull of higher paid jobs in towns.

In New Zealand, for example, numerous studies have indicated that it has been males and females in their late teens and early twenties who have been the most mobile (Frazer, 1971; Heenan, 1968a; 1968b; 1979; 1988). Also, those with higher levels of educational attainment and higher aspirations have been more likely to be migrants (Vellekoop, 1968). Similarly, there is evidence that occupations emphasising educational attainment show higher rates of migration (Forrest and Johnston, 1973; Keown, 1971). White-collar workers and those in the professions have been more migratory, reflecting, perhaps, the nature of the labour market in which demand for a particular skill can extend from the local to the national level. However, given the higher propensity of young people to migrate, there has been a bias towards single persons in the 15 to 24 year age range being most migratory.

In terms of push factors, poor services and facilities in rural areas have often been highlighted as being very important. However, Bracey's (1953; Brush and Bracey, 1966) study in Somerset, England, in the late 1950s showed only a mild correlation between the incidence of depopulation and poor public utilities. This apparent anomaly led him to conclude that other, social, criteria needed to be investigated to help explain spatial patterns of depopulation.

Other studies have shown that feelings and attitudes towards a place and locality play an important role in fostering or deterring the wish to migrate. However, these attitudes are closely related to critical social and economic factors, which Woods (1985) argues need to be built into a coherent general theory for migration, that is 'specifically designed to deal with motivations together with their ultimate causes and expressions' (Woods, 1985: 4). Meanwhile, Figure 3.6 shows how the various factors considered above might be built into such a general theory.

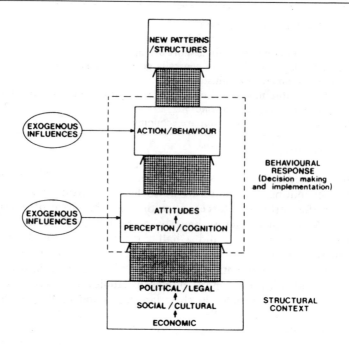

Figure 3.6 Framework for a geneal theory of migration (*Source:* Woods, 1985.)

3.3.2 Push and pull factors – a case study

The example of school-leavers migrating from a small New Zealand town can be considered as an illustration of the ways in which general push and pull factors translate into a range of decisions concerning out-migration from a rural area (see also Johnston, 1971).

In a study of school-children in the small town of Temuka (population 3,771 in 1981) in rural South Canterbury South Island, New Zealand (Plate 6), for those wishing to leave the town, higher education and/or further training were cited as the chief reason for leaving (Table 3.8) (Robinson, 1986a). This stated reason was closely followed by employment, with the wish to travel and to broaden experience ranked in third place. Marriage and family reasons were rarely singled out as important in their own right. However, amongst those wishing to remain in the town, 'family' closely followed by 'friends' were cited as the main reasons underlying this intention.

The expressed desire to leave the town after completing secondary education was associated with the age of the student, the intended school-leaving age and whether it was thought that the type of job or further education desired was available in the locality. Older children expressed the strongest desire to migrate, as for them migration was directly related to the intention of pursuing further education not

Plate 6 Rural Depopulation – 2. Temuka, South Island, New Zealand: Although there is a slow rate of population increase, the better-educated of its schoolchildren express a strong desire to leave the local community

Table 3.8 Reasons for leaving or staying in Temuka, New Zealand, as expressed by secondary school children

Average rank	Reasons	% Placing this first	Average rank	Reasons	% Placing this first
Prefer to stay			*Prefer to leave*		
1.	Family	33.71	1.	Higher education/training	38.89
2.	Friends/social	24.27	2.	Employment	35.12
3.	Employment	23.65	3.	Experience/travel	23.30
4.	Recreation	1.63	4.	Marriage	0.18
5.	Pleasant location	12.33	5.	Family	2.33
6.	Others	4.41	6.	Others	0.18

Source: Robinson (1986a: 225)

available locally. In addition to the desire for educational advancement promoting out-migration, 91 per cent of those intending to leave Temuka felt they could not obtain the job they desired locally. This desire was linked to the occupational status of the children's parents – those children whose parents had manual or semi-skilled jobs were more likely to have expressed a desire to remain in the town after leaving school. Higher social status correlated with the intention to remain at school for

Table 3.9 Views of Temuka expressed by school-children at Temuka High School, New Zealand

(a) Views of those wishing to migrate		(b) Views of those wishing to remain in Temuka	
Good points	% of respondents citing factor	Bad points	% of respondents citing factor
Small/quiet	55	Small/quiet	58
Friends	43	Friends	58
Family	30	Family	32
Sports	28	Sports	30
Location	28	Location	26
School	3	Employment	8
Employment	2	School	4
Shops/services	1	Shops/services	4
Bad points	% of respondents citing factor	Bad points	% of respondents citing factor
Employment	73	Entertainment	47
Entertainment	52	Employment	45
Lack of privacy	20	Shops/services	17
Shops/services	17	Location	11
Small/quiet	17	Small/quiet	11
Location	5	Lack of privacy	9
Sports	2	Sports	4
Friends	2	School	2
Family	1		

Source: Robinson (1986a: 226)

a longer period. This indicated the desire to obtain better qualifications and hence the greater likelihood of at least temporary migration in search of further education or training.

Additional influences upon an expressed desire to migrate were stated in the children's views of the locality. Lack of employment and entertainment were felt to be the main drawbacks by both potential stayers and leavers, but the former was much more clearly recognised to be a disadvantage by those wishing to leave (Table 3.9). This group also gave greater emphasis to problems associated with the lack of privacy in the small community and the fact that it was small and quiet. In these stated views and preferences no significant difference was found between the sexes nor an association with location of residence (farm or non-farm) and length of residence in the locality. Yet, when the migration patterns of recent school-leavers were examined, there was a clear distinction between male and female. This was because a higher proportion of females had left Temuka to obtain further education upon leaving school. In contrast, the local economy (incorporating a large local manufacturer and industrial/service employment in nearby Timaru, a port 11 miles to the south with 29,000 population) had been able to absorb a substantial proportion of the male school-leavers. One-quarter of these stated they had not been able to obtain the job they had desired and there was a suggestion that they might subsequently migrate. Yet, several of those in higher education away from the locality expressed the intention of returning to work in or near Temuka upon obtaining their qualifications. Thus migration amongst school-leavers was not a once and for all decision, and there were a series of temporary movements to and from the locality.

The findings from the study of Temuka compare with similar work carried out elsewhere in New Zealand, for example, Vellekoop's (1968) work on Westport and Glendining's (1978) on Eketahuna. The latter community, in the northern Wairarapa, is less than a quarter the size of Temuka and experienced depopulation in the 1970s. The emphasis upon migration in order to obtain better job opportunities is common to the three studies as are the general dissatisfaction with local employment opportunities and some of the qualities used to describe the communities (e.g. close-knit, importance of friends and family ties).

The national context for these studies is the growth of a northward drift of population, i.e. towards Auckland and the northern half of North Island, New Zealand, a feature of internal migration well documented since the early 1960s (Forrest and Johnston, 1973; Heenan, 1968a; McCaskill, 1964; Roseman and Crothers, 1984). However, the greatest volume of internal migration in recent years has occurred within the country's thirteen Statistical Areas, emphasising the importance of local mobility. Emigration has also been important, especially in the 1970s (Farmer, 1979: 39–40). For both internal migration and emigration, the high mobility age groups for both males and females have been from ages 18 to 35 years, the ratio of movers to non-movers peaking at an earlier age for women than for men.

3.4 Depopulated regions

The geographical impact of migration has been studied from a variety of standpoints (e.g. White and Woods, 1980a: 44), two of these being considerations of the area of origin and the character of emigrants. Examining such areas and their emigrants, the prevailing structural context must not be overlooked. In the case of rural depopulation the migrant streams of the last two centuries have altered the pattern of rural population growth, broken down barriers between 'national' social systems and old-established 'local' ones (e.g. Ogden, 1980; Saville, 1957). Thus the depopulation has been both a product of existing structural features of the economy and society and a contributor to significant alteration of these social and economic structures at a national level (Grigg, 1980).

One part of Western Europe which provides a good illustration of the problems confronting areas experiencing long and continued depopulation is Mid-Wales where some parishes have suffered from loss of population since the mid-19th century. Overall, there has been a 19.2 per cent decrease in population since 1901. The process of depopulation in Mid-Wales has consisted of four chief elements. One has been the push factor of the labour shedding character of hill farming, the mainstay of the local economy. Not only have the number of agricultural labourers diminished, but also the number of agricultural holdings has declined through the process of farm amalgamation. Hill farming has become increasingly reliant upon government subsidies for its very survival, and smaller farms have tended to lose their viability as the economic size of operation has favoured the large producers. Thus the number of farms in Mid-Wales has fallen from 10,793 in 1885 to 5,325 in 1985. With no alternative local opportunities within farming, both farm amalgamations and the shedding of farm labourers have contributed to out-migration.

The loss of labour from the chief form of employment in the area has contributed to both a change in the age structure of the population and a decrease in population density. The dispersed distribution of an aging population makes it increasingly unattractive for new industrial development despite government incentives to encourage this. So, a reliance upon agricultural employment has been retained, contributing to an overall income level below the national average and a relatively high level of unemployment. These help to accelerate out-migration, further impoverishing local life and a sense of deprivation, especially promoting greater feelings of dissatisfaction amongst the young (Taylor, 1979; White, 1980). The loss of population also means that the provision of public services becomes more and more expensive per head of population. This then entails a higher per capita subsidy from government (Ogden, 1980).

White's (1980) study of Lower Normandy, France, extends some of the points made for Mid-Wales, as he highlighted the crucial aspect of the selectivity of migration by age, sex and occupation during a 13.2 per cent loss of population from sample communes between 1962 and 1975. Young adults were the main emigrants, though with some differences between male and female reflecting educational and occupational variations. There were low rates of out-migration from agriculture, though this still meant a high absolute outflow because of large numbers employed on the land. The highest proportions of workers emigrated from the agricultural processing and professional sectors, though for the latter this was more than balanced by inmigration. An important result of the trends in migration was an aging of the population which reduced the financial efficiency of local government whilst raising its cost to the remaining residents. Some local facilities, schools for example, were becoming redundant and there seemed to be several 'negative' consequences to the depopulation. The way in which these can be self-

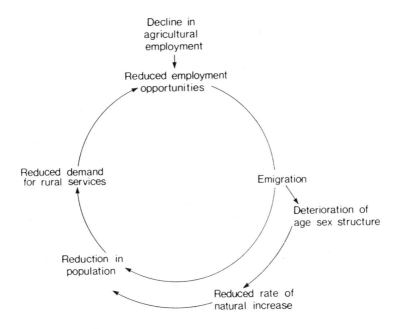

Figure 3.7 The cyclic nature of rural decline (*Source:* Gilg, 1983: 95)

reinforcing, leading to a 'vicious circle of decline' (Gilg, 1983: 94), is illustrated in Figure 3.7.

The result of long continued depopulation is often a society that represents the 'hard core' of a once larger farming community. It is lacking in financial resources, is supported by a minority secondary population and isolated miscellaneous groups such as craftsmen and foresters suffering from the same deprivation. Adaptations in the community's social provisions and organisations have been necessary, though in the past two decades the growth of return migration and second home ownership have introduced new elements to the population.

Some of the ways in which the various push and pull factors described above translate into rural out-migration and depopulation at regional and national level are considered below as case studies for two sizable areas where rural depopulation has had a significant influence upon changes within rural society.

3.4.1 Southern Europe

Population outflow from rural areas of southern Europe has occurred on a large scale in both the 19th and 20th centuries. In some cases this exodus has been focused beyond the region itself and has been part of a general migration away from poverty both in the countryside and the towns. For example, one of the major sources of rural migrants in southern Europe has been southern Italy. Excluding internal migration, this region had a combined net emigration of nearly 4.5 millions between 1871 and 1951. And even this figure underestimates outward movement as it ignores the widespread practices of repatriation and re-expatriation, which might increase the numbers migrating abroad to 9 million. In Italy, and in Spain too, greater contacts between rural inhabitants and a different world beyond the village have not only increased material well-being in rural areas, but have opened the floodgates of out-migration (King, 1976).

The spatial pattern of this depopulation has been determined frequently by variations in the rural economy. White (1985), for example, showed that for the Cilento district, 75 miles (120 km) south of Naples, the presence or absence of tourism was a major influence upon population change in the 1960s. Other factors were the degree of local accessibility, as measured by the quality of roads to particular settlements, and altitude of settlements. In effect, two separate sub-systems could be distinguished in the area, one near the coast which was dominated by tourist development, and the other of declining inland settlements characterised by a high degree of local autonomy and self-sufficiency yet with links through migration to a more distant world (e.g. Lopreato, 1967). There was a suggestion in this latter area that the larger settlements were experiencing the most rapid depopulation, a strong contrast with the pattern of rural population loss experienced in northern Europe.

Pitt-Rivers (1976: viii) likened the social changes occurring in rural southern Europe as a whole from the 1950s to that occurring in more recognisably 'primitive' and peasant societies in Africa and South America. Similar patterns of depopulation, out-migration and the introduction of new economic forms were characterising rural Mediterranean Europe as those changes that had begun earlier in parts of northern Europe and North America. In the Spanish context, Aceves and Douglass (1976: xi) referred to this as the 'opening up' of Spanish rural society: demographic change contributing to and forming part of rapid changes in life-style and in the value systems of rural people (e.g. Douglass, 1975).

Brandes (1975), in a study of a farming community 125 miles (200 km) west of Madrid, showed that the rural exodus has not been merely a reflection of the peasants reacting to events beyond their control. Rather than fleeing the countryside because of its poverty and the seductive pull of the prospect of greatly increased prosperity in the growing cities, he showed that out-migration has continued in spite of a significant increase in farm incomes in the village. The first phases of out-migration in the 1940s and 1950s brought increased wages for farm labourers, secure employment, land control through renting and share-cropping plus direct economic aid from migrant relatives. These changes brought clearly observable changes in the village through house improvements and material possessions plus better diets and general health. Although the difficult terrain limited the effectiveness of technical innovation in agriculture, land reforms brought greater wealth and freedom from reliance upon low and uncertain wage labour. Money from migrant relatives also added to incomes, whose increases were apparent in major house improvements, purchase of consumer durables, improved diets and a sharp fall in infant mortality. To these changes can be added others associated with community structure, in which village society has become more egalitarian, leisure activities have increased and inter-personal behaviour has altered as what was essentially a peasantry emulated the 'urban elite' of Madrid and the urban emigrants from the village (e.g. Friedl, 1964).

A more recent study of the eastern Montes Orientales of Granada by Beck (1988) reveals a similar pattern of rural exodus in the 1970s and 1980s, mainly directed to Barcelona and West Germany. The introduc-tion of labour-saving techniques in agriculture greatly reduced local income opportunities, thereby promoting both long-term and seasonal migration. There was some growth in 'petty agrarian capitalism' through land purchases, with money earned outside local agriculture. Also, with the development of the welfare state from 1977, living conditions in the villages increased, reducing the gap between urban and rural standards of living.

The significance of Beck's study and others on depopulating communities in southern Europe is that they indicate that despite increased wealth and a changing life-style, these do no necessarily impede further out-migration:

Villagers aspire not only to greater material welfare, but also to direct participa-tion in the bustling life of Madrid. Radio and television, though partially alleviating the boredom of country life, are scarce compensation for the lack of movies, theatre, parks, and other centers of diversion found in the city (Brandes, 1975: 15).

Thus once an experience of urban life is gained, by an occasional visit or indirectly through contact with friends and relatives from the city, the lure of the city is often far more powerful than the increasing material benefits developing in rural communities. Former patterns of localism and autonomy are broken down, and the changing structure of the population, consequent upon large-scale out-migration, prevents a return to old social patterns (e.g. Behar, 1986; McNeill, 1978: 138–55).

Redclift (1973) showed that once remote villages become dependent on forces originating outside the village then changing population composi-tion makes it increasingly difficult for groups of villagers to regard themselves as part of a collective entity with shared traditions and institu-tions. In the case of his study of Gema in the Spanish Pyrenees, the forces were the influences of industrial migrants, tourists and land developers.

Emigration from many rural areas of southern Europe has often been a reflection of the response of population to limited natural resources. A good example of this has occurred in Malta where agricultural productivity has been limited by thin, stony limestone soils, low rainfall and high rates of evapotranspiration. Also lacking in indigenous sources of power and industrial raw materials, Malta has been a representative of other parts of the Mediterranean countries in microcosm. For example, in the early 1920s, Malta exported significant numbers of migrants to the United States, and there was also large-scale out-migration post-1945. Overall, between the 1948 and 1967 Maltese censuses, 90,000 people emigrated, representing 30 per cent of the 1948 population (Jones, 1973: 103). Assisted passages schemes to Australia, Canada and the United Kingdom were an important factor in determining the final destination of many emigrants. However, most of these migrants were not leaving the land. Only in the flow of migrants to Australia was the proportion of migrants who were farmers and agricultural workers greater than the proportion of this occupational group within the Maltese population as a whole (Jones, 1973: 111). A frequent pattern was for farm labour shedding to promote greater pressure in urban areas in Malta, a growth in unskilled urban manual labour and then a high rate of emigration from this group.

Between 1946 and 1979 over 140,000 Maltese emigrated through assisted passages schemes, 57.6 per cent to Australia, 22 per cent to the United Kingdom and 13 per cent to Canada (King, 1979). However, there was also a steady stream of *return migrants* who returned to their original village or urban district. This return stream seemed to be independent of opportunities in the original departure area and reflected a family life-cycle control: 'an initial emigration when single for a couple of years, a return for a year or so to find a wife, and then a more protracted family migration of around 10–15 years' (King, 1979: 247). Thus out-migration cannot be regarded as a once-and-for-all decision, but may be counterbalanced at a later date by return migration or even a series of departures and returns by the same individuals.

3.4.2   Depopulation on the Prairies

Despite their more recent history of occupation, many rural areas in North America still exhibit a temporal pattern of depopulation not very dissimilar to their European counterparts. In parts of North America outmigration has been a dominant demographic characteristic since the late 19th century (e.g. Pierson, 1973). Photiadis (1965; 1974; 1976) demonstrated several similarities between rural Appalachia and Greek mountain villages in terms of the ways in which communities had been affected by long-term out-migration. He argued that the socio-psychological linkages developed with the larger society, combined with limited economic potential in the rural areas themselves, brought out-migration and, eventually, a mass exodus and collapse of village social organisation. Yet,

Table 3.10 Population growth of the major cities on the Canadian Prairies

	1901	1921	1941	1961	1981
Prairie provinces	419,512	1,956,072	2,421,935	3,178,811	4,438,454
Rural districts	316,277	1,252,604	1,498,300	1,210,605	1,173,647
% Rural	74.4	64.1	61.9	38.1	26.4
Main five cities (1)	54,270	361,384	505,953	1,243,745	2,469,281
% in five main cities	12.9	18.5	20.9	39.1	55.6

(1) Calgary, Edmonton, Regina, Saskatoon and Winnipeg

Source: Population Censuses

there are other areas, in Australasia too, where depopulation in the mid 20th century has followed very rapidly upon pioneer settlement, and brought a distinctive pattern of depopulation to the countryside. One of the best examples of this is the Canadian Prairies.

For Canada, Fuller (1985) distinguishes four consecutive phases of population mobility: initial rural settlement, labour migration, depopulation and return migration. The dominant pattern of early mobility was westwards towards the frontier of settlement. Subsequently, the labour migrations of rural workers were also westwards, e.g. the movements of workers who manned the threshing crews on western grain farms in the early 20th century. So, some of the main foci of these migrations were in the Prairie provinces where in-migration was followed very quickly by depopulation and the rapid growth of the major metropolitan areas, especially during the past three decades. This growth has been fuelled partly through out-migration from rural communities and the small service centres that developed during the first few decades of European settlement (Table 3.10).

Given that pioneering agricultural occupation only occurred after 1870 and was still under way on a large scale in certain areas in the 1920s, the Peace River district for example, the movement from a simple agricultural society to one caught up in the metropolitanisation process has been extremely rapid. This speed of change, coupled with the type of farming characteristic of the Prairies, has produced a highly distinctive social and economic response (Carlyle, 1988). The predominantly large-scale, family-operated farms remain, but they have become even larger and more reliant upon the substitution of machinery for labour. The sparsely settled nature of the Prairies has been accentuated and the loss of hamlets and small villages has placed severe strains upon the maintenance of services and utilities. Whilst this is a familiar story in other parts of the Developed World, in the Prairies the element of large distances between farms and service centres adds an extra dimension to the picture. The farm population has fallen by three-fifths from its peak in the 1930s so that in 1981 only 12 per cent of the Prairie population lived on farms. Less than one-third of the total population was in rural townships whilst over half were in the five major cities.

This transformation has been marked by an uneven pattern of depopulation, closely associated with the changing basis of the farm
economy. Whilst there have been pockets of population growth, especially in the Edmonton-Calgary corridor (Smith, 1988), in areas dominated
by extensive agricultural production there have been losses of over half
the population post-1930. Such losses have been most pronounced in
areas reliant upon extensive wheat production, and hence the highest
rates of depopulation have been in southern Saskatchewan (Figure 3.8).
Here depopulation has brought a consolidation of rural services and a
number of farm mergers, more than doubling the average farm size over
a period of just 45 years. Yet these very large holdings often exist
alongside much smaller holdings, demonstrating two distinct processes at
work. The larger farmers have had to become more efficient and more
business-like to combat the forces exerted by high costs of suppliers on
the one hand and competitive retailers and wholesalers on the other.

So, agribusiness and rising output per unit of labour have accompanied
depopulation as have increased indebtedness and reliance upon outside
support. The agribusiness developments have involved movements away
from the norm of single family operation of farms through the growth
of corporate structures and co-operatives whilst outside support has often
taken the form of subsidised shipment of grain. Meanwhile, if smaller
holdings have been retained it has often been through the growth of part-
time farming on a seasonal basis, further eroding the traditional pattern
of rural communities.

Todd (1980), in a study of southern Manitoba, showed that, despite
agricultural improvement and enhanced profitability, negative employment and income multipliers ensued. Labour shedding and some reductions in income have hastened out-migration, further exacerbating the
'depressed income standing of rural communities' (Todd, 1981: 464).
Even close to the major metropolitan centres there has been a loss of
population and services, and a mixture of depopulation, suburbanisation
and repopulation. For example, Todd and Brierley (1977) noted that
suburbanisation of rural census sub-divisions around Winnipeg and Bran-
don had brought especially high growth in general services, trade, finance
and public administration and defence. The sub-divisions around these
two urban centres also showed dramatic reductions in people without
formal education and growth in numbers in receipt of high-school education. In contrast, under-performers in education in the rural areas of
southern Manitoba were explained by out-migration, consequent upon
agricultural intensification, which chiefly involved the better educated
(Todd, 1979a; 1979b).

During the 1980s the numbers of farms and farm population have
continued to decline, and the rationalisation of the network of grain
elevators and rural railways has progressed further. The destruction of
whole communities, including those founded by specific ethnic and
cultural groups, has meant that a great deal of the cultural distinctiveness
of the Prairies is disappearing along with the modification in the rural

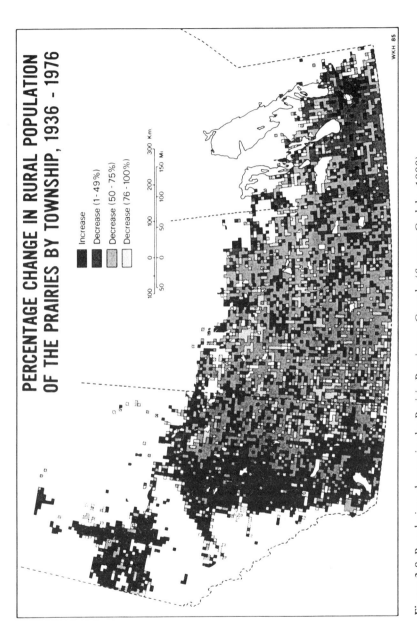

PERCENTAGE CHANGE IN RURAL POPULATION OF THE PRAIRIES BY TOWNSHIP, 1936 - 1976

Increase

Decrease (1 - 49%)

Decrease (50 - 75%)

Decrease (76 - 100%)

WKH 85

Figure 3.8 Population change in the Prairie Provinces, Canada (*Source:* Carlyle, 1988)

settlement pattern. The rural population has become a shrinking minority of the Prairie population, with a cycle of decline continuing as its driving forces still favour farm amalgamation and centralisation of goods and services. The 'wide open' Prairie landscapes remain as a prevailing image of Prairie life, but the realities of that life now reside in the big cities and not in the scattered hamlets and service centres that were dominant in the early decades of the century.

3.5   Rural depopulation and counter-urbanisation

In the 1970s and 1980s the extent of rural depopulation has not been so widespread. Many areas that had experienced depopulation for several decades have started to regain population once more as urban influences have translated into a variety of factors promoting new settlement in rural areas: in-migration has replaced out-migration even in many parts of the rural periphery. Conversely, depopulation has continued to affect both lowland areas with prosperous agriculture as well as more remote uplands and the periphery. Throughout the Developed World, depopulating villages can be found in close proximity to villages increasing population through in-migration (as implied in Figure 3.3). Whilst the traditional factors promoting depopulation are still operating in many depopulating villages and townships, e.g. reduction in the farm labour force, inadequate housing, poor services, Weekley (1988) argues that new factors are also at work. It is these that increasingly must be considered if the pattern of 'pockets of depopulation within overall growth' is to be understood, especially within prosperous farming areas.

Two of the most crucial factors promoting continued depopulation amidst general growth have been two aspects of change already closely identified with the urbanisation of the countryside. These are the increased proportion of elderly in rural areas – geriatrification; and a progressive increase in middle-class residents in rural areas – sometimes referred to as rural gentrification (e.g. Parsons, no date). These developments can promote depopulation by contributing to declining size of households, property amalgamation and replacement, and conversion of permanent dwellings into second homes. Especially in situations where there is no private or local authority housing development, these can lead to progressive housing decline. Hence, a modern form of the 'sub-urbanisation' of the countryside can produce depopulation. Weekley's (1988) study in the English Midlands notes that this has been the case largely because rural planning policies have restricted new housing developments in some villages that are most threatened by the forces of urbanisation. Given this key role of planning policy, the effects of geriatrification can then produce the apparent paradox of depopulation within overall growth (Robert and Randolph, 1983).

It is this phenomenon of population growth in the countryside that is considered now, in terms of the reversal of the long continued migration flow from rural areas to the towns.

4

Counter-urbanisation

4.1 The nature and extent of counter-urbanisation

4.1.1 Definitions

By the 1970s the reversal of internal migration flows in the developed countries which had started in the United States in the 1960s had spread to several parts of Western Europe. Instead of the towns decanting population from the countryside, the opposite trend began to occur. Not only did this change involve the growth of commuter villages, it also included some urban to rural migration affecting more remote rural communities which had experienced depopulation for many decades (Figure 4.1). This process, often referred to as *'counter-urbanisation'* (e.g. Champion *et al.*, 1989), has involved small but noticeable movements away from large towns: the larger the city the greater the net out-migration. In part, the movement has been towards peripheral rural areas, though in some cases it has involved migration from large urban centres to smaller towns (Fielding, 1986; Townsend, 1986). Where the chief movement of population has been from urban to rural, the term *'population turnaround'* has sometimes been employed (e.g. Dean *et al.*, 1984a).

The transformation of population movement is graphically illustrated for the United States by Eberle's (1982) account of her own family's move from Evanston, Illinois, to a small town 'in the country'. She likens the reversal of the traditional patterns of migration to 'thousands of iron filings being drawn irrevocably toward a magnet suddenly revers(ing) their direction.' The migrants to small town America she describes as people seeking life 'closer to the natural world' and escaping some of the disadvantages of urban life. She also refers to the existence of 'community' in small towns, with many people's ideal sizes tending towards 500 for a 'neighbourhood' and 5,000 for a 'community'.

Mumford's (1973) *The fall of megalopolis* gave a vivid portrayal of the problems affecting the modern metropolis and the growing difficulties experienced by the citizenry to the extent that they might wish to

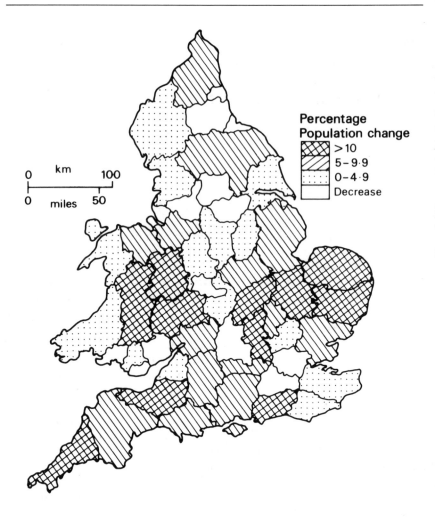

Figure 4.1 Population change in England and Wales, 1971–81 (by county)

overcome these by seeking a different residence. Indeed, he prophesised a distinct, almost natural, limit to the growth of the metropolis:

. . . beyond a certain point, which varies with regional conditions and culture, urban growth penalises itself. Too large a part of the capital outlays and annual income of the city must be spent in devices for increasing congestion and mechanically relieving its worst results (Mumford, 1973: 22).

These 'negative' aspects of urban living plus a range of other factors promoting urban out-migration have been part of a series of processes all of which have contributed to the 'wave' of counter-urbanisation so prominent in the past two decades. Hall (1986: 65), for example, refers to four processes exhibited by urban areas in the late 20th century. These are:

(a) *Suburbanisation*: the reduced growth of population and employment in central areas of cities compared with outer suburbs, followed by a decline in the inner districts.

(b) *De-urbanisation*: the decline in the inner city spreads to affect the whole functional urban region. Eventually the aggregate population and employment of urban areas may grow more slowly than in non-urban areas. This is effectively 'deglomeration', but, essentially, is what is referred to in most uses of the term 'counter-urbanisation'. Even beyond metropolitan areas, it has often been rural rather than urban districts that have experienced the most rapid growth. For example, Lichter *et al.* (1985) pointed out that within the turnaround in the United States it was often the rural parts of the non-metropolitan area rather than the urban parts that had grown most rapidly since 1970. This is an indication of *deconcentration* occurring within the non-metropolitan sector. Indeed, rural areas accounted for over 80 per cent of aggregate United States non-metropolitan population change in the 1970s (see Table 4.1). In the south and west a rural population decline in the 1950s had been converted to strong rural growth (Wardwell and Brown, 1980).

(c) *Negative returns to urban scale*: these are most manifest in major metropolitan centres in the form of slower than national average growth followed by decline in population and employment.

(d) *Regionally structurally-induced effects*: in older-industrialised regions the above trends may be accentuated by the structural decline of employment in basic industries (*de-industrialisation*). The converse may be associated with 'growth' regions containing higher level service functions.

These various trends and processes have been identified for the United States (Berry, 1976; Vining and Strauss, 1977; Sternlieb and Hughes, 1977), Western Europe (Hall and Hay, 1980) and even worldwide (Vining and Kontuly, 1977; Vining *et al.*, 1982). However, in both Canada and Australia there is the suggestion that whilst suburbanisation is still occurring, there is also a tendency for a general concentration in metropolitan centres as opposed to outright de-urbanisation. For New Zealand, Hall (1986) shows that whilst Christchurch and Wellington have experienced suburbanisation followed by de-urbanisation, the latter has not occurred in Auckland where growth of employment has continued, confirming its position as New Zealand's dominant metropolis.

The term 'counter-urbanisation' has proved difficult to define without ambiguities. Fielding (1982) refers to it as a process in which population change through migration is inversely related to settlement size. Robert and Randolph (1983) extend this by giving the process two prerequisites: *decentralisation* and *deconcentration*. The former describes movement

Table 4.1 Growth of metropolitan and non-metropolitan counties in the United States by census division, 1960–80

Census division	Metro			Non-metro		
	Per cent of region's population 1980	Per cent change 1960– 70	Per cent change 1970– 80	Per cent of region's population 1980	Per cent change 1960– 70	Per cent change 1970– 80
New England	76.6	12.6	1.0	23.4	13.3	16.5
Mid Atlantic	87.7	9.2	– 2.7	12.3	6.6	11.7
East North Central	78.1	12.7	1.9	21.9	5.7	9.8
West North Central	53.3	14.6	5.6	46.7	– 2.3	4.9
South Atlantic	70.4	25.9	20.8	29.6	3.2	19.6
East South Central	51.9	11.9	13.9	48.1	0.8	15.1
West South Central	70.4	21.0	26.3	29.6	1.3	15.4
Mountain	63.1	34.1	41.3	36.9	30.5	38.8
Pacific	89.4	27.0	18.6	10.6	10.9	30.7

Source: Bohland (1988)

out from a central city but not beyond the functional urban system. Deconcentration refers to movement down the urban hierarchy, either between city regions or into rural areas. However, separation of deconcentration and decentralisation is problematic given that the boundaries of functional urban systems are elusive. Yet, much of the research upon counter-urbanisation sees the term as more than just the sum of these two 'fuzzy' processes. Instead, it is presented as a tangible force that is causing a redistribution of population in many countries. This view of counter-urbanisation as a force is criticised by Dean *et al.* (1984a) who also criticise the many parochial 'local' studies of what is essentially an international phenomenon. They also dislike the positivist mode in which nearly all the studies have been set (see also Vartiainen, 1989a).

Returning to the terms employed by Robert and Randolph, Keinath (1982) described population decentralisation in the United States as part of the decentralisation of American economic life. He viewed this as part of a move from the 'centre' to the economic periphery, though with certain parts of the periphery favoured more than others. The greatest increases in economic activity have occurred in the middle Sun belt and north-western USA, characterised by growth in the 'six pillars' of services, durable manufacturing, mining, wholesale trade, construction and transportation (Sale, 1975). However, within these growth regions significant variations occurred, sometimes reflecting greater expansion in non-metropolitan areas. Overall, the high growth periphery has developed a more diverse economic base with a dynamic tertiary sector. This growth in services and a general lack of large, integrated industrial complexes has characterised the so called 'post-industrial society' (Keinath, 1985).

At a local level, in northern Lower Michigan, for instance, it appears that one major effect of post-industrialism has been to provide more white-collar jobs, i.e. *labour-market infusion*, but to exacerbate unemployment amongst blue-collar workers, i.e. *labour-market overload* (West *et al.*, 1987). This implies that counter-urbanisation and the population turnaround are part of new patterns of economic growth in which differentiation of labour is not only favoured by new forms of industrial development, but also the location of that labour then is a substantial element in reversing previous patterns of population dynamics. Freed from the need to work in urban and industrial centres by new 'post-industrial' ex-urban development, and also seeking an ex-urban residence, the white-collar workers have been at the centre of the turnaround.

4.1.2   The extent of the turnaround

As part of the population turnaround, during the 1970s suburban growth was accompanied by falls of 15 per cent or more in the population of inner areas of cities in many parts of the Developed World. Such diverse cities as Frankfurt, Montreal, Stockholm and Paris experienced such losses. But counter-urbanisation implies a stage of development beyond this: the overall decline in population of the metropolitan area. In the 1970s this was the case for several European cities. Those experiencing up to a 10 per cent loss in population included Basle, Duisberg, Amsterdam, Copenhagen, Vienna and Birmingham. In the same decade the level of urbanisation in the United States stopped increasing for the first time since censuses were first instituted there, and the growth of small towns was recorded in areas previously well removed from the main centres of population growth, for example the Piedmont Triangle of South Carolina.

Kontuly *et al.* (1986) suggest that in West Germany the turnaround, which they refer to as *deglomeration*, has been under way since the early 1960s. For Italy, Dematteis (1987) identifies the late 1960s as the time when counter-urbanisation started, with an acceleration from 1974, after the 'oil crisis', and following a diminution of the outflow of migrants from southern Italy. The range of factors he presents to explain the spatial and temporal patterns of counter-urbanisation indicate the complexity of the process: the housing market, de-industrialisation of urban areas and industrialisation of peripheral areas, socio-historical and cultural factors, regional and urban policies, and special policies for the Mezzogiorno (Coombes and Dallalonga, 1987):

In France counter-urbanisation has represented the first net out-migration from Paris and its region since records of migration were first kept. There has been a drift of population towards Mediterranean France for some time, but the out-migration from Paris during the 1970s was also associated with movement to rural areas in western France (Ogden, 1985). In Brittany, for example, the long period of net migration outflow has been reversed from the late 1960s. Whilst a substantial element of this in-migration has been retirement migration, over one-quarter of migrants have been children arriving with economically active parents. Within the in-migrant stream, professional, managerial and entrepren-eurial groups have been important, moving to the major urban centres within Brittany, e.g. Rennes, Brest and Quimper. However, the largest flows of in-migrants have come from neighbouring regions and, as yet, there has not been a turnaround in net flows between core (the Paris Basin) and periphery (Dean, 1987).

The reasons for such movements are complex, but in many cases may be linked to government encouragement of decentralisation, for example schemes for new towns and aid to peripheral regions. This 'push' from government may be combined with easily recognised factors of migration relating to the diseconomies of living in large cities and the perception of other places as more environmentally attractive (Keeble, 1984; Lewis, 1988). For example, in Eire, Horner and Daultrey (1980) attributed rural population growth in the 1970s to the stimulus of the greater prosperity of Irish agriculture under EEC membership. Other positive influences were rural electrification, subsidised post-primary education, provision of rented accommodation for tourists and subsidies for small farmers (Attwood, 1978; Horner, 1986; McDermott and Horner, 1978).

Attempting to quantify the process of counter-urbanisation is extremely difficult because of the inadequacy of official definitions of the term 'urban'. In Western Europe, according to World Development Reports, only the Netherlands recorded a lower proportion of urban population in 1981 compared with 1960, yet the majority of countries recorded falls over the same period for the proportion of urban population living in the largest city. In this respect out-migration from Oslo and Vienna was most dramatic, followed by Copenhagen and London. The proportion of the Norwegian population living in Oslo fell by 18 per cent from 1960 to 1980. The corresponding figures for Vienna, Copenhagen and London were 12 per cent, 8 per cent and 4 per cent. In contrast, proportions for Athens, Rome and Madrid rose by 8 per cent, 4 per cent and 4 per cent, reflecting not only the degree of variation in counter-urbanisation, but also problems with suitable definitions of 'urban' and boundary problems for individual cities.

A more detailed picture for the Netherlands is presented in Table 4.2. This example was selected because in the Netherlands the overall popula-tion of the 17 major municipalities declined throughout the 1970s. In 1970 29.3 per cent of the population lived in 15 municipalities with more than 100,000 inhabitants; in 1980 26.8 per cent lived in 17 such municipalities.

Table 4.2 Population change in the major Dutch cities, 1971–81

Municipality	Population in 1971 000s	Population change 1971–5 %	Population change 1975–81 %	Population in 1981 000s
Amsterdam	831.5	− 8.8	− 6.1	712.3
Rotterdam	686.6	− 9.6	− 7.2	576.3
The Hague	550.6	− 12.3	− 5.4	456.7
Utrecht	279.0	− 8.2	− 7.7	236.2
Eindhoven	188.6	+ 1.8	+ 1.9	195.7
Haarlem	172.2	− 3.7	− 5.0	157.6
Groningen	168.8	− 2.4	− 1.0	163.0
Tilburg	152.6	− 0.3	+ 0.7	153.1
Nijmegen	148.8	− 0.7	− 0.3	147.3
Enschede	139.2	+ 1.4	+ 2.3	144.3
Arnhem	132.5	− 4.8	+ 2.0	128.7
Apeldoorn	123.6	+ 7.4	+ 6.0	140.8
Breda	121.2	− 2.6	− 0.8	117.1
Zaanstad	116.1	+ 7.3	+ 4.1	129.7
Leiden	101.2	− 3.3	+ 5.4	103.2

Source: Bevolking der Gemeenten van Nederland

There is also a great deal of variation with regard to the amount of counter-urbanisation within Western Europe. Apart from the United Kingdom and the Netherlands, the European urban systems have not reached the same stage of urbanisation as the United States. This is particularly true for Southern Europe where urbanisation is actively increasing in Greece, Italy, Portugal and Spain. Countries with relatively low levels of urbanisation, such as Austria, Eire, Finland and Norway, are experiencing growth in some of their urban areas, but not in others (e.g. Nicholson, 1975; Vartiainen, 1989b). Of the more urbanised countries, several cities in France, Sweden and West Germany are exhibiting the characteristics of counter-urbanisation. Some of these changes are illustrated in Table 4.3, categorising population changes between 1960 and 1980 in some of the major European cities.

Kontuly and Vogelsang (1988) argue that counter-urbanisation can be measured in the form of a negative relationship between regional population size and net migration rate. This relationship has actually strengthened in the early 1980s, partly because the tendency for out-migration from urban areas filtered down from older age groups to younger ones. Demographic projections predicting an aging population suggest that the strong counter-urbanisation trend may well continue.

The process is more clearly defined in North America, although here there has been some re-urbanisation in which population in the central areas of cities has stabilised or even started to increase, reversing the absolute decline in the population of city centres evident in the 1960s and 1970s. This reversal of the urban to rural movement has occurred

Table 4.3 Types of population change in selected west European cities, 1960–80

Decline in metropolitan area and inner city 1960–80	Decline in inner city, 1960–80, decline in metropolitan area, 1970–80
Bristol	Amsterdam
Charleroi	Basle
Covilha	Berne
The Hague	Birmingham
Liverpool	Copenhagen
London	Duisburg
Newcastle	Evora
	Hanover
	Lausanne
	Manchester
	Mirandela
	Oxford
	Rotterdam
	Saarbrucken
	Stuttgart
	Vienna
	Zurich

Decline in inner city, 1960–80	Decline in inner city, 1970–80
Bordeaux	Antwerp
Frankfurt	Bologna
Geneva	Brussels
Ghent	Dortmund
Hamburg	Dublin
Lille	Gothenburg
Lyons	Graz
Paris	Groningen
Stockholm	Helsinki
	Larvik
	Leiden
	Liege
	Linz
	Malines
	Malmo
	Milan
	Nantes
	Naples
	Nuremburg
	Oslo
	Utrecht
	Vienna

Source: OECD, 1983

in many North American cities, often as part of the process of *gentrification* or 'landscape of improvement' in which particular groups of people are attracted to inner city areas where they alter the urban fabric (London, 1980). Those areas that have proved to be the most susceptible to gentrification have been those inner residential areas in close proximity to the city centre, especially those near the presence of an existing 'elite' social group (Jackson, 1985; Ley, 1984; 1986; Smith, 1982).

In Canada the 1970s were the first decade since the 1930s to register a population growth rate in rural areas superior to that of urban areas. As elsewhere this 'turnaround' can be attributed to the expansion of the rural non-farm population, though the picture is confused by definitional changes within the national population census. The analysis by Joseph *et al.* (1988) showed significant regional variations in this pattern of growth, with 'urban spillover' growth most important in Quebec, Ontario and the Prairie Provinces.

Much of the literature on counter-urbanisation emphasises that it is a process associated with the movement of people who are making a 'clean break' with the city, i.e. people migrating beyond the commuter belt to live and work in the countryside. This process is now examined more closely.

4.2 Forces of counter-urbanisation and the counter-urbanisers

An example of counter-urbanisation in a British context is Cornwall, well removed from major metropolitan areas (excepting Plymouth). 'Non-retirement counter-urbanites' can be divided into six categories as follows (Dean *et al.*, 1984b; Dobson, 1987a; 1987b; Spooner, 1972):

(a) career transients or 'spiralists'. People moving into the region for reasons associated with their career, but whose residence is only temporary;
(b) commuters;
(c) professional self-employed and small businessmen, choosing to live in an area they perceive offers a high quality of life;
(d) small manufacturers, people seeking a 'new start' and people involved in permanent job transfers;
(e) job specific and non-specific migrants. These represent the highly skilled migrants with the ability to find work in various locations, and who may be changing careers and moving to an area with pleasant surroundings;
(f) partial employment migrants. People who choose to be employed for only part of the year and who can derive benefit from the seasonal nature of the Cornish tourist industry.

It is clear from this categorisation that it is the white-collar workers, and especially particular groups within this broad category, that have dominated the turnaround. More specifically, Thrift (1987a) argues that the main driving force of the shift of industry in Britain from urban to rural areas is the emergence of a *'service class'* which has sought rural locations. Hence, he identifies social class change as a central mechanism and, in particular, the growing importance of a service class of professionals and managers (see also Abercrombie and Urry, 1983; Massey, 1983). The image of this service class is rural, with a predilection for their rural idyll which lies at the heart of their particular lifestyle (Thrift, 1987b). This interpretation emphasises choice and a deliberate seeking out of attractive rural locations, i.e. the 'pull' factor within in-migration (e.g. Joseph *et al.*, 1989).

Exactly how this 'pull' has been extended to more remote or peripheral rural regions has been disputed. For example, in the United States, McCarthy and Morrison (1977) have argued that a major factor in the growth of non-metropolitan areas far removed from principal urban centres has been the way in which the interstate highway network has broken down the accessibility barrier to remote rural regions. This has been disputed by Briggs and Rees (1982) in an examination of economic trends in non-metropolitan areas in the 1970s. The presence of interstate highways was seen to be a permissive factor rather than a direct stimulus to growth. Instead, emphasis was placed upon the role in employment creation of *branch plants* as these are the dominant type of manufacturing establishment in rural areas. Their importance in the economy as a whole is indicated by the fact that non-metropolitan areas in the United States added 56 per cent of the country's growth in manufacturing jobs from 1962 to 1978 (Haren and Holling, 1979). Half of this gain had occurred by 1967, and it was the service sector that dominated non-metropolitan employment growth in the 1970s. Another factor acting as a direct and growing stimulus to non-metropolitan development in the 1970s may be the higher proportions of unearned income (e.g. investment income and transfer payments) to total personal income in non-metropolitan areas, and especially those more removed from urban areas.

Other studies have attempted to summarise the full range of factors underlying recent rural growth. For example, according to Carlson *et al.* (1981: 13) major factors producing growth in rural regions in the USA include:

(a) growth of rural-based recreation activities;
(b) increased student populations of higher education institutions located in rural areas;
(c) dispersion of manufacturing, business activity, and services;
(d) rural residential development adjacent to metropolitan areas;
(e) new resources developments, often energy related;
(f) continuing higher birth rates in rural areas;
(g) retirement of older persons to rural areas (see Wardwell, 1977).

In summarising the causes of the population turnaround in Australia, Hugo and Smailes (1985) listed eight non-mutually exclusive causes, as indicated in Table 4.4. With respect to Australia, three general causal mechanisms were cited as filling key roles: the expansion of urban fields;

Table 4.4 Causes of the population turnaround in Australia

1. The turnaround is only a temporary fluctuation in the general trend toward urban concentration in response to the economic recession of the 1970s.
2. The turnaround is a demographic effect caused by changes in the particular age and life-cycle population mixes of metropolitan/non-metropolitan populations.
3. The turnaround is a result of successful public regional development and decentralization policies, particularly those relating to deconcentration of manufacturing industry from large cities.
4. The turnaround is an area-specific effect traceable to employment growth in particular, localised industries in favoured non-metropolitan regions (e.g. mining, defence), rather than a general broad-scale phenomenon.
5. The turnaround is a result of the gradual emergence of scale diseconomies in large urban areas, which combine with growing social problems to increase the push factor in migration streams from urban areas.
6. Reduced distance friction associated with new transport and communication technology has allowed a further rapid extension of urban commuting fields into widely dispersed but still metropolitan-focused economic networks.
7. There has been a basic change either in people's values and life-style preferences or in their ability to act on such preferences, acting in favour of residence in rural or small-town environments and against large cities.
8. The turnaround is primarily a result of structural change in modern Western economies as the proportion of tertiary and quaternary employment increases relative to secondary employment, while the decline in primary employment has almost run its course.

Source: Hugo and Smailes (1985)

changes in people's residential choices (e.g. De Jong and Sell, 1977); and changes in economic structure producing an employment led turnaround (e.g. Jarvie, 1981).

In terms of population growth in Australia between 1976 and 1981, the largest increases were recorded in areas outside the major metropoli, along the coasts of New South Wales (Burnley, 1988), Queensland and Western Australia (Figure 4.2). However, many of the most rapidly growing small towns and rural areas were distant from metropolitan centres so that mere overspill cannot explain their rapid growth (e.g. Hugo, 1983). It seems that not only has the direction of migration been reversed, but, more importantly, the propensity for migration from the countryside has been reduced.

In a detailed study of South Australia, Hugo and Smailes (1985) found a variety of factors at work to support the three general causes referred to above (Figure 4.3). In particular, structural change was present, related to growth in the public tertiary sector, plus other increases in retailing and services (Smailes, 1979). However, these structural changes were reinforced by several behavioural influences, e.g. entrepreneurs desiring to locate in rural areas and children of farmers choosing to stay in rural localities (Zuiches, 1981). These two processes operated

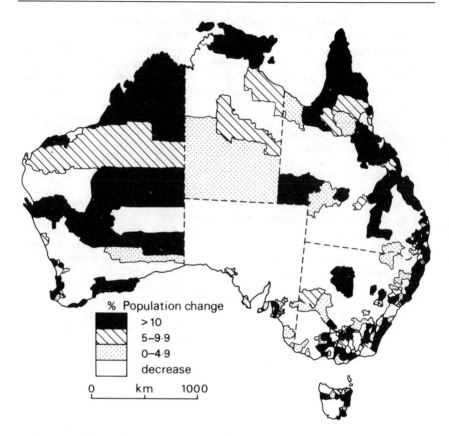

Figure 4.2 Non-metropolitan local government areas in Australia recording net migration gains 1976–81 (based on Hugo and Smailes, 1985)

alongside a general widening of the commuter belt around Adelaide, bringing a general growth to new areas. Overall, they concluded that the turnaround could not be interpreted as 'employment led' but was multi-causal to the extent that their explanatory sketch of the turnaround incorporated 26 different 'boxes' (Figure 4.3) (Smailes and Hugo, 1985).

The significance of the range of economic and social factors referred to by Smailes and Hugo has also been examined in work in Canada. Todd's (1983) analysis of the population turnaround in rural Manitoba showed that, whilst the rural farm population was declining in the late 1960s and early 1970s, the rural non-farm population was growing, though at a slower rate from 1971–6. He examined several of the factors presumed to be influential in these changes: quality-of-life indicators, objective measures of economic activity and subjective indices expressing people's preferences for small town living. Whilst economic variables such as the level of manufacturing activity and cutbacks in primary economic activity figured in his models, social factors were highly

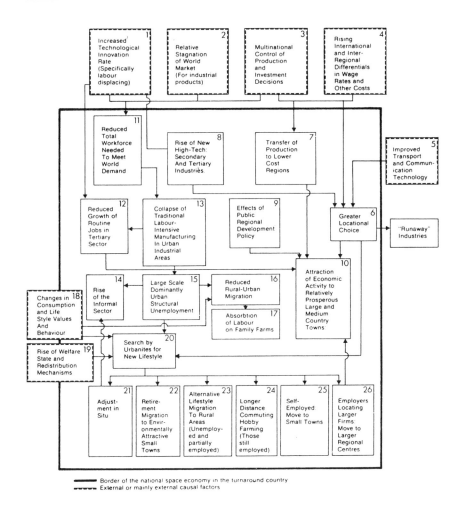

Figure 4.3 The population turnaround: an explanatory sketch (*Source:* Hugo and Smailes, 1985)

significant. In particular, emphasis was placed upon availability of medical services in small towns, reflecting the way in which communities were attracting retirement migrants (see also Barsby and Cox, 1975). The patchy nature of the turnaround on the Prairies is illustrated by Stabler's (1987) work on rural retailing. This showed that the growth of population in non-metropolitan areas in the Prairies did not result in a corresponding increase in retailing in these areas. Indeed, of 1,027 small communities surveyed, over half actually declined in terms of retail functions. Furthermore, each functional category provided fewer consumers' and producers' services in 1981 compared with 1961. Hence, 'while an increasing number of people prefer to live in non-metropolitan areas, their shopping patterns, nevertheless, reflect a preference for the options provided by larger urban places. Their presence has not revived the trade centre status of the small communities they have chosen to live in' (Stabler, 1987: 43).

These conclusions are extended by the work of Johansen and Fuguitt (1979), in a sample survey of villages in the United States. They found that the greatest declines in retailing had tended to occur in rural areas with the easiest access to urban areas. Villages with lower levels of urban accessibility tended to retain traditional roles of retail provision for a longer period. For north-west Wisconsin, McGranahan (1980) suggested that loss of commercial functions in small communities had been accompanied by retention or even gains in 'residential and social functions', e.g. church membership. This notion of community attachment affecting geographic mobility was confirmed in a survey of residents in Washington State by Fernandez and Dillman (1979), showing that community attachment influenced mobility in an age specific context, with attachment retarding mobility for individuals aged over 45 years.

4.3 Examples of counter-urbanisation

As indicators of the way in which urban to rural migration has affected a range of rural areas beyond the confines of the immediate urban commuter belts, two examples are cited below: one at a regional level for an area previously characterised by a long period of substantial depopulation, and one national situation in which the pattern and processes of counter-urbanisation can be recognised as complex. These examples are followed by consideration of three key aspects of counter-urbanisation: part-time farming, second homes and retirement migration.

4.3.1 The Highlands and Islands of Scotland

Between 1971 and 1981 the English- and Welsh-born population in the Highlands and Islands of Scotland increased by one-third and represented 30 per cent of the region's increase in resident population (Jones, 1982). Work by Forsythe (1974; 1980; 1982a) suggested that the all-important consideration in moving was found to be 'to live in a nicer area', with responses referring to scenic beauty, tranquility, space, remoteness and outdoor recreation being pre-eminent. A critical push factor underlying the migration was the desire to leave behind 'the rat-race', for the majority involved in moving had a background in professional, managerial and allied occupations. This migration was a conscious social distancing from metropolitan work structures, consumption patterns and life-styles. The moves were made primarily by people over a wide age spectrum, though with few in the 16 to 24 years age range, and with a high proportion of self-employed heads of household. Business openings,

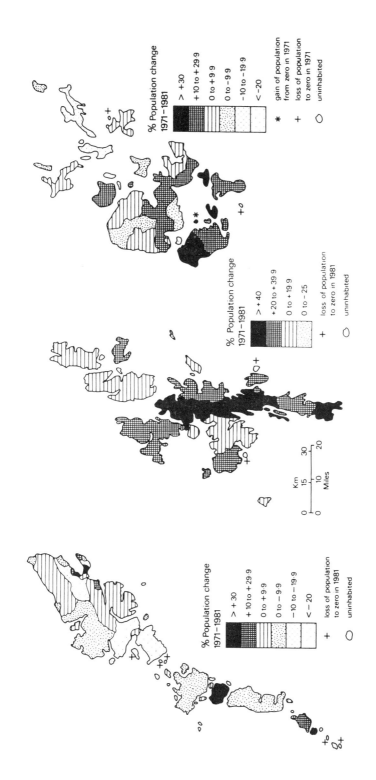

Figure 4.4 Population change in Scotland's Northern and Western Isles, 1971–81 (based on Nurminen and Robinson, 1985)

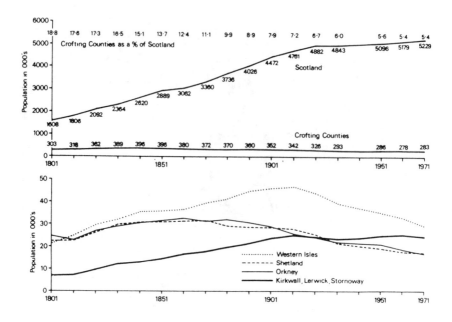

Figure 4.5 Population change in Scotland, the Crofting Counties and the Isles, 1801–1971 (*Source:* Nurminen and Robinson, 1985)

often associated with the primary sector or construction industry, had been created by the entrepreneurial enterprise of the newcomers. This pattern of employment was associated with high job satisfaction, self-fulfilment and provision of work opportunities for the migrants' children (Jones *et al.*, 1984).

Even in remote areas, where depopulation has occurred for over one hundred years, there have been gains in population during the past two decades. These gains have been associated with oil-related growth, retirement migration and tourist development. This growth has even been extended to the outlying island groups, the Orkneys and Shetlands (the Northern Isles) and the Outer Hebrides (the Western Isles), though with a great amount of spatial variation (Figure 4.4). A significant characteristic has been the continued growth in the proportion of the population in the three main towns in the Isles, Kirkwall, Lerwick and Stornoway, meaning that the Orkney mainland, Shetland mainland and Lewis respectively have had greater long-term stability in population than other islands (Figure 4.5). The three towns now account for 37 per cent of the population of the Isles compared with 19 per cent in the 1880s.

In the Orkneys, the one island in the north of the group to experience a percentage gain in population of over 10 per cent was Rousay where there has been an influx of migrants from England. Numbers are small, but Forsythe (1974; 1980; 1982a; 1982b) recorded an in-migration of over fifty families from the late 1960s onwards, most of them from southern England. The unspoilt environment, the 'better way of life' and the availability of relatively inexpensive farms and crofts have been the most important pull factors for in-migration (OIC, 1983). Similar in-migration has occurred on other islands in the Orkneys, e.g. Eday and Sanday, but on these it has not counterbalanced population loss (Lumb, 1980: 66; 1981). The causes of continued out-migration include lack of job opportunities; concentration of oil and service employment in Kirkwall, Stromness and Flotta; young people leaving for training and further education; families leaving to be nearer their children at secondary school in Kirkwall; persons wishing to be nearer health and social services in Kirkwall; and the popularity of Kirkwall and Stromness as places of retirement (Nurminen and Robinson, 1985; Wilson, 1981).

The effects of return migration, in-migration from England, the oil industry, and also continued loss of locals in their teens and twenties have been to produce three very different types of area in the Isles:

(i) The rapidly growing island capitals and their dormitory settlements, where much of the new employment has been oil-and services-related;

(ii) Dispersed growth in the Northern Isles, which includes population growth around oil terminals plus areas of in-migration for retirement or to lead a 'better way of life'. In the Western Isles this includes communities affected by growth in service industry and the retention of traditional cottage industries;

(iii) Stabilisation or continued decline, which have typified the more remote rural areas.

The extent to which growth will be maintained when oil-related employment ceases is a question now well to the fore in local thinking, but with no obvious answers.

The uneven pattern of population growth throughout the Highlands and Islands, and especially in the Northern and Western Isles, is indicative of the range of forces promoting growth, some of which comprise the processes of counter-urbanisation. Despite the growth of oil-related employment, a significant component of population increase has comprised retirement migrants and return migrants (Jones et al., 1986). For example, in several communities in the southern part of the Western Isles the returning population is that generation which, at the end of the 1940s, was 'tending to leave the island to seek employment on the mainland or abroad, or at sea' (Hobson, 1949: 80). These people are returning to take up small crofts which can provide some subsistence and limited additional revenue to old age pensions or other income. However, it has been the limited alternatives to crofting and fishing that have reduced the attractiveness of the region for people moving to seek new employment. In this respect the importance of governmental activity must not be overlooked in its attempts to promote the growth of industrial and service employment.

In this example, the chief agency involved in job creation has been the Highlands and Islands Development Board (HIDB) which, since its establishment in 1965, has been empowered to carry out its own projects, acquire land and buildings, promote development through the provision of grants and loans as well as giving advice and training. In particular, it has attempted to stimulate new private enterprises, especially in the areas of fishing, craft industry, tourism and large-scale capital intensive industry (Spaven, 1979), with a focus on the three 'growth areas' of Fort William, Inverness and Caithness (Hughes, 1984). These growth areas have been associated with the largest numbers of new jobs, but in the Isles there has been a higher proportion of jobs created and retained per head of population. Even so, unemployment rates in rural areas have largely remained above the national average, mainly because of continued structural weaknesses in the economy. For example, in the Western Isles the reduction in male unemployment in the mid-1970s through the creation of jobs in the tertiary sector was not associated with the development of continuing growth of employment in manufacturing.

In-migration for retirement, oil-related employment, some service employment (e.g. in the tourist industry at Aviemore) and for the 'good life' in remote communities has brought the population turnaround to the Highlands and Islands. However, this may represent only a temporary phase within the longer-term pattern of out-migration. Rapid population growth in certain areas can be set against continued slow decline elsewhere and the limited life-span of the oil boom. The economic problems of this periphery may act as a significant deterrent, so that the process of counter-urbanisation is replaced by depopulation and the re-urbanisation seen already in parts of North America.

4.3.2 The United States

The dynamics of the population turnaround have been investigated for the United States by Johansen and Fuguitt (1984) who based their findings on a survey of 572 villages. A basic result of the surveys was the elimination of metropolitan growth as an explanation of village growth, i.e. growth was not simply the product of metropolitan sprawl swallowing up adjacent settlements. Important factors affecting the pattern of the turnaround were regional location, county growth, village accessibility to larger centres and the initial size of the village. For example, there were increasing regional differences in growth levels, with the highest growth in the West, followed by the South, Plains and the North.

Village growth was associated closely with county growth whilst proximity to cities was much less important in the 1970s than in the previous two decades. Smaller villages, which had lagged behind with regard to growth in the 1950s and 1960s, attracted more rapid rates of population increase in the 1970s. Village growth was part of a process of deconcentration related to factors such as recreation, energy development and

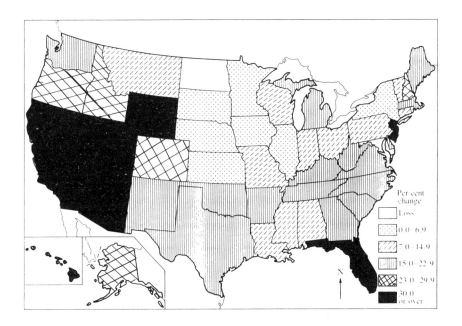

Figure 4.6 Change in non-metropolitan population (by state), 1970–80 (based on Bohland, 1988: 42)

extended commuting, especially in non-agricultural counties. Hence, there has been a major shift in the settlement hierarchy, symbolised by the way in which the increase in manufacturing and related economic activities is following a pattern more independent of the urban hierarchy than before (Lonsdale and Seyler, 1979; Till, 1981).

The highest proportional gains in non-metropolitan counties in the 1970s occurred in the south and west (Figure 4.6) as a result of gains of migrants from the cities. Williams and Sofranko (1979) suggested that the motives for this migration may not be tied as closely to employment and work-related factors as suggested by earlier work on migration. Indeed, Roseman and Williams (1980) found that issues such as environmental quality were a major factor in persuading people to leave the cities, with social ties to a non-metropolitan location also influencing moves. Economic motives remained important, but with 'quality of life' assuming increasing importance as a motivation for migrating (Goldstein, 1976: 424).

Bohland (1988: 45), summarising findings from various studies (e.g. Beale, 1977); Long and De Are, 1980; Long and Frey, 1982; McCarthy and Morrison, 1977; Williams and Sofranko, 1979), identified several attributes associated with out-migration to non-metropolitan counties which have affected the changing patterns of both depopulation and counter-urbanisation:

(a) Proximity to metropolitan centres;
(b) Dispersion of economic opportunities;
(c) Changes in the American life-style;
(d) Expansion of public institutions in non-metropolitan counties (see Bohland and Treps, 1981);
(e) Movement of urban elderly to rural retirement areas;
(f) Growth in the importance of rural recreational areas (e.g. Hart, 1984).

These attributes represent more than just the simple expansion of the urban fringe, as many counties remote from urban centres have experienced positive net migration (Sternleib et al., 1982). However, migration to the urban fringe rather than to 'deeper' countryside has itself often been associated with a change in the location of employment rather than just with extended commuting (Morrison and Abrahams, 1982).

Despite the turnaround of the 1970s, during which non-metropolitan counties increased their population by 25 per cent, in 1980 three-quarters of the population of the USA lived in metropolitan counties, and in the southern states the metropolitan counties were still growing more rapidly than the non-metropolitan counties. Indeed, in the early 1980s the indications were that the turnaround was slowing or had even been reversed. For example, between 1980 and 1984 the metropolitan population grew by 5 per cent whereas non-metropolitan growth was only 3.8 per cent. Furthermore, Engels and Forstall (1985) noted that, especially in the North-West and Mid-West, central cities within metropolitan areas were growing more rapidly than non-metropolitan counties. This reversal may be due to weaknesses in rural economies, loss of rural services and higher costs imposed by deregulation of airlines and telephones.

In effect, the recessionary period of the early 1980s in the United States coincided with a slowing in the growth rate of non-metropolitan areas so that they began to grow more slowly than metropolitan areas (Elgie, 1984). It is possible that changes in non-farm self-employment and the incidence of female participation in the labour force have been influential in this change (Cook, 1987). In Oregon and Washington State, for example, areas providing opportunities for both male and female employment have begun to be more influential in directing migration flow whilst the presence of high levels of small businesses and self-employment have acted as a deterrent to migration.

In attempting to predict the pattern of counter-urbanisation over the next decade or so, it seems that continuing diseconomies of urbanisation will conflict with desires for centralisation and the attractions of urban life as revealed in the emergence of urban gentrification in the 1980s. Diseconomies such as the high costs of land development, labour and housing, and increased congestion and pollution should continue to make small towns and non-metropolitan communities attractive for new industrial development, especially of the large space-using type. In this respect, in Canada the recent siting of new car plants in Quebec at Bromont (by Hyundai), and in Ontario at Alliston (Honda) and Cambridge (Toyota), seems instructive (Thraves, 1988: 198). Thus, despite a slowing of growth in non-metropolitan areas in the mid-1980s, growth prospects still seem good for small- and medium-sized communities, especially within the still expanding commutersheds of the large metropolitan areas (Bryant *et al.*, 1982: 219–22).

4.4 Part-time farming

Whilst one theme within literature on the 'turnaround' and counter-urbanisation has been the role of services and industry seeking rural locations, another has been the influence of agricultural change, and especially the growing importance of *part-time farming*. This phenomenon has been associated traditionally with urban fringe locations. However, it has become more widespread in recent decades, often representing an outlet for former urban dwellers seeking a rural retreat. So, it has become a part of counter-urbanisation, whilst also retaining a particular place in urban fringe locations.

4.4.1 The nature of part-time farming

The rural-urban fringe may be recognised as a distinctive area in a physical sense with characteristic land use associations where urbanisation impinges directly upon rurality in some recognisable form. It may also be recognised in sociological terms. Pahl's studies in the rural-urban fringe of north London drew attention to these sociological aspects whereas previously much geographical study of such areas had actually been based on the physical delimitation and definition of land uses. Typically these had revealed a wide variety of uses, often with a high degree of inter-mixing, reflecting the haphazard growth of the city at different rates in different directions. The nature of urban growth has produced an incoherent land use pattern comprising a wide range of uses, including agriculture, rural settlements, modern residential estates, industry, out-of-town shopping centres, derelict land, cemeteries and sewerage works.

At the city's edge the proximity of the city has had a major influence upon agriculture in four ways:

(a) through the loss of farmland for urban development;
(b) through the purchase of land for development, producing smaller farming units and the fragmentation of holdings;
(c) through speculation in anticipation of development, creating a deterioration in agricultural standards, possibly likely to occur on all

farmland in the vicinity of towns and through vandalism;

(d) through the spread of what Carter (1981) refers to as 'rural retreaters' or *hobby farmers* who farm on a part-time basis.

Part-time farming was included under Lewis and Maund's (1976) term 'population' and was seen by Mignon (1971) as a recent development in the long process of industrialisation and urbanisation of the countryside. He recognised three types of farmers whose farming activity is not their sole form of income:

(a) In the pre-industrial period there were farmers who supplemented their income from farming with that derived from work in craft industry;

(b) From the early 19th century there was the development of the worker-peasant in which factory work provided an additional income to farming. As discussed in Chapter 3, this latter combination of farm-work and factory-work still persists in certain parts of Western Europe. In 1976 in West Germany, Italy, the Netherlands and Eire, of all types of employment, the holding of second jobs was most prevalent within agriculture, though only for Italy (8.4 per cent) was the proportion of second job-holders above 5 per cent. In West Germany of those with a second job whose main employment was in agriculture, nearly half had their second job also in agriculture whereas the figure for the other countries was close to one-third. In West Germany 70 per cent of all second job-holders were engaged in agriculture compared with 39 per cent in Eire, 20 per cent in the Netherlands and only 4 per cent in the United Kingdom (Alden and Sacha, 1978; Frank, 1986; Schmitt, 1984). Lack of data prevents comparisons with France and Denmark.

(c) Farming as a hobby: essentially a post-1945 development (e.g. Wilhelm, 1977).

It must be stressed, though, that *hobby farming* is not synonymous with part-time farming, which actually includes a number of different groups of farmers. For example, for Wales, Aitchison and Aubrey (1982) distinguished six types of part-time farmers:

(a) Those moving into part-time farming from full-time farming because their holdings were too small for full-time operation;

(b) Those similar to (a) but representing a more advanced phase of retreat from full-time farming;

(c) Farmers always operating their holdings on a part-time basis, but with an approximately even balance of income between on-farm and off-farm activities;

(d) Farmers always operating their holdings on a part-time basis, but deriving little income from farming;

(e) Those who before becoming part-time farmers had no previous occupational experience of farming;

(f) Marginal part-time farmers using their holding not so much as a

Table 4.5 Characteristics of part-time farming in Wales

Farm type	Number of farms	Scale (ha)	Arithmetic means Commitment On-farm (hours)	Arithmetic means Commitment Off-farm (hours)	Dependency %	Career context frequencies (%)* (a)	(b)	(c)
a	18	68	56	14	76	5	78	17
b	26	26	23	46	16	—	100	—
c	24	47	39	32	54	100	—	—
d	49	25	23	45	16	100	—	—
e	23	22	36	38	30	—	—	100
f	71	11	19	46	7	—	—	100

Source: Aitchison and Aubrey (1982). Descriptions of each farm type are given in the text.
(a) Always engaged in part-time farming.
(b) Entry from full-time farming.
(c) Entry from non-farm sector.

work-place but as a place of residence, but with sufficient land to pursue a range of rural activities. As shown in Table 4.5, these formed the largest group of part-time farmers, though they operated the smallest holdings, devoted the smallest amount of time to farming activities and produced the smallest agriculturally derived incomes.

Part-time farmers in the United Kingdom include at least five distinct groups:

(a) crofters in the Highlands and Islands of Scotland;
(b) farmers who are able to make a satisfactory living from less than the Ministry of Agriculture, Fisheries and Food's (MAFF) standard man-days definition;
(c) farmers who, under pressure from changing economic circumstances, develop a secondary activity based on their farm but taking labour away from strictly agricultural work, for example horse-breeding, farm shops or farm holiday schemes;
(d) institutional control of land which may or may not involve farming in a manner likely to violate the MAFF's definition of full-time farming;
(e) the development around towns of farming by people who derive an income from employment unrelated to their farm.

Hobby farming forms part of this last mentioned category in which the typical farmer comes from the professional, administrative and managerial classes. A common occurrence has been for a prospective hobby farmer from this group to purchase a farmhouse and a little land whilst selling off the remaining land to neighbouring farmers, thereby contributing both to farm sub-division and farm enlargement. With

Table 4.6 Differences between full-time and part-time farming in Kent, 1960s

(a) *Livestock output contributed by different enterprises*

	Livestock output contributed by each enterprise (%)	
Type of enterprise	Part-time farms	Full-time farms
Milk production	27	48
Rearing dairy cattle	4	6
Beef	10	8
Sheep	7	5
Pigs	19	14
Poultry	33	19
Standard output (£) per acre from livestock enterprises	28.3	35.6

(b) *Number of enterprises*

	Number of enterprises (%)	
Number	Part-time farms	Full-time farms
1	28	15
2–3	49	38
4–5	22	34
more than 5	1	30

(c) *Variation with distance from London*

Distance by road from London (miles)	Number of part-time farms (%)	Acreage in part-time farms (%)
13	46	30
18–20	60	27
30–35	54	32
36–41	50	40
50–52	48	44
56–62	29	21
Area surveyed (on London-Hastings transect)	49	35

Source: Gasson, 1966

profits of less importance than for the farmers whose livelihood depends solely upon agriculture, activities involving limited amounts of time assume greater importance. This can include the keeping of beef cattle, sheep, poultry or more exotic livestock such as rabbits and goats. Cropping using contract labour may be a possibility, with some forms of horticulture also capable of fulfilling this time criterion.

Some of the differences between full-time and part-time farming in the 1960s were investigated by Gasson (1966; 1967) in a study of the area between London and Hastings on the East Sussex coast. Of all the farm holdings in this area 40 per cent were run on a part-time basis. The majority of the part-time farms were owner-occupied, with three types of people dominant. These were the higher professional group, those of independent means (often retired) and people from a lower professional group. Between them these three categories accounted for over two-thirds of the part-time farmers. There was a definite correlation between distance from London and a decrease in the incidence of part-time farming, and, similarly, a correlation between size of part-time farms and distance from London.

Gasson's study also highlighted certain differences between part-time and full-time farming. The former derived a larger amount of its income from cropping, and especially cereals. Its modal size of holding was much smaller, lying in the 5 to 20 acres (2 to 8 ha) range in contrast to 150 to 300 acres (60 to 120 ha) for the full-time farmers (Table 4.6). The main conclusions from this study were that part-time farming was essentially farming for recreation by people not previously connected with agriculture. The farming with which it was associated was more specialised and yet more simplified than under full-time systems.

Similar findings were revealed in Layton's (1978; 1979; 1981) studies in the rural-urban fringe of London, Ontario. His study area was in the Windsor-Quebec axis, where urban pressures were exerting great influence upon the adjacent countryside (Plate 7). Around London, Layton identified two clearly recognisable groups of farmers: those completely reliant upon farming for their income and those who were either part-time or hobby farmers with non-farm employment to supplement their income. The hobby farmers had a more urban background and purchased farms as a place of residence and for personal recreation. Yet, even 44 per cent of this group of farmers farmed on a strongly commercial basis rather than maintaining an interest in just 'farming for pleasure'. The hobby farmers had tended to purchase land in the aesthetically more attractive areas and were inflating land prices. This was preventing the amalgamation of commercial farm units, with the result that the family farmer was gradually disappearing in favour of either very large agribusinesses or small part-time and hobby farms.

By 1978, 44 per cent of American farmers worked off their farm for 100 days or more; and 92 per cent of US farm families had some form of non-farm income in 1979 (Albrecht and Murdock, 1984). The numbers of small farms (under 20 ha), many of them part-time, have grown subsequently. Explanations for this growth advanced by Daniels (1986), include:

(a) Preference for living in small towns or rural settings (e.g. Healy and Short, 1981);
(b) Lack of restrictions on creation of hobby farms;
(c) Major tax incentives have made the purchase of a small farm attractive, e.g. deduction of mortgage interest payments and local property taxes from federal taxable income, farm use-value property taxation, and incorporation of farm operations for investment tax credits and tax sheltering;

Plate 7 Part-time Farming in Canada – Numerous former full-time farms, such as this one near Tottenham, Ontario, have been occupied by former urban residents commuting to major metropolitan centres but still utilising some of the farmland for limited agricultural activity

(d) Relative cheapness of rural land and houses;
(e) Improvement of transport and communications networks, enabling greater separation of workplace and residence, plus ease of access to urban amenities for rural inhabitants;
(f) Ruralisation of industry has increased the number and range of job opportunities in non-metropolitan USA, helping to create higher disposable incomes which can be used to support hobby farming (e.g. Dillman, 1979).

Daniels' own work on the Willamette Valley, Oregon suggests that, in the rural-urban fringe, hobby farmers are increasingly competing with commercial farms over the same land base. This competition raises land prices and fragments holdings, and therefore can prevent expansion of commercial operations.

4.4.2 Part-time farming in England and Wales

Gasson's (1984; 1986; 1988) recent studies of part-time farming in Britain and Western Europe show that the phenomenon is by no means confined to the small sub-viable holding and is not dominated by hobby farming, even in Britain where more commercial part-time farming has

Table 4.7 The proportion of farmers who are part-time – selected countries

Country	Year	Per cent part-time	Country	Year	Per cent part-time
Japan	1975	87	Sweden	1971	39
Norway	1979	69	Italy	1970	38
FR Germany	1975	55	Finland	1969	37
Austria	1973	54	Canada	1970	31
Switzerland	1975	51	Australia	1972	27
Spain	1972	48[†]	United Kingdom	1979	27
United States	1974	45	Netherlands	1975	25
Belgium	1970	43	France	1970	23
			Ireland	1972	22

Source: Gasson, 1988: 2

Table 4.8 Distribution of part-time farms in England and Wales, 1983, by size of holding in standard man-days (smds)

Size of holding in smd	Per cent of part-time holdings	Part-time holdings as per cent all holdings
Under 100	47.2	44.9
100–249	23.2	32.7
240–499	13.5	20.5
500–999	9.6	14.1
1000 and over	6.5	13.8
All holdings	100.0	30.7

Source: Gasson, 1988: 14

frequently been dismissed as of little significance. This dismissive attitude probably reflects the lesser importance of part-time farming in Britain when compared with other developed countries (Table 4.7). However, part-time farming has increased in recent years and is likely to continue to do so if some suggested reforms to the Common Agricultural Policy (CAP) are carried out, e.g. the EEC's Green Paper, *Perspectives for the CAP* (Commission of the European Communities, 1985).

   Using data from a labour input enquiry, Gasson (1988) has shown that nearly half the part-time holdings in England and Wales are under 100 standard man-days (Table 4.8). The south-east, East Anglia and south-west have the highest concentrations of part-time farming, perhaps reflecting the importance of access to London. Other aspects of her studies have extended the knowledge of how part-time farming operates in England and Wales and of the characteristics of part-time farmers and their families.

   In terms of the latter, despite the heterogeneity of part-time farmers, two contrasting types can be distinguished, the one more closely associated with farming as a business and an occupation, and the other with a farming life-style. It is the former group who have large holdings and who generally come from a background in which farming has been the main activity. These are in the minority though, and tend to farm in areas where agribusiness has fostered interests in farming-related business. More common have been the small part-timers moving into farming for motives of investment, residence and amenity. These households in which farming is a secondary role are in the majority, and they account for around one-fifth of the entire farming population of England and Wales (Gasson, 1988: 156). They tend to operate smaller holdings, and often both farmers and their spouses hold jobs off the farm. It is also common for both farmer and spouse to work on the farm.

Gasson's studies demonstrate the wide variations between part-timers despite her recognition of a certain duality amongst them (see also Munton *et al.*, 1989). Nearly two-thirds of part-time farmers derived their main source of income from non-farming activity. This seems to imply that the attraction of part-time farming might lie primarily outside any strict financial considerations. Yet, in her farm survey carried out in 1984, over half of her sample cited financial and broadly related reasons as the main benefits of part-time farming. The additional income provided, plus the farm's ability to utilise available farming labour, are often significant considerations over and above those relating solely to 'life-style' and the farm as a rural retreat. Nevertheless, the latter were of greater significance where farming was a secondary activity (see also McQuin, 1978).

In terms of the type-of-farming practised, the following summary by Gasson (1988: 96) is an excellent description:

Part-time farmers in general operate on a smaller scale than full-timers, they have less time available for farming and they are not under the same pressure to maximise or even to maintain their farm incomes. These tendencies are reflected in their choice of enterprises. Part-time farmers tend to favour simpler and less demanding systems than those who depend on farming for their livelihood.

Thus, labour-extensive crops, such as cereals, have been favoured rather than labour-demanding ones in many examples of part-time farming in the USA, West Germany and Norway (OECD, 1978: 11). The high labour demands of dairy cattle have also led to their avoidance by part-time farmers in Wisconsin (Kada, 1980), Canada (Bollman, 1982), Norway (Symes, 1982), southern Sweden (Persson, 1983) and Eire (Higgins, 1983). However, other livestock enterprises have proved popular with part-time farmers, especially beef production and sheep (Table 4.9). Dairying has only tended to be important for part-timers with no other job (e.g. Cawley, 1983).

Gasson's survey also demonstrated that 10 per cent of all part-time farms in England and Wales had little or no commercial farming activity.

Table 4.9 Type-of-farming and farmers' main activity on part-time farms in England and Wales

Type of farming	Farmers with other jobs major	minor	Farmers without other jobs
	per cent of holdings		
Dairying	4.1	7.3	23.5
Cattle and sheep	59.7	52.9	37.5
Pigs and poultry	4.1	3.9	3.6
Cropping	21.0	20.6	21.5
Horticulture	6.1	8.0	5.2
Mixed	5.0	7.3	8.7
All types	100.0	100.0	100.0

Source: Gasson (1988)

Typically, on such holdings, land would be let on short-term grazing tenancies or the commercial activity would be the grazing of horses. Other activities included non-agricultural purposes, such as riding schools or caravan parks, or subsistence smallholdings with no predominant enterprise (Gasson, 1988: 98). It was also common for part-timers to use land extensively, with high proportions of grassland and unproductive land. The latter may reflect poorer maintenance standards too on those part-time holdings where farming is largely a hobby or for limited economic return (e.g. Munton, 1983: 126–31).

A variety of statistics can be used to indicate the significance of part-time farming within the overall pattern of agricultural development. For Britain, just three 'facts' can serve the purpose:

(a) Over 40,000 families are being enabled to remain on the land by virtue of other sources of earned income;

(b) About 10 per cent of part-time farming households (or 5,000 families) made less than £4,000 from farming and other activities combined in the financial year 1983/4. One-third of these families were in the South-West Region;

(c) Four-fifths of these low income families derived over half their household's income from farming and frequently had a farm-based second activity. The commonest of these were the provision of accommodation, sporting and recreational activities, and adding value to produce through processing or retailing.

These suggest that whilst most part-time farmers are 'attached' to the land and a certain life-style, they are not really a part of the 'maintenance of the fabric of rural society' as implied in the EEC Green Paper, *Perspectives for the CAP* (Commission of the European Communities, 1985). They are least common in areas of depopulation, being most prevalent in the prosperous south and south-east where they often

represent a further example of former urban residents seeking a rural
residence. Some of this group of part-timers is playing a role in preserving a particular type of rural environment which is in accordance with
generally recognisable 'conservationist' views. However, there is also
evidence for poorer maintenance of land amongst a section of part-timers
plus low productivity which could be seen as 'inefficient farming' by
those concerned with maximising output per hectare or per unit of
labour (see also Buttel, 1982b).

4.5 Second homes

One aspect of counter-urbanisation evident for over one hundred years,
albeit on a small scale in most countries pre-1945, is the ownership of
second homes in the countryside by people who had a permanent urban
residence. This practice, termed 'seasonal suburbanisation' by Clout
(1970), was common amongst the urban elite of many continental cities
in the 19th century. It involved the ownership of a second house,
cottage, farmhouse or flat outside the city, primarily for use at week-ends
or during vacations (and hence the term 'summer suburbs'). More
recently, the type of accommodation involved has also included house-
boats, caravans and purpose-built chalets or cottages (Thissen, 1978).
Once the preserve of the very wealthy, the ownership of second homes
is still associated with higher income groups, but the numbers of owners
involved has grown, reflecting the growth of wealth, mobility, increased
leisure time and the rise in the desire for recreation and living in rural
surrounds. Countries where a tradition developed for the urban elite to
own second homes, such as France, have continued to have a larger
proportion of their population owning or making use of such residences.
So in France one-quarter of all households have access to a second home
whilst in Britain, where the tradition has been much less well developed,
the corresponding figure is less than 4 per cent of households. The
maintenance of this difference over a long period of time may reflect the
variation in urban structure between continental Europe and Britain. In
the former, the middle classes, and often also the wealthiest sectors of
society, have occupied flats and apartments, and so may have had greater
desire to seek the open spaces of the countryside for leisure than
comparable groups in Britain who had private gardens. For example,
two-thirds of Swedes owning second homes have an apartment as their
primary residence (Bielkus, 1977). Unfortunately, in Britain the reliability
of statistics on second home ownership is highly questionable, though
Shucksmith (1983) suggests there were around 200,000 in the early
1980s, using data from market research organisations and the 1979
National Dwelling and Household Survey.

Amongst other differences between countries in the distribution of
second homes is the distance between primary and secondary residence.
Whilst longer distances tend to be associated with the larger cities, there

is a contrast between continental Europe and North America/Australasia. For example, in Denmark and Sweden over two-thirds of second homes are within 30 miles (50 km) of the main residence whilst in North America this distance is commonly over 60 miles (100 km) and frequently up to 200 miles (320 km). This reflects differences in the availability of fast roads, cheapness of travel, attitudes to journey times and distances, and varying distances from urban centres to areas of scenic beauty which tend to be the main areas of second homes. The contrast between the distance required to travel from the sprawling megalopolis of north-eastern USA and that from Stockholm to a 'comfortable' location is a good example of this type of variation.

As illustrated by the title of Coppock's (1977) edited volume on this topic, *Second homes, curse or blessing?*, it is a phenomenon associated with a variety of advantages and disadvantages for the people and places it affects. At one level it can be regarded as economically advantageous in being instrumental in introducing money to rural areas. Thus it may help create employment, as suggested by Clout (1972) for Mid-Wales where he referred to six second homes creating the equivalent of one full-time job. In this same area of Wales, Bollom (1978) estimated that although second homes were used on average for only 90 days per annum, they were responsible for an additional income to the local area of between £400 to £600 per second home (see also Davies and O'Farrell, 1981; Sarre, 1981). On the island of Arran in the Firth of Clyde, Pacione (1979) estimated that nearly one-third of the housing stock was second homes and that 70 per cent of the second home owners employed islanders on a part-time basis as cleaners, gardeners and caretakers. Also, seven out of ten building firms on the island obtained income from building second homes. The creation of employment, coupled with injections of new income for the local economy and support of services, can clearly be advantageous in rural areas which might otherwise be losing population. In depopulating areas, surplus land and buildings can be taken by second home owners, so preventing their deterioration and even abandonment. However, this interpretation ignores several problems that can be produced (Albarre, 1977) (Plate 8).

As with the ex-urban residents in commuter villages, the very presence of second home owners alters existing situations. Certain aspects of the character of the host communities are altered, even if in some cases the most readily apparent alteration only takes the form of new specialist retailing (e.g. antiques, crafts, country produce). In the case of new employment created, the new temporary residents can take a place towards the top of the social hierarchy. Yet, they can also be in direct competition for property with local first-time house buyers or renters. So locals can be out-bid for property, or prices can be increased by the new influence on the market, again to the detriment of the locals (Shucksmith, 1981). In some cases this has provoked conflict between the newcomers and local inhabitants, some of the most extreme examples being in Wales where it has been exacerbated by the cultural factor of the perceived threat felt by locals to their Welsh language and culture. Whyte (1978) reports a similar conflict on Skye between Gaelic-speaking locals and English-speaking second homers. Other problems may include the greater costs of provision of utilities in areas of second home ownership, these costs being partly transferred to permanent residents possibly not as affluent as their new neighbours. At one time such problems were held to be so great, with the high incidence of second homes causing a destruction of 'communities', that a proposed plan for England's Lake District contemplated a ban on further second home buyers (Clark, 1982a). However, this was deemed unworkable, and several studies have suggested that in areas experiencing long and continued depopulation the influx of second home owners confers several benefits, especially the retention of services and the stimulus to the local economy (Clout, 1969; 1971; Shucksmith, 1983).

The greatest increase in the number of second homes post-1945 has occurred in Spain which had just under 2 million such residences in the early 1980s. This has given perhaps nearly 20 per cent of Spanish households access to a second home, though a substantial proportion of such homes are either owned by foreigners or used principally by foreign visitors. The growth of tourism has fostered a concentration of foreign-owned and -used second homes on the Balearics where the numbers of these residences increased five-fold in the 1960s and 1970s. Coastal locations and concentrations in and around tourist centres dominated the initial expansion, but, subsequently, Spanish second home ownership in the Balearics has grown. This has been associated with the growth of hobby farming which has brought the development of second homes in rural areas away from the coast. Land in such locations has been considerably cheaper than on the coast, encouraging purchase by people who derive their principal income from the tourist industry (Barke and France, 1988; Garcia-Ramon, 1987).

4.6   Retirement migration

There is evidence to suggest that a significant number of second home owners eventually retire to their temporary residence so that second home ownership can be seen as a stepping-stone in this process. For example, in Pacione's (1979) survey of second home owners on Arran, 18 per cent of owners referred to purchase of a second residence as being a preliminary to retiring there. This can be viewed as just one aspect of a growing phenomenon of *retirement migration* to rural areas, with the elderly being attracted by cheaper accommodation and the perceived benefits of rural life (Biggar, 1984).

The growth in this form of migration in England and Wales in the 1960s was shown by Law and Warnes (1975; 1976; 1982) in a map of 'retirement areas' (Figure 4.7). These they defined as areas with the proportion of their population of over 60 years old more than 25 per

Plate 8 Second homes in Wales – Former farmhouses and farm labourers'
cottages, especially those in scenic areas such as these in Gwynedd, have become
major attractions for English retirees and second home owners. The resultant
buoyant housing market has tended to put such properties beyond the reach of
local inhabitants

cent in excess of the national mean plus an increase in the number of
retirement cohorts between 1961 and 1971 (or a decrease at less than
half the national rate). This highlighted coastal areas, especially the south
coast resorts, and rural areas in the South-West and East Anglia. More
recently the rising cost of housing on the south coast has increasingly
deflected elderly migrants towards the countryside. Here the influx has
placed a growing strain upon public transport and local facilities. The
demand upon such services as home help, meals-on-wheels, health
visiting, chiropody, regular visits by general practitioners and geriatric
care in rural hospitals has created major difficulties in certain areas, as
revealed in the United Kingdom, for example in several surveys of social
services carried out after the Chronic Sick and Disabled Persons Act of
1970.

In the United Kingdom in the 1970s, the growing differential between
property prices in urban and rural areas encouraged new retirement
migrants to move to the latter, especially around major resorts, e.g.
Brighton, Torquay, Colwyn-Rhyl, Blackpool and Bournemouth. Meanwhile, the resorts themselves have experienced declines in their
pensionable population. In addition, more remote rural areas also
received a rising influx of retirement migrants, e.g. central Wales, the
Welsh borders and north-east Scotland (Warnes and Law, 1984; 1985).
Similar trends have been apparent in France and the United States whilst

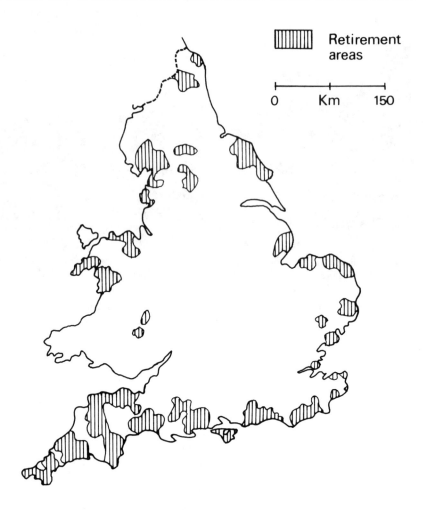

Figure 4.7 Retirement areas in England and Wales (based on Warnes and Law, 1975)

in Australia the resorts of Queensland have been a major attraction (Warnes, 1983).

In the United States, Ullman's (1954) pioneering article on 'amenity' areas drew attention to the importance of a physical setting conducive to recreation as being a factor in inter-regional population shifts. Johansen and Fuguitt (1984: 46) linked these amenity areas with retirement migration as well as the creation of employment in the tourist industry and allied services. This phenomenon has been a significant factor within population growth in the 'Sun Belt', especially Florida, California and Colorado (Biggar, 1984). Perhaps the picture of new urban communities in sprawling extensions to Miami Beach and other Florida cities is the one most closely associated with retirement migration in America, but non-metropolitan areas have also been affected (Freudenburg, 1982). Rural retirement communities are now being planned throughout the Sun Belt states, often with residence limited to individuals of above fifty years of age, and with the physical and social design geared specifically to the interests of the elderly (Broschen and Himminghofen, 1983).

The affluence of that sector of the population participating in retirement migration dictates that the development of desirable retirement services if often viewed as a potential growth industry (Lassey *et al.*, 1988: 152). This has certainly been the case recently in New Zealand where the most rapid rates of rural population growth in the early 1980s have been associated with retirement migration. Thus Willis (1988: 232–3) quotes rates of increase above 25 per cent and even above 50 per cent for several rural districts on the east coast of the North Island between 1981 and 1986: 'As the population of New Zealand ages and as people increasingly desire the warm climates and beach lifestyles these centres will continue to grow.' Similar attractions are encouraging high levels of retirement migration throughout the Developed World, with growing impacts upon rural life, especially through the way in which services and retail businesses in some communities have become reliant upon this section of society (e.g. Hunt *et al.*, 1984; Krout, 1988; Li and MacLean, 1989; Rohr-Zanker, 1989).

PART TWO
RURAL ECONOMY

5

The new agricultural revolution

Agriculture is at the centre of rural economies. Not only is it still the dominant user of land throughout most of the Developed World, its economic and social contributions to rural areas are far more significant than its frequently small share of gross domestic product or employment would suggest. For many people 'rural' and 'agricultural' have become vital symbols of a different type of environment beyond the complexities of urban living. And whilst smaller proportions of the overall population have been economically active in farming, the sheer amount of land devoted to agriculture has served as a kind of bulwark against urbanism in the minds of many urban dwellers. However, just as these urbanites have increasingly sought residences closer to their rural idylls, so some of the problems of urban, industrial life have affected agriculture, involving it and the farming community in complex issues influencing economy and society as a whole (e.g. Cox *et al.*, 1986).

There have been dramatic changes within the farming industry itself during the past four decades and these plus dynamic external forces have precipitated a number of key conflicts central to the nature of post-war agricultural development and the changing directions of the rural economy. To understand these changes and the range of crucial issues affecting modern agricultural development, it is necessary to examine the main features of farming's transformation post-1945: the constituents of what has been termed the *Third (or New) Agricultural Revolution* (e.g. Grigg, 1989).

5.1 The agricultural transformation post-1945

The changes affecting agriculture throughout the Developed World since 1945 can be aptly described as 'revolutionary' in that farming in the 1980s is a vastly different proposition from that just forty years earlier.

A transformation covering virtually all farming systems has been effected through the interaction of numerous factors:

(a) In some cases, notably the EEC countries and the United States, commitment of government to aid and subsidise agriculture has been vital in helping to stabilise the industry by achieving both technical progress and an economic stability within farming (Friedmann and McMichael, 1989);

(b) The growing wealth of consumers has increased demand, especially for 'luxury' foods;

(c) There has been a 'green' revolution in which new techniques have supplanted old traditions, with mechanisation becoming commonplace;

(d) The decline of the general agricultural labourer has continued, being replaced by fewer skilled workers operating new machinery or working as a specialist in a particular branch of farming;

(e) The heavy investment in plant and machinery, especially on arable farms, has produced a highly capital- and energy-intensive industry whilst also fostering the demand for land, as land has remained the essential capital input in farming;

(f) Market forces and the need for high levels of capitalisation have promoted closer links between farming and both its suppliers (backward linkages) and retailers/wholesalers (forward linkages).

An illustration of the effects of these changes is the way that agricultural productivity in Britain has quadrupled post-1945. There have been some spectacular increases in yields and overall production, total output in Britain doubling between 1945 and 1985, e.g. wheat production rose four-fold between 1950 and 1985, partly through a more than doubling in yields (Figure 5.1a and b), whilst the average lactation yield per cow also doubled between 1938 and 1985 (Figure 5.1c) (Soper and Carter, 1985). Yet, these rises were produced from diminishing amounts of agricultural land and labour (Figure 5.1d), phenomena repeated throughout the Developed World.

The six aspects of the New Agricultural Revolution described above have been present to varying degrees throughout the Developed World. Different combinations of the constituent factors have been present in individual countries, though the general effect has been towards significant increases in production, output per worker and output per unit area. In the United States, for example, Buttel (1980: 45) has highlighted five principal structural changes in agriculture during recent decades. These trends are interrelated and mutually reinforcing:

(a) a trend towards large-scale, specialised farm production units;
(b) increased mechanisation;
(c) increased use of purchased biochemical inputs;
(d) a trend towards regional specialisation of production;
(e) an increased level of food processing and inter-regional marketing.

The results of these trends are the 'revolutionary' increases in output, reliant especially upon machinery and chemical inputs:

Mechanisation, the use of artificial fertilisers and chemical methods of pest control, the adoption of improved irrigation techniques, the breeding of new crop varieties that are more responsive to fertiliser, more resistant to drought, more rapidly maturing and less vulnerable to disease than their natural ancestors, have combined to produce rapid and continuing increases in output per acre – a yield 'take-off' in mid-latitude farming regions (Manners, 1974: 182).

This description of modern agriculture refers to a system of farming in which a high level of technical skill has been allied with energy subsidies in an operation involving high productivity. It is this energy subsidy derived from fossil fuels that has enabled developed countries to reduce the proportion of population directly employed in agriculture to less than 10 per cent.

The drive towards greater productivity has also given rise to several significant shifts in agricultural land use (Briggs and Wyatt, 1988). For example, in Britain, partly as a result of government support policies, one of the long-term changes has been an increase in the acreage of cereals. The barley acreage nearly trebled between 1950 and 1970 as increased yields, the development of barley-based feeding systems and favourable guaranteed prices stimulated its cultivation (e.g. Dawson, 1980). Wheat, too, returned acreage levels last recorded in peace-time in the 1890s, and new crops, such as oilseed rape, became important (e.g. Wrathall, 1978; 1986a; 1988a; Wrathall and Moore, 1986). In contrast, the horticultural acreage fell as a result of increased foreign competition under the auspices of the European Economic Community's (EEC) Common Agricultural Policy (CAP) (see Best, 1979).

Certain common elements of the 'revolution' can be found throughout the Developed World, but before focusing upon some of these elements, it must be remembered that key differentiating features of rural areas remain in the form of variations in farm structure, organisation, and type-of-farming practised. Despite the labour shedding that has been experienced, the Mediterranean countries still have a larger farm labour force, and here too there tends to be a greater predominance of small-holdings and fragmented farms than in other parts of Europe. Major differences still exist between the structure of farming in the largely mono-cultural grain producing regions of the New World and the mixed farming regions of Western Europe. Indeed, significant agricultural variation exists within most countries of the Developed World despite the modifying influences of increased specialisation, widespread restructuring, the growth of agribusiness, loss of agricultural land, and the emergence of new aims and goals for modern farmers (Table 5.1).

a)

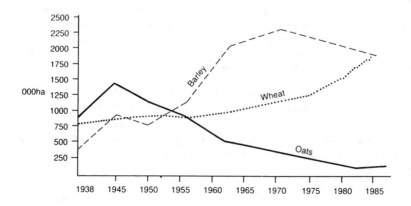

c)

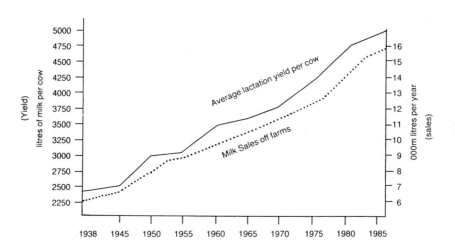

Figure 5.1 Selected agricultural changes in Great Britain
(a) The acreage of barley, oats and wheat, 1938–85
(c) Milk yields and sales, 1940–85

b)

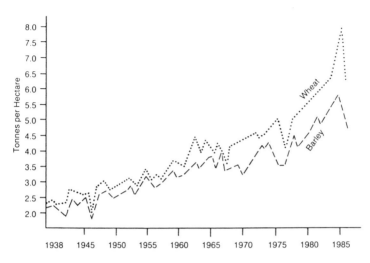

d)

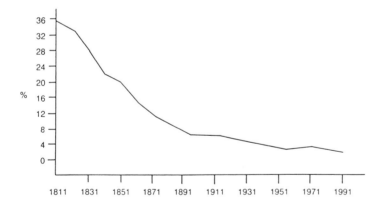

Figure 5.1 Selected agricultural changes in Great Britain
(b) The yields of barley and wheat, 1938–85
(d) Agricultural employment as a proportion of all employment, 1811–1991
(based on Robinson, 1988a: 152)

Table 5.1 Agricultural indicators in the developed world, 1985

	Average of size of holdings ha	% of work-force in agriculture/ forestry	Land area 000ha	Arable %	Permanent Pasture %	Forest %
Australia	2800.0	5.5	761793	6.4	57.3	13.9
Austria	11.6	6.5	8273	17.3	24.0	38.9
Belgium/Lux.	15.4	3.0	3287	24.5	21.3	21.2
Canada	213.5	3.8	922097	4.6	3.4	38.1
Denmark	25.0	7.6	4237	62.3	5.7	11.6
Eire	22.6	17.1	6889	14.1	70.5	4.9
Finland	13.3	9.1	30547	7.8	0.4	76.0
France	25.4	7.9	54563	34.3	23.0	26.7
Greece	4.3	29.9	13080	30.4	40.2	20.0
Italy	7.4	12.4	29402	41.6	17.2	21.7
Netherlands	15.6	5.0	3394	25.5	33.7	8.7
New Zealand	267.2	9.7	26867	1.9	51.7	26.8
Norway	10.3	6.1	30786	2.8	0.3	27.1
Portugal	n/a	23.5	9164	38.7	5.8	39.7
Spain	n/a	18.0	49940	41.1	21.4	30.1
Sweden	31.7	4.3	41162	7.2	1.4	64.2
Switzerland	17.4	4.4	3977	10.4	40.5	26.5
United Kingdom	68.7	2.7	24160	28.9	48.4	8.9
United States	180.1	2.6	916660	20.7	26.3	28.9
West Germany	15.3	5.6	24428	30.5	18.9	30.0

Sources: FAO Production Yearbook, 1986; Eurostat, Basic Statistics of the Community, 1986, Yearbooks of individual countries

5.2 The drive towards increased specialisation

A dominant theme within rising agricultural output has been increased specialisation. The greater concentration and regional specialisation of agricultural production that has occurred pre-dates World War Two but has accelerated in recent years (e.g. Winsberg, 1980). Capital-using technological change has favoured certain types of farming, e.g. irrigated agriculture and pastoral farming in the United States, which, in the case of the Great Plains, has reduced the influence of market access upon location (Visser, 1980).

Mixed farming, involving the harmonisation of crop and livestock production, has greatly diminished to be replaced by a specialisation which has put an end to the operation of joint economies and which has incurred higher environmental costs (e.g. Lawrence, 1987). For example, one of the commonest forms of specialisation in eastern Britain has been that of intensive arable farming. Intensification has involved heavy invest-ment in machinery which could only prove profitable given a larger area under cereals. The result has been a dramatic rise in the acreage of

Table 5.2 Changes in United States agriculture, 1960–87

	1960	1987
Arable (000ha)	188,309	187,881
Permanent pasture (000ha)	255,927	241,467
Forest (000ha)	259,363	265,188
Irrigated area (000ha)	11,959	18,102
Agricultural population (000)	22,158	7,211
Economically active in agriculture (000)	7,331	3,143
% employed in agriculture	12.0	2.6
Tractors (000)	4,770	4,676
Harvesters-threshers (000)	1,065	645
Farms (000)	2,173	3,963
Average farm size (acres)	297	461
Farms < 10 acres (000)	244	188
Farms > 500 acres (000)	336	366
Tenants as a % of all farmers	13.0	11.5
Farm output per hour (1977 = 100)	37	155

Source: US Yearbooks

cereals and of remunerative break crops such as sugar beet and oilseed rape (Robinson, 1988a: 185–205). Elsewhere, other farmers have specialised in livestock enterprises, and the same trend towards intensification as pursued by highly capitalised pig and poultry farming has been followed by similar methods associated with cattle husbandry.

Examples of increased specialisation within the United States include the greater concentration of cotton, tobacco, peanuts, rice and sugar cane in environmentally favoured 'islands' in the southern states as these areas have eclipsed less competitive areas with poorer quality crops (Hart, 1978). Hart (1980) cites Carroll County, Georgia as one such example of an area that has suffered, losing its cotton lands either to timber for paper manufacture or for beef/poultry production. In the Piedmont area of Georgia and the Carolinas as a whole nearly 40 million ha (10 million acres) of land passed out of farm ownership between 1939 and 1974.

In the United States new agricultural technology has transformed the Corn Belt, turning it from a mixed farming area to one specialising in cash grain production. Although family farming has remained the norm, the size of holding has doubled and the rural landscape has been greatly modified to accommodate the new farm machinery. The number of tractors and combine-harvesters has risen by over 50 per cent since 1960 whilst yields have doubled during this period (Table 5.2). Hart (1986) dates the transformation of the Corn Belt to 1933 and the introduction of hybrid seed maize, but 'take off' did not occur until after World War Two and the use of agricultural chemicals and machinery to raise yields (e.g. Kenney *et al.*, 1989).

The increased size of holdings in the Corn Belt has been brought about by farmers renting land as small family farms have been sold, thereby releasing land onto the market. However, family farming has remained dominant albeit in a different guise. Estate tax laws have encouraged incorporation and more elaborate corporate structures to be developed, but these have largely been devices to facilitate inter-generational transfers of assets within a family. Survival has necessitated specialisation, usually by combining maize and soyabeans in rotation. The latter crop has become the second crop of the Corn Belt, being especially important in north-central Iowa, the Grand Prairie and the Maumee Plain.

Rising output has necessitated a greater reliance upon export markets, but global recession and the strength of the US dollar in the early 1980s reduced these exports, depressed prices below costs of production and created major problems within the Corn Belt. In 1980 34.5 per cent of the maize crop and 40.4 per cent of the soyabean crop were exported: American farmers had developed a critical over-reliance upon the world grain market.

This conclusion, reflecting one of the negative concomitants of specialisation, illustrates that whilst agricultural change post-1945 can be viewed as a direct product of technological success, this technological basis can also be seen as 'mis-directed, based on false, out-of-date assessment of market demand and a narrow economic evaluation of production methods' (Munton, 1988). In effect, farmers are now on a treadmill driven by technological advance. Innovations have been adopted enthusiastically, but have generated economic, environmental and political problems compounded by support policies (e.g. Body, 1982; 1984; Dexter, 1977).

The radical developments in agriculture since 1945 have brought profound changes both in production and in the rural landscape. Amongst the *environmental problems* caused have been 'the destruction of representative ecosystems and individual species, the loss of the amenity value of the landscape, nitrate pollution of water supplies, eutrophication of water-courses, nuisance from intensive livestock production and straw-burning, and soil erosion' (Hodge, 1986: 180). This revolutionary alteration of farming in recent decades, which has turned food production into a large-scale industry run very differently from pre-war agriculture, has promoted much concern over its effects on the countryside. In particular, the rise of environmental concern has been related closely to the way in which the technological transformation of farming has wrought changes in the countryside affecting well loved components such as hedgerows, hedgerow trees, woodland, areas of rough grazing, downs, moors and wetlands (Cox and Winter, 1986; Shoard, 1980).

Long-term features of rural landscapes have been threatened by the growth of farm mechanisation, in Western Europe especially through the transformation of the traditional picture of horses pulling machinery in small, hedged fields. The use of machines has promoted field enlargement and the removal of obstructions such as hedgerows and stone walls. Movable field boundaries of post and wire fences have been a common replacement. This has been accompanied by new farm buildings of concrete and corrugated iron to house animals or store feed, produce and machinery. The new wave of distinctive farm buildings has been associated with the growth of intensive animal husbandry. New style hen houses for battery-produced birds, large milking parlours for dairy cattle and, more recently, tower silos have become familiar features of the landscape.

A simple illustration of the environmental consequences of the New Agricultural Revolution can be given from the Welsh uplands where there has been a great reduction in rough grazing inversely related to increased sheep numbers. The latter have been encouraged by headage payments under the EEC's Less Favoured Areas scheme. This scheme has promoted increased stocking rates beyond the carrying capacity of the semi-natural vegetation of the rough grazing, and so farmers have then carried out grant-aided improvement under grassland conversion projects. This has reduced rough grazing and dwarf shrub vegetation whilst promoting new pastures (Anderson and Yalden, 1981; Hodge, 1978; Wathern *et al*, 1986; 1988). Further examples from arable districts of eastern Britain are considered in Chapter 10.

Although the high level of specialisation within 'agriculture would probably have resulted from the combined effects of rising labour costs and falling real costs of capital and technical progress, the process has often been led by government agricultural policies. These policies, both directly and indirectly, have given rise to particular types of agricultural change, of which the various forms of *restructuring* have attracted particular academic attention.

5.3   Agricultural restructuring

Agriculture in the Developed World post-1945 has been an important arena for the twin processes of accumulation and concentration of capital and wealth. For example, farming has become a more attractive target for financial institutions which can combine profits from farming with capital gains from appreciating land values. Various pressures upon farmers too have dictated the increased size of operation needed to enable viability (see Britton and Hill, 1975). Economies of scale have divided farming businesses into the 'haves' and 'have nots' in terms of ability to obtain benefit from these economies. Indeed, Kantsky argues that the disappearance of the 'have nots' is effectively a precondition for the formulation of large enterprises (Banaji, 1976). This simple division of farming into two types of business, whilst being overly simplistic, is an illustration of the way in which farm businesses have been polarised by various forms of agrarian change, involving size of holding, capitalisation, the labour force and type-of-farming. These changes can be considered under the umbrella heading of 'restructuring'.

Fitzsimmons's (1986) study of speciality crop production in the United States revealed two particular characteristics of agricultural restructuring: *vertical integration* by the concentration of control over the production process, at the same time as production activities within the farm are *vertically disintegrated* through a complex network of capacity and specialty sub-contracting.

These characteristics were amplified in Marsden *et al.*'s (1986) examination of the restructuring process in British agriculture which highlighted the diversity of experience of farmers and the growing disparity between small farmers and large *agribusinesses* produced by post-1945 changes. Restructuring itself they saw as having four principal characteristics:

(a) Overproduction of agricultural goods in the face of stable demand. Technological improvements and state support have accelerated tendencies towards concentration and centralisation of capital in different parts of the food production industry (Busch *et al.*, 1989). For farmers this has meant increased indebtedness (Burrell *et al.*, 1985: 66).

(b) 'Downstream' environmental consequences have led to major landscape modifications (e.g. Lowe *et al.*, 1986; Shoard, 1980; Westmacott and Worthington, 1984).

(c) Increased differentiation has occurred between farms and also between regions. A 'dualism' has been created between marginalised family farms on the one hand and large heavily capitalised businesses on the other. This in turn affects the direction and nature of capital penetration.

(d) The transformation of the family farm has created a wide range of different types of farm under this umbrella category. In particular, the farm household has had to seek alternative sources of income and capital. This has contributed to the growth of part-time farming (Buttel, 1982b; Gasson, 1988), especially within the Metropolitan Green Belt (Marsden *et al.*, 1986: 274–5); Munton, 1983), and the more recent increase in farm-based tourism (Bouquet and Winter, 1987).

The process of restructuring can be seen to have given rise to a series of different strategies by farmers, depending upon the centrality of revenue from farming to their overall income. Marsden *et al.* (1986) identify three broad strategies. The first is one which regards farming as only a minor contributor to income, i.e. hobby- or part-time farming. The second is a survival strategy in which diversification plays a significant role. And the third is an accumulation strategy through corporatisation or agribusiness development involving complex relationships with finance and industrial capital (Marsden, 1984; Marsden *et al.*, 1989).

Marsden *et al* (1987) also recognise four scales at which the process of uneven development of agriculture has occurred.

(a) regional divisions of capital and labour;
(b) sectoral divisions of capital and labour, e.g. between arable and livestock;
(c) internal organisation of individual production units;
(d) differing divisions between the units of production and other agencies involved in the food chain, e.g. manufacturers of inputs, food manufacturing, retailers and financial institutions.

They highlight the fact that, increasingly, industrial capital has played a part in determining the nature of agricultural development. They also suggest that whilst a 'dualist' farm structure has developed, separating highly capitalised, large-scale agribusinesses from small, part-time quasi-subsistence farms (Buttel, 1982b; Wallace, 1985), this ignores the complexities within the family farming sector. It is clear, though, that

the intensive capitalisation of farm production in some regions at the expense of others, provides a contemporary example of the unevenness of capital penetration and the superimposition of new spatial divisions of labour (Marsden *et al.*, 1987: 299).

Whilst both Britain and the United States provide good examples of the duality created in the farm structure by agricultural modernisation, less modernised economies, such as that in Eire, also exhibit an accentuated *dualism*. For example, income differentials between small and large farmers have increased, and rural out-migration and farm abandonment have increased the under-use of land in remote areas. Commins (1980) argues that because problems within Irish agriculture have been dealt with at an aggregate level, relying upon market and technical solutions, the trend towards a dualistic farm economy has actually been fostered by government, creating similarities with the richer countries of the Developed World. This has been true of the influence of the CAP since Eire joined the EEC in 1973.

In Britain the average size of farm holding in 1935 was 28.1 ha (69.5 acres) of crops, fallow and grassland plus 14 ha (34.5 acres) of rough grazings. In 1985 the average was 75.9 ha (187.5 acres) plus 11.1 ha (27.5 acres) of rough grazings. This is a simple illustration of scale economies operating post-1945: the number of holdings has fallen and the size of holdings has risen. In part this has been fostered by government policies aimed at reducing the number of small farmers, but it also reflects what Newby (1987b: 197) terms the change from 'farming landowner to landowning farmer.' More efficient, larger farm units have been formed and these have usually been owner-occupied. At one end of the scale has come the large agribusiness and at the other the small dairy farmer. Throughout, owner-occupation has proved attractive, e.g. in 1985 70.2 per cent of farm holdings in Britain were owned or mainly owned by the farm operator, with the highest proportion of ownership occurring in the Home Counties and Wales.

Table 5.3 An ordinal scale of subsumption in the internal organisation of farm businesses

Score	Farm capital	Land rights	Internal relations Business management structure	Land relations
1	Farm family, individual head of household	Simple owner-occupation, sole owner-operator	Single unit farm managed by head of household	Farm labour farm, 1 full-time family worker
2	Farm family, shared and nominal corporate family ownership	Farmily owner-occupation	Single unit farm, joint family management	Family labour farm, more than 1 full-time family worker
3	Farm family, corporate ownership of single farm unit	Corporate family ownership (nominally renting)	Single unit farm, family owned but employing manager.	Family labour farm, + casual hired labour only
4	Farm family, corporate ownership of multi-farm unit	Mixed family ownership and tenancy (where both are significant)	Multi-unit/ farm, family owned and managed	Mixed labour farm, 1–3 full or part-time hired worker (hired < family)
5	More than family corporate, involving non-family capital	Simple family tenancy, (renting from 1 non-family owner)	Multi-unit/ enterprise business, family owned employing manager	Mixed labour farm, over 3 full or part-time hired labour (hired < family)
6	Family and landowner corporate business	Complex family tenancy (renting from 2 + non-family owners)	Multi-unit/ enterprise business, non-family owned and managed	Mixed labour farm employing over 50% contract hired labour
7	Non-family, single corproate owner	Non-family corporate inhand (with manager)	Multi-unit/ enterprise business, non-family owned, employing manager	Wage labour farm, all hired labour
8	Non-family, multi-corporate owner	Non-family corporate tenancy	Multi-unit/ enterprise business, non-family owned, contract co. managed	Wage labour farm, all contract labour

Source: Marsden *et al.*, 1987

Given the importance of capital accumulation, concentration of owner-ship, and major disparities between farm businesses, it is not surprising that studies of agricultural change and restructuring have attracted work from a Marxist perspective (e.g. Massey and Catalano, 1977). Marxist interpretations of the growth of capitalisation of agriculture are based on ideas relating to the need for capital to switch from one 'circuit' of accumulation to another to permit higher returns or the maximisation of surplus value (e.g. Smith, 1984). Yet, such notions have not been properly tested with respect to agriculture despite some supporting evidence from American studies (e.g. Edel, 1981; Herdt and Cochrane, 1966). Also, the nature of agriculture, with its close links to physical factors and variability of land characteristics, presents a 'special case', differentiating it from other industries. The inherent qualities of agriculture have acted towards the creation of stable production patterns and a strong element of inertia in the organisation of farm businesses, especially when relatively small producers have been dominant. Thus, despite the growth of agribusiness, throughout the Developed World families rather than corporate bodies have remained the predominant owners and occupiers of farm businesses.

In recognising this, Marsden *et al.* (1987) attempted to demonstrate the degree of capital penetration (or *subsumption*) in agriculture in selected areas by using a numerical scale. This ranged from one, for the 'standard' owner-occupier family farmer managing the farm and using only family labour, to eight, for the non-family, multi-corporate owner with non-family corporate tenancy and using only contract labour (Table 5.3). When applied to different sample areas, this index suggested that increased marginalisation was occurring upon small dairy farms, partly as a result of milk quotas introduced in 1984. It also revealed the increased exploitation of family labour extending the involvement of farm women on the smaller farms, and the growth of 'agribusiness'.

Increasingly, research in a variety of disciplines has recognised the various aspects of restructuring, especially the 'forward' and 'backward' linkages developed by farming into the agricultural supplies industry and marketing of farm produce (e.g. Bowler and Ilbery, 1987). Foci upon agriculture alone have been replaced by ones referring to the 'food and fibre industry' which combines supply, production and marketing (Le Heron, 1988; 1989). In addition, new approaches to this broader field have been developed, notably in the political economy field (e.g. Bradley, 1981; Buttel, 1982a; 1982b; Cloke, 1989b; Marsden, 1988). For future work by geographers, these studies have picked out three types of restructuring within the food and fibre industry which have potential for a spatial focus in investigations:

(a) the capital accumulation process;
(b) competitive movement of 'fractions' of capital within the food chain;
(c) the micro-level of farm production.

Across a broad inter-disciplinary spectrum, though, there have been

Plate 9 Intensive Agriculture – 2. One of the most appropriate applications of the term 'agribusiness' is in the growing and marketing of flowers. This flower auction at Aalsmeer, the Netherlands, covers over 100 ha (250 acres), sells 3 billion flowers and 300 million plants annually, has an annual turnover of 1.5 billion Dutch guilders and accounts for 43 per cent of all flower sales in the Netherlands

several studies focusing upon that aspect of restructuring involving the growth of agribusiness.

5.4 Agribusiness

Using Carlson *et al.*'s (1981) definition, *agribusiness* refers to all organisational structures associated with the supply of inputs to farming, such as fuel, seeds, chemicals and equipment, the farm production process, the marketing and processing of farm products (Plate 9), the distribution of these products and their marketing and sales (see also Barlow, 1986a; Bowers, 1985; Kinsey, 1987). The term 'agribusiness' is commonly used to refer to the growth of corporate structures, the development of forward and backward linkages in agriculture and the growth in scale of farming enterprises. Throughout the Developed World there has been a tendency for these larger farms to become concentrated spatially. Two examples of the development of agribusiness are given below as illustrations, one summarising developments in the United States and the other a description of the growth of a single business enterprise in England.

a.

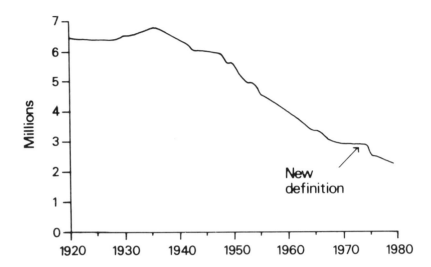

b.

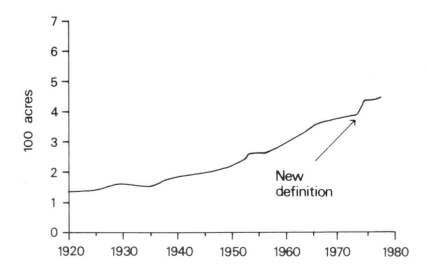

Figure 5.2 Trends in the number and average size of farms in the United States, 1920–80; (a) Numbers of farms; (b) Average farm size (based on Bohland, 1988: 174)

5.4.1 Agribusiness in the United States

Farming in the United States is the best exemplifier of the development of agribusiness, where it has been associated with increases in the size of farms and reduction in their numbers. For example, the number of farms fell by 4.6 million between 1935 and 1984, a decline of 77.6 per cent whilst average farm size rose two-and-a-half times to 174 ha (430 acres) (Figure 5.2). The larger farms have a greater proportion of their production under contract, are more capital-intensive, receive more funds from government in support of agricultural commodities, place greater reliance on cash-wage labour, have higher fixed costs but also have higher level debts.

In terms of annual sales, the major concentration of large farms extends from the Mid-West to the Great Plains, running from the Canadian border to Texas (Plate 10). These farms are concerned mainly with cash grain or livestock production. Smaller concentrations of large volumes of annual sales occur along the eastern coastal plains and in the irrigated basins and valleys of the South-West. Using other measures of size, though, slightly different concentrations appear. For example, Gregor (1982) referred to the level of capital investment in production as a good indicator of 'industrialisation' of agriculture. Using this plus labour investment, he developed two indices of *agricultural industrialisation*: *scale* or the size of operation, and *intensity* or measures of input. As shown in Figure 5.3 the highest intensities were recorded closest to the major eastern markets and in the cash-grain, mixed farming and dairy areas of the Mid-West. Other high intensity production included citrus production in Florida, tobacco- and peanut-growing in the Carolinas and Georgia, and the irrigated agriculture in parts of Arizona, California and the Pacific North-West. Gregor's scale factor tends to be complementary to his intensity factor, and emphasised the more extensive production systems in the western USA plus more limited areas of high inputs in Florida, a few parts of the Mid-West, southern Texas and parts of the Mississippi Valley.

Vogeler (1981) showed that both vertical and horizontal integration have occurred within American agriculture, often jointly, creating an 'agribusiness system' 'where input resources, production and marketing are spatially and administratively integrated under single ownership to maximise profit' (Bohland, 1988: 179). Typically, such operations are owned by corporations or large family partnerships. Vertical integration, in which production from inputs to distribution of the finished product is controlled, has proved attractive, partly because corporations have been able to spread costs across several phases of an operation and so undercut competitors in any particular sector if necessary. Also, higher-profit margins in certain sectors, e.g. wholesaling, can compensate for lower margins on actual farming operations. Horizontal integration, when firms take over others engaged in a similar business, has become common for firms producing inputs to agriculture or processing/

Plate 10 Intensive Agriculture – 1. The United States: One characteristic of increased capitalisation of farming has been investment to extend the irrigated area, demonstrated near Corpus Christi, Texas, by a mobile sprinkler system

marketing agricultural commodities. For example, after a series of mergers and purchases in the 1970s, in 1980 fifteen firms accounted for over 60 per cent of all farm inputs, forty firms for two-thirds of all food processing and forty-five companies for over three-quarters of wholesale and retail food revenues in the United States (Vogeler, 1981). The underlying role of government policy in promoting this situation must not be overlooked:

Agricultural policy is an instance of state policy in general which operates to aid and legitimate control by large corporations over the forces and relations of production (Frundt, 1975: 2).

To a remarkable extent a few large corporations have come to dominate food processing and wholesaling in the United States and, through vertical integration, farming enterprises too. In many cases their influence extends beyond the USA as they are multi-national conglomerates in which food processing is just one of several areas of interest. Notable examples include Unilever, Shell, Boeing, Getty Oil and insurance companies like Connecticut General (Cortz, 1978). But despite large corporations' growing involvement in farming, they owned less than

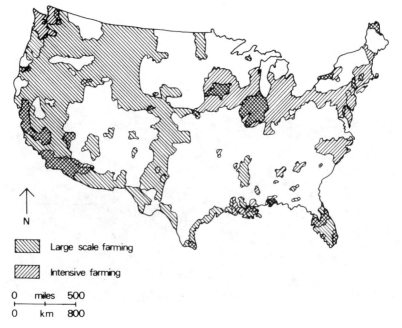

Figure 5.3 The industrialisation of agriculture in the United States (based on Gregor, 1982)

1 per cent of farms in the mid-1980s according to official statistics. In fact, agriculture continues to contain several deterrents to direct corporate involvement. Hoggart and Buller (1987: 69) highlight three:

(a) the length of time taken by production cycles. This ties up capital, reduces profit rates and restricts responsiveness to price fluctuations;
(b) the seasonal pattern of work produces inefficiencies in the use of machinery, management and labour;
(c) the presence of uncontrollable and unpredictable variables such as disease, pests and the weather.

Non-farm holdings in agriculture have only tended to rise when other sources of profitable return have entered periods of reduced profitability (e.g. Lyson, 1984), or incentives such as tax relief on farmland have proved attractive (Lapping and Clemenson, 1984).

Some Marxist theories lend support to the growth of agribusiness, and also to its 'patchy' and uneven advance. For example, one view of this uneven development of capitalism in agriculture has been in terms of Marx's distinction between *labour time* and *production time*. The former consists of the periods when labour is actually applied, whilst production time covers the entire production cycle. If there is a distinction at the farm level between these two types of time, Mann and Dickinson (1978: 466) have argued this will have 'an adverse effect on the rate of profit, the efficient use of constant and variable capital, and the smooth functioning of the circulation and realisation process.' This favours larger units with 'more efficient' use of labour. Yet Mooney's (1982) examination of this thesis with respect to the development of capitalism in US agriculture between 1944 and 1974 did not reveal non-coincidence of labour time and production time to be a limitation for capitalist penetration in the production of eight commodities examined. There is the suggestion that capital can profitably 'penetrate' a wide range of farming enterprises, even if those enterprises operate on only a small scale. However, the limitations of scale may inhibit agribusiness development in some sectors (see Newby, 1987a).

It is clear, though, that increased capital input has introduced more integration of production, wholesaling and retailing via more elaborate corporate structures. Such changes have provoked academic interest in the 'food and fibre industry' rather than on agriculture alone (e.g. Le Heron, 1988). An example from the British poultry industry illustrates the type of integration occurring, and also the growing involvement of multi-national corporations in agricultural production.

5.4.2   The Sun Valley poultry enterprise

Sun Valley Poultry Ltd was formed in Herefordshire, England, in 1960–1 as a co-operative scheme started by several farmers seeking to take advantage of the growing demand for poultrymeat (Figure 5.4a). Consumption of poultrymeat per head in the United Kingdom rose by 60 per cent from 1957 to 1965 (Dutch MAF, 1967), and it was this increased demand that was exploited by the new co-operative. The individual farmers originally each produced poultry very much as a subsidiary enterprise, but were able to increase output through their co-operative venture. A factory for broiler production was built and the co-operative was developed as one overall organisation with both vertical and horizontal integration. The initial target was the production of 50,000 broiler chickens per week from 14 farms producing hatching eggs and 17 intensive production units. Their factory, on the outskirts of Hereford, then prepared the birds for market, with the majority being sold on contract to the main supermarket chains as frozen chickens (Figure 5.4b). The original target output was exceeded within three years of operation as other farmers in the district began producing for Sun Valley. So, by 1965, 150,000 chickens and 5,000 turkeys per week were being shipped from the factory. These were produced from 20 hatcheries, 65 chicken-feeding farms and six turkey-feeding farms. Most of these units were small, but, with an assured outlet in the co-operative's factory and a growing market for the finished broilers, returns were high despite the halving of the prices of poultrymeat between 1955 and 1965.

The principal sales outlet was Marks and Spencer who were receiving over 100,000 chickens and turkeys per week by the mid-1970s. By this time Sun Valley were averaging a throughput of 280,000 chickens and

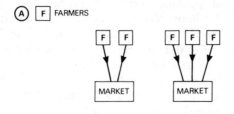

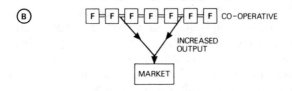

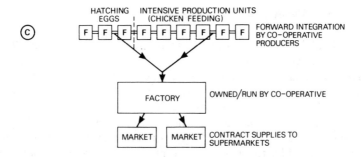

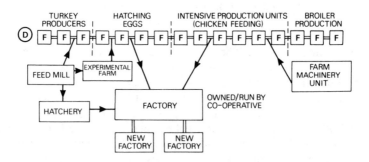

Figure 5.4 The expansion of Sun Valley Poultry Ltd in the English Midlands
(a) The early-1960s
(b) The late-1960s
(c) The early-1970s
(d) The 1980s
Source: Robinson, 1988a: 246

15,000 turkeys per week. Their factory had been enlarged and covered nearly 3 ha (7.5 acres), employing 650 people. Sales of poultrymeat trebled between 1965 and 1973 to £12 million per annum, by which time further expansions of the initial farm-based production had been made. This expansion of activities illustrated the ramifications of this type of integrated enterprise: a subsidiary firm, S.V. Chickens Ltd, was established for broiler production, housing birds in purpose-built accommodation; Sun Valley established their own hatchery, feed mill, farm maintenance service and poultry-growing farms; there were over 100 farmers participating in the co-operative; and the feed mill consumed over 1,220 tonnes of cereals per week, stimulating cereal-growing locally and also boosting pig production as pigs could utilise the surplus feed (Figure 5.4c).

Further expansion in the 1980s has been as part of the Union International Group of companies. Production has expanded to over £50 million per annum of chickens, turkeys and associated produce such as processed meats. New factories have been opened to tap new markets. Exports have grown, with the firm operating its own aircraft for rapid transport of breeding stock to Europe. New developments have included other forms of intensive livestock production, initially investigated on a 120 ha experimental farm (Figure 5.4d). Expansion of production has meant that over 25 million chickens and over 1.5 million turkeys were produced annually in the mid-1980s and the firm's labour force had risen to 2,500. Local farmers in Herefordshire provided over 12,000 tonnes of cereals for feed, or around 15 per cent of the company's annual cereals intake (Robinson, 1988a: 244–6).

5.5 Farming on the urban fringes and the protection of farmland

5.5.1 The rural-urban fringe

As agriculture has become more susceptible to capitalist penetration, it has also faced the relentless challenge of competition for the land it occupies from other parts of the 'arena of capital'; especially the housing market and industrial development. The dispersed nature of the urban form has grown post-1945 as the increased ownership of cars by private individuals has permitted the extended separation of work-place and residence. The enlargement of the maximum commuting zone between household and work-place has brought with it the growth of suburbs attached to the urban core as well as the growth of both existing and new settlements in the area beyond these suburbs. Within this simplistic division of core, suburbs and outlying settlements, Russwurm (1975; 1977) amongst others has recognised a more complex mixture of different land uses characterising the area beyond the suburbs. This mixture includes ribbon development, uses requiring large amounts of space (e.g. stock-yards), and various types of residential property belonging to commuters and other ex-urbanites. Other more subtle forms of urban influence on land use also occur by way of increased non-farm ownership of land and modifications to the type of agricultural land use. Using data from the Valuation Department of central government, Moran (1978) explored the changing influence of distance from the central business district (CBD) on the value of land in the urban periphery of Auckland, New Zealand, between 1955 and 1970. He showed that the gradient of unimproved value with distance from the CBD had steepened through time as the conversion of land to urban uses and sub-divisions of rural properties into smaller holdings increased the unimproved value of land, though with significant local variation. This local variation could also be seen around Auckland in terms of spatial patterns of agriculture (Moran, 1974; 1979). Factors such as the postponement of rates, the direct marketing of produce, the historical legacy of land use patterns and the locational control of town supply dairy farming had created distinctive areas of intensive production close to the city, e.g. town supply dairying in the vicinity of Te Hihi and vineyards/orchards in the Henderson Valley.

Similar examples to the one for Auckland can be found for farmers in most rural-urban fringes. In effect, the pockets of intensive production close to urban areas distort Sinclair's (1967) generalisations that the value of land for agriculture increased with distance from the expanding metropolis and that a gradient of agricultural uses will develop from more extensive near the edge of the built-up area to more labour and capital intensive at a greater distance. Sinclair's proposals are the antithesis of the pattern of land uses depicted in models of agricultural land use based on considerations of economic rent, and they reflect influences such as the negative effects of proximity to urban centres (e.g. pollution, vandalism), speculative landholding awaiting the next surge in urban expansion, and the farmer's ability to obtain high returns from land without intensive production (e.g. 'horseyculture' and sod farming) (Figure 5.5) (Mather, 1986: 132-9; Parenteau, 1980).

Idle or under-used farmland on the fringes of cities can be attributed to:

(a) the influence of urban intrusion, e.g. trespass and pollution, mitigating against efficient farming;
(b) the presence of large numbers of hobby farmers who do not wish to make full agricultural use of their land;
(c) the effects of high municipal taxes combined with the uncertainties of farming act as disincentives to invest or go beyond low intensity farming (Blair, 1980; Brown *et al.*, 1981; Bryant, 1982; Munton, 1983; Sinclair, 1967).

Recent work by Munton *et al.* (1988) has suggested that references to distinctive types-of-farming within the rural-urban fringe have been something of an oversimplification. It should be recognised that farmers

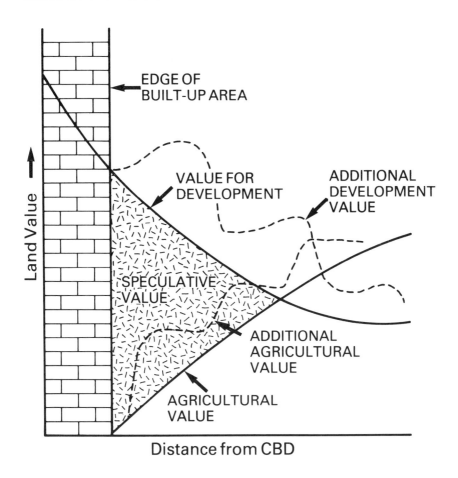

Figure 5.5 A model of land values around a city (based on Sinclair, 1967 and Mather, 1986: 133)

operate within a range of 'technological and financial imperatives' which can actually be quite unrelated to a general urban-fringe location. Consequently, it is possible to find major differences between farms around major cities within relatively close proximity (e.g. Bryant and Greaves, 1978), or even within particular urban fringes at comparable distances from the urban edge.

This demonstration of significant variations within a fringe area is certainly possible within the London Green Belt where Munton *et al.* (1988) recognised four general characteristics:

(a) Great diversity of capitalisation over a small area, i.e. from small family to corporate concerns to interests not primarily concerned with agricultural production, e.g. hobby farming and 'horseyculture';

(b) A highly competitive land market encouraging speculation and subsequent fragmentation of agricultural holdings;
(c) Diversity of income sources and business ventures on and off the farm;
(d) Pressure upon agricultural land from planned developments such as motorways, housing development and amenity land use.

They demonstrated the way in which locally specific responses were being made to the wider structural processes affecting British agriculture as a whole, e.g. new patterns of capital investment, spreading of family interests in the farm business, increase in off-farm activity (e.g. Goodman and Redclift, 1985). Thus 'urban-fringe farming' cannot really be considered to be a clearly recognisable 'morphological type'. Farmers in the fringe are under the same pressures as farmers elsewhere; they respond in different ways according to the range of local opportunities and constraints presented to them, and these vary even within the context of an individual fringe area. Two major differences between the fringe and 'deeper' countryside would appear to be the greater extent of hobby farming in the fringe as a reflection of particular circumstances present in fringe areas (e.g. Brown et al., 1981), and the threat posed to agricultural land by urban sprawl.

5.5.2   Farmland protection and loss

The loss of farmland to urban sprawl has prompted different reactions in the constituent countries of the Developed World, largely related to the perceived importance of prime farmland and the presence or absence of an agricultural surplus (Bryant and Russwurm, 1982). Four examples of contrasting reactions illustrate the diversity of national attitudes.

5.5.2.1   CANADA

In Canada changes in land use can be studied by a computerised information system, the Canada Land Use Monitoring Program (CLUMP) (Lands Directorate, 1980). Established in 1978, it provides information on changing land use for researchers from various disciplines, planners, land resource managers and policy-makers. It was designed to complement the land capability surveys of the Canada Land Inventory and so monitor land use and change in land use both spatially and sequentially. Amongst the types of area selected for special attention are *urban-centred regions (UCRs)*, comprising the main labour market area of an urbanised core or continuous built-up area having 100,000 or more population (Lands Directorate, 1985a).

In the UCRs between 1966 and 1986 just over 300,000 ha (714,000 acres) were converted from rural to urban use, the most rapid rate of conversion occurring from 1976. The largest losses of land to urban use

Table 5.4 Total Rural Land (TRL) and Prime Agricultural Land (PAL) converted in the Canadian UCRs for four monitoring periods, 1966–86

Province (No. of UCRs)	1966–71 TRL ha	PAL ha	% of TRL	1971–76 TRL ha	PAL ha	% of TRL	1976–81 TRL ha	PAL ha	% of TRL	1981–86 TRL ha	PAL ha	% of TRL	1966–86 TRL ha	PAL ha	% of TRL
B.C. (7)	7 515	1 154	15	7 665	1 690	22	23 372	5 272	23	6 778	1 244	18	45 330	9 360	21
Alta. (5)	14 698	8 911	61	12 279	8 936	73	11 077	6 821	62	13 637	6 761	50	51 691	31 429	61
Sask. (4)	1 487	951	64	2 410	2 090	87	4 507	2 509	56	2 209	1 368	62	10 613	6 918	65
Man. (2)	5 199	4 733	91	1 441	1 356	94	3 975	3 433	86	2 431	1 925	79	13 046	11 447	88
Ont. (26)	36 952	29 125	79	21 260	16 558	78	27 070	20 276	75	20 670	17 081	83	105 952	83 040	78
Que. (19)	15 632	8 409	54	11 082	5 486	50	17 609	7 346	42	6 264	3 671	59	50 587	24 912	49
N.B. (3)	1 803	292	16	2 798	868	31	4 830	892	18	1 917	373	26	10 848	2 425	22
N.S. (2)	1 810	663	37	1 143	582	51	3 928	1 481	38	1 162	321	28	8 043	3 047	38
P.E.I. (1)	307	309	99	414	414	100	1 523	1 463	96	34	13	38	2 280	2 197	96
Nfld. (1)	685	—	0	672	4	>1	1 085	10	>1	608	1	>1	3 050	15	>1
70 UCRs	86 090	54 545	63	61 164	37 984	62	98 976	49 503	50	55 210	32 758	59	301 440	174 790	58

Source: Warren *et al.*, 1989

Table 5.5 Increases in urban areas in Canadian UCRs, 1976–81

(a) *Relative increases*

Top 20 cities	A	B	Bottom 20 cities	A	B
Moose Jaw (Sask)	50.67	44.7	Windsor (Ont)	8.87	94.6
St-Jerome (Que)	41.29	14.2	St Thomas (Ont)	8.79	90.0
Nanaimo (BC)	38.93	10.9	Newcastle (Ont)	8.57	88.1
Kelowna (BC)	37.43	44.9	Vancouver (BC)	8.55	18.5
Chilliwack (BC)	34.55	28.7	Winnipeg (Man)	8.49	92.8
Sarnia (Ont)	33.10	86.9	North Bay (Ont)	8.17	12.1
Red Deer (Ala)	32.08	82.3	Ottawa-Hull (Ont/Que)	8.07	48.5
Victoriaville (Que)	30.45	43.9	Hamilton (Ont)	7.86	89.7
Joliette (Que)	29.55	42.9	Alma (Que)	7.55	52.3
Prince George (BC)	28.71	4.1	Halifax (NS)	7.14	20.6
Prince Albert (Sask)	28.48	46.1	Calgary (Ala)	7.02	67.7
Charlottetown (PEI)	26.00	96.1	Edmonton (Ala)	6.86	54.0
Lethbridge (Ala)	25.37	91.3	Chicoutimi-Jonquiere		
Sherbrooke (Que)	24.00	22.3	(Que)	6.03	52.2
Granby (Que)	23.72	27.8	Sault Ste Marie (Ont)	5.93	9.8
Belleville-Trenton			Kitchener-Guelph (Ont)	5.70	72.0
(Ont)	22.67	73.8	Oshawa (Ont)	5.68	93.8
Saskatoon (Sask)	20.92	59.0	St Catharines-Niagara		
Rimouski (Que)	20.85	27.1	(Ont)	5.54	85.6
Sorel (Que)	20.67	18.5	Toronto (Ont)	5.00	95.1
Fredricton (NB)	20.35	25.3	Thunder Bay (Ont)	4.50	13.0
			Montreal (Que)	4.20	74.8

A = per cent; B = per cent of increase on Capability Classes 1, 2, & 3

(b) *Absolute increases*

Top 20 cities	A	B	Bottom 20 cities	A	B
Vancouver (BC)	6073	1125	Kamloops (BC)	549	14
Chilliwack (BC)	5520	1585	Peterborough (Ont)	424	324
Toronto (Ont)	4985	4738	Oshawa (Ont)	420	394
Edmonton (Ala)	4789	2585	Sorel (Ont)	410	76
Ottawa-Hull (O/Q)	4179	2027	Thunder Bay (Ont)	391	51
Nanaimo (BC)	4136	451	Rimouski (Que)	388	105
Montreal (Que)	4010	2998	Brockville (Ont)	367	154
Quebec-Levis (Que)	3805	1310	Chicoutimi-Jonquiere		
Calgary (Ala)	3420	2315	(Que)	366	191
Winnipeg (Man)	3258	3023	Cornwall (Ont)	359	277
Victoria (BC)	2602	922	Sault Ste Marie (Ont)	336	33
Saint John (NB)	2462	147	St-Hyacinthe (Que)	330	210
Kelowna (BC)	2428	1091	North Bay (Ont)	322	39
Halifax (NS)	2206	454	Vallyfield (Que)	306	251
Prince George (BC)	2063	84	Woodstock (Ont)	286	251
Hamilton (Ont)	1876	1682	Chatham (Ont)	255	255
London (Ont)	1815	1800	Stratford (Ont)	231	231
Syndey (NS)	1722	1027	St Thomas (On)	220	198
Saskatoon (Sask)	1609	949	Newcastle (On)	194	171
Sarnia (Ont)	1584	1374	Alma (Que)	151	79
			Rouyn-Noranda (Que)	132	7

A = ha; B = ha of increase on Capability Classes 1, 2, and 3 ˙

Source: Lands Directorate (1986)

were in the major cities, Calgary, Edmonton, Montreal and Toronto where a 1 per cent growth in the urban population produced a 1.5 per cent increase in land under urban use (Yeates, 1985: 6–7). In British Columbia the rate of loss for 1976–81 was almost three times that for 1966–71 and 1971–6, but in all provinces the amount of loss of prime agricultural land fell in the 1980s (Warren et al., 1989) (Table 5.4). Relative increases from 1976 tended to be greatest in small towns, especially in British Columbia and in the Central St Lawrence Lowlands of Quebec (Table 5.5) (Robinson, 1988b).

Counter-urbanisation and centrifugal forces diffused from the larger Canadian cities during the 1970s to smaller urban areas. Previously, with less pressure upon land in the rural-urban fringe around these smaller cities, there has been more land available for conversion and, possibly, less attempt made to control urban sprawl through the application of specific planning controls. Yet, in absolute terms, it is still the major cities that are dominating the conversion of rural land. The larger UCRs had the smallest relative and largest absolute expansions in urban areas whilst smaller centres had greater relative increases and urbanised more land per capita.

This difference between small and large UCRs is demonstrated clearly in Alberta where the two cities dominating the urban hierarchy, Calgary and Edmonton, increased their urban populations by 26 per cent and 17 per cent respectively between 1976 and 1981. Yet, their urban areas expanded at a rate well below the national average of 11 per cent. This contrasted with the experience of smaller towns in the same province, such as Lethbridge and Red Deer, which had smaller proportional increases in population, but experienced greater relative growth of their urban areas.

These differential changes in the amount of land converted to urban areas may be related to variations in population growth by measuring the number of hectares of rural land converted per 1,000 increase in urban population (Table 5.6). Thus in the Maritimes, rural to urban land conversion has been associated with relatively small increases in population, suggesting low density use of the land. Elsewhere, and especially in Alberta and Ontario, much higher density use of land has been associated with the conversion. This pattern tends to reflect the distribution of large and small UCRs and the quality of land affected by urban expansion. In terms of the former, there has been an inverse relationship between size of the UCR and the rate of conversion of rural land per 1,000 increase in population. So the nine UCRs with a population in excess of 500,000 have the highest population densities and accounted for 78 per cent of the population increase in the UCRs between 1981 and 1986, and 66 per cent of the total rural land converted whilst only increasing their urban area by seven per cent.

Between 1981 and 1986 prime agricultural land accounted for 59 per cent of the land converted in Canada as a whole, illustrating the conflict between urban growth and the retention of the best agricultural land.

Table 5.6 Rates of land conversion in the Canadian provinces, 1976–81

(a) By province

Province	Ha land converted from rural to urban/ 1000 change in population	Urban area increase %	Population increase %
New Brunswick	1,657	15.9	1.2
Prince Edward Isle	1,580	26.0	3.0
Nova Scotia	1,193	9.7	1.0
Manitoba	791	9.4	0.8
Quebec	334	10.3	1.3
Newfoundland	109	9.6	6.5
Saskatchewan	96	19.3	10.7
British Columbia	84	16.6	9.2
Ontario	80	8.1	5.4
Alberta	46	8.5	21.4

(b) By UCRs

Population of UCR			
25,000–50,000	341	24.2	7.4
50,001–100,000	367	18.2	4.4
100,001–250,000	202	11.7	4.3
250,001–500,000	159	7.2	3.8
> 500,000	61	7.0	6.0

Sources: Census of Canada 1981; Lands Directorate (1985a)

More than 55 per cent of land in Agricultural Capability Classes One, Two and Three lies within 100 miles (160 km) radius of major metropolitan areas (Neimanis, 1979). The concentrated nature of urban pressure is further demonstrated by the UCRs around Toronto. The region bounded by Toronto, Barrie and Windsor converted 14,500 ha (35,830 acres) of agricultural land to urban uses between 1981 and 1986 of which nearly 94 per cent were prime agricultural lands, with 65 per cent in Class One. Toronto UCR absorbed 10,047 ha (24,825 acres) of prime agricultural land during this period, compared with 4,036 ha (9,973 acres) in Edmonton, 2,665 ha (6,585 acres) in Montreal and 498 ha (1,230 acres) in Vancouver.

Given the very small area of high quality farmland in Canada, it is not surprising that great attention has been devoted to preserving this land (e.g. Hoffman, 1982; Smit *et al.*, 1983). Two areas have received particular attention because they are both associated with some of the most intensive agricultural production in the country: the Niagara Fruit Belt and the Okanagan Fruit Lands.

The *Niagara Fruit Belt* (Figure 5.6) is the centre of Canadian wine and soft fruit production, a critical threat to which is posed by urban sprawl onto the areas of most intensive fruit-growing. Agricultural land use is

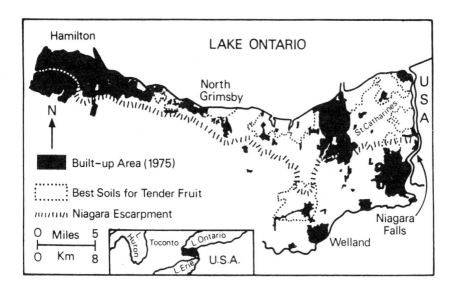

Figure 5.6 The Niagara Peninsula, Ontario, Canada (*Source:* Robinson, 1988b: 236)

threatened by growth of internal centres such as St Catharines and Niagara Falls, external pressures from Hamilton and Toronto, the routing of a main trunk road, the Queen Elizabeth Way, through land of high agricultural capability and the development of recreation and mineral extraction (Kreuger, 1977; 1978).

The amount of land devoted to fruit and vines in the Fruit Belt in 1951 was 21,450 ha (53,000 acres). Nearly forty years later less than half the acreage of tree crops remains and the total agricultural acreage has fallen by 20 per cent. Although the area under vineyards has increased (by 29 per cent) through new plantings along the Niagara Escarpment, the competition for prime fruit-growing land is great. The high rate of loss of farmland in the 1960s was one of the factors which brought about the establishment of regional government in 1970 in the form of the Regional Municipality of Niagara (Jackson, 1986). Its Policy Plan of 1973 stated a basic objective as being 'to protect the agricultural industry and its resources.' After ten years of public hearings and enquiries, boundaries of permitted urban development were established that were in broad agreement with the wishes of the local citizen's conservation group, PALS (Preservation of Agricultural Land Society).

The ten years represented a period of major political activity related to the demarcation of the boundaries of urban areas. Many felt the Policy Plan proposed urban boundaries far in excess of future requirements (Gayler, 1982), but, generally, municipal and regional politicians were unwilling to reduce the area designated for urban development. From

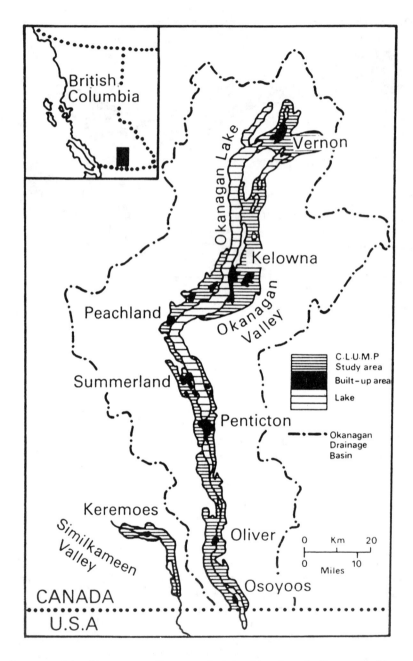

Figure 5.7 The Okanagan Valley, British Columbia, Canada (*Source:* Robinson, 1988b: 238)

1978 boundary disputes were brought before the Ontario Municipal Board (OMB), with PALS making use of a Green Paper on *Food Land Guidelines* produced by the Ontario Ministry of Agriculture and Food (1977; Johnston and Smit, 1985). This provided specific criteria to assist local governments in designating agricultural land within their official plans (Furuseth and Pierce, 1982: 58–60). As a result of the hearings, over 2,000 ha of land were removed from urban designation and the OMB advocated strict control policies outside the urban boundaries based on soil capability rather than on existing levels of production (Jackson, 1982; Kreuger, 1984).

As in the Niagara Region, in British Columbia the few pockets of land suitable for intensive food production are in close proximity to concentrations of population (Figure 5.7). Prime agricultural land is very limited in the province: *the Fraser Valley and the Okanagan*, the two most important areas of fruit production, having only 109,459 ha (270,475 acres) of Capability Classes 1 to 4 land (BC Select Standing Committee on Agriculture, 1978). The pressure of urban expansion upon this land has been great, with one estimate suggesting that in the 1960s and early 1970s over 6,000 ha (14,825 acres) of prime farmland were being lost annually to this competing use (Lands Directorate, 1980: 19). The scale of this conflict and the severe shortage of agricultural land prompted an early response from the provincial government which pioneered legislation to establish a comprehensive agricultural land protection policy.

In 1973 a land-zoning programme was introduced to protect farmland, paying special attention to the need to maintain and develop family farming (Bray, 1980; Stupich, 1975). Zoning and regulatory controls over agricultural resources and Agricultural Land Reserves (ALR) were placed in the hands of a Provincial Lands Commission. The ALR, covering 4.9 per cent of the province, acts as an area zoned exclusively for farming so that its designation supersedes all other land planning and zoning decisions (Wilson and Pierce, 1984). To remove land from the ALR, local authorities have to make special petitions to the Commission or the Provincial Government. Various surveys in the late 1970s suggested that the integrity of the ALR was being preserved and that there was widespread support for the farmland protection policy (Malzahn, 1979; Manning and Eddy, 1978; Raynor, 1980). Indeed Pierce's (1981a) survey of the rates of conversion of farmland in the province showed that land transfers out of the ALR were much slower than before their creation.

Despite these encouraging findings, high rates of loss of agricultural land are apparent for several of the smaller towns in British Columbia. This may indicate some localised relaxation of the strict enforcement of the ALRs, but it also reflects a noticeable change in the distribution of population, accompanied by increased low-density dwelling activities. For example, the Victoria Metropolitan Area experienced a growth in its urban population of only 6 per cent between 1976 and 1981 whilst its rural counterpart experienced a 32 per cent increase (Lands Directorate, 1985b). This was associated primarily with conglomerates of low-density housing. Some of this development involved infilling on existing sub-divided land, but good quality farmland was also affected, and it is clear that further plans may be needed to guide and control future urban and suburban growth more effectively despite an overall reduction in the rate of loss of farmland during the 1970s and early 1980s (Pierce, 1981b).

5.5.2.2 SWEDEN

In Sweden efforts to retain farmlands have been part of a more comprehensive policy to enhance the economic viability of family farming. This policy was established in the 1977 Agricultural Act which stressed the need for cultivating all land suitable for agriculture and for maintaining existing production capacity. This reversed previous legislation under Acts in 1947 and 1967 in which Sweden pursued a programme to reduce both the number of farms and farmers (Lapping and Forster, 1982). From 1960 to 1973 over 8,000 holdings ceased to exist each year (OECD, 1978), much of their land falling idle rather than being incorporated in remaining holdings. The new concern with family farming, which has continued in the 1980s, was prompted by a concern for maintaining economic development in peripheral regions, previously strongly susceptible to farm closures, plus evidence showing family farms were more efficient than large-scale commercial operations (e.g. Granath, 1978; O'Hagan, 1978). Aesthetic and environmental considerations also came into play with government recognising that abandoned farmland was 'unattractive' and detracted from the image of the countryside.

Through a system of County Agricultural Boards, proposed farm sales are vetted and some amalgamations effected. The Boards have comprehensive powers to regulate land markets, extend credit, provide technical advice to farmers, and organise and manage farm amalgamation. They have exercised these powers to prevent loss of farmland to urban and speculative development whilst building up land blocks for redistribution. The high degree of municipal ownership of urban land has helped to reduce the pressure upon farmland in the rural-urban fringe (e.g. Brasch, 1979).

5.5.2.3 BRITAIN

Conservationists in Britain have often focused upon loss of farmland to urban sprawl (e.g. Best, 1977; 1981; Coleman, 1978; Wibberley, 1959). However, in the face of overproduction of cereals in recent years, especially in North America and Western Europe, government policy has begun to encourage greater diversity of land use through promotion of farm forestry and preservation of special environments on farms, e.g. ancient woodland and species-rich meadows (Brown and Taylor, 1988).

Edwards (1986) has estimated that at present rates of growth of productivity, and with a small increase in demand, the United Kingdom could maintain its present level of self-sufficiency to the year 2000 even with a reduction in its agricultural area of 4 million ha (9.9 million acres) (23 per cent). If this reduction was of poor farmland then even more land could be released. So a crucial question for policy-makers has become not one of sustaining output but of limiting it so that surpluses are cut or avoided. The policies concerned are likely to be continuations of quotas on potatoes, sugar beet and milk; the use of price cuts to reduce the output of other commodities, especially cereals; and programmes to encourage the transfer of land out of agriculture (EDC for the Agriculture Industry, 1987).

By early 1989 1,816 farmers in the United Kingdom had opted to take land out of production under the *set-aside scheme*. This would affect over 56,650 ha (140,000 acres) or 1 per cent of all farmland and on 1.2 per cent of full-time holdings. Of this land set-aside, 2 per cent has been earmarked for woodland, 7 per cent for non-agricultural use and most of the remainder for fallow.

The concept of 'set-aside' has now been incorporated in government proposals in a £28 million per annum scheme to cut food surpluses. Farmers will be encouraged to take arable land out of production and put it under an approved alternative use, e.g. golf courses, riding schools, but not grazing or sale for development. Farmers able to do this for at least 20 per cent of their arable land for five years will receive up to £200 per ha (£81 per acre). This scheme, effective from late 1988, will work in tandem with cash penalties for overproduction to limit EEC cereal production to 160 million tonnes, and to keep total CAP spending around £20 billion. Within the set-aside scheme, payments will be based on the farmers' arable crop pattern in 1987–8, with payments varying according to the new land use adopted. For example, the payments will be £200 per ha (£81 per acre) per annum for land use conversions to fallow and woodland, but with a £20 per ha (£8 per acre) reduction in poorer agricultural areas (e.g. the Less Favoured Areas) and a £20 reduction if the fallow is rotated around the farm. For non-agricultural use the set-aside payment will be reduced by £50 per ha (£20 per acre). The government's argument is that the funding for set-aside recognises the farmers' environmental services to the community in keeping such land in attractive condition and capable, if necessary, of returning to agricultural use. The initial uptake of set-aside comprised 2,000 applications affecting over 50,000 ha (125,000 acres). Nearly one-third of applicants was in East Anglia.

In 1987 the Ministry of Agriculture, Fisheries and Food (MAFF) launched a £25 million package intended to encourage farm diversification through alternative uses of farmland, and especially the expansion of 'environmentally friendly' farming (e.g. MAFF, 1987). Under the acronym, ALURE (Alternative Land Use and Rural Economy), these proposals have attracted tremendous interest, mainly because they may signal a definite break with the long-established pre-eminence of policies supporting increased production. Not surprisingly, though, the four main policy proposals (summarised in Table 5.7) have been controversial.

Table 5.7 MAFF's ALURE Policy Proposals, 1987

1. £10 million per annum for the development of on-farm woodlands. Farmers to be offered variable payments to reflect the loss of income from land being afforested;
2. An additional £3 million per annum to be given for the expansion of the traditional private sector forestry programme, primarily in lowland areas;
3. £7 million for a doubling of the number of Environmentally Sensitive Areas (ESAs) which were first introduced with the 1986 Agricultural Act;
4. £5 million per annum to encourage diversification of farm businesses by providing grants to assist the establishment of on-farm ancillary business.

Although they represent the promotion of *extensification* of agriculture, they have also been interpreted in some quarters as likely to encourage building development on good farmland by reducing MAFF's input to planning controls (Blowers, 1987). Reaction from the farming community has also been quite unfavourable as it has been felt that insufficient financial incentives were being offered for diversification. Yet in terms of their first eighteen months of operation, the proposals have had little impact upon farming. From January to October 1988 there were 896 applications for the farm diversification grant scheme, and Cloke and McLaughlin (1989) argue that it is largely only a small group of entrepreneurial farmers who are likely to take up the scheme. The funding of the ALURE policy must be set against the £1,800 million that is paid in agricultural support, and it is clear that further financial encouragements for diversification will be required if farm forestry, farm tourism, new and unconventional products and value-added systems (e.g. pick-your-own schemes) are to develop further in British agriculture.

5.5.2.4 THE UNITED STATES

The debate on the conversion of farmland to other uses in the United States has focused on three issues (Fisher, 1982):

(a) the magnitude of farmland conversions;
(b) the nature and severity of the costs which these land losses impose upon society;
(c) the appropriateness and efficiency of various public policies.

The first of these issues has been addressed by the National Agricultural Lands Study using data prepared by the Soil Conservation Survey (Fischel, 1982). This study estimated an annual rate of loss of between 0.8 and 1.2 million ha (2 to 3 million acres) from 1967 to 1977. However, Simon and Sudman (1982) challenged this as being too prone to error, and, using a broader database, produced an estimate of 0.4 million ha (one million acres) loss per annum.

Brown *et al.* (1982) argued that such disputes, based on lack of adequate data, permitted widely differing perceptions of the problem. At one level the introduction of payments to farmers to take land out of growing crops, because of overproduction, suggests there is no need to be worried about farmland loss in terms of its effect on agricultural production. For example, 25 per cent of land in the United States is in farms; even if the area of cities doubled, they would only occupy about 4 per cent of all land and 10.8 per cent of the non-federal land (Andrews, 1979). But not only is the land lost often of high quality, especially that to urban sprawl (e.g. Dillman and Cousins, 1982), but non-economic arguments are also important. Thus, increasingly, protection of farmland has become an issue taken up by environmental and conservationist groups concerned with the changing appearance of the countryside (Stopp Jnr., 1984).

In the United States 'set-aside' has been used both as a way to ease overproduction of certain crops and as a means of reducing the risk of soil erosion by encouraging farmers to plant trees on the most erodible farmland. Following bumper harvests in the early 1980s a Payment In Kind (PIK) programme was introduced in 1983, encouraging farmers to take their land out of production in exchange for payment. Over one million farm holdings participated in the programme, representing one-third of the country's cropland. This resulted in a 38 per cent decline in the maize harvest, a substantial Treasury saving in storage costs, but an actual increase in farm income (Goodenough, 1984). Subsequently, as part of the 1985 Farm Security Act, a Conservation Reserve Program (CRP) was introduced setting aside 18.2 million ha (45 million acres) of land for a ten year period. However, the Program has held little attraction for many farmers, who look upon set-aside as the antithesis of 'proper' farming. Instead, it has often been speculators who have helped to establish the Program, buying up farms from which farmers have been forced out by the price-cost squeeze, and taking advantage of the available government funds for set-aside, at an average of US$118.50 per ha (US$48 per acre): as much as a 15 per cent net return per year on investment. The negative aspects of the scheme are the reduction of land available for young farmers to rent, the potential loss of money to the rural economy, and the effect upon the farm supply industry.

The first legislation to protect farmland was introduced in Maryland in 1956, though it was in the 1970s that the growing concern for preservation was shown in the passing of similar legislation in 44 states. Also, in 1976 the federal Environmental Protection Agency started to require a statement of agricultural impacts for major federal projects. The preservation programmes, now practised in 49 states, have three main strategies: financial incentives, enforced land use policies and voluntary programmes (see Table 5.8). Of these, Volkman (1987) argues that the voluntary schemes, as represented by the creation of Agricultural Districts, offer one of the most effective ways of slowing intensive development. Such schemes are operated in nine states (California, Illinois, Maryland, Minnesota, New Jersey, New York, Ohio, Pennsylvania and Virginia) and have involved the creation of an Agricultural

Table 5.8 Agricultural land converted to urban, built-up, transportation and water uses, by former agricultural uses, United States, 1967–75

Census region	Cropland's share of region's rural land, 1977 (%)	Quantity of cropland consumed (per million acres)	Cropland's share of rural land converted (%)	Pastureland and rangeland consumed (per million acres)	Other agricultural uses consumed (per million acres)	Total
Northeast	20	0.6	20	0.1	2.3	3.0
South	22	2.5	21	2.1	7.4	12.0
North Central	35	1.6	31	0.8	2.8	5.2
West	18	0.7	23	1.3	1.0	3.0
Total	37	5.4	23	4.3	13.5	23.2

Source: Volkman (1987)

District in which an area of farmland is set aside from non-farm development for a fixed period of time. Farmers meeting certain requirements can join the scheme in return for various benefits, e.g. reduced property taxes, reduced payments for special utility assessments. Although Furuseth and Pierce (1982: 55) note that incentives have rarely been sufficiently strong to withstand the pressures of major urban development, they may be more useful in 'deeper' countryside and are amenable to modifications which can strengthen limitations on non-agricultural development (Volkman, 1987: 30; see also Ervin, 1988; Furuseth, 1985; Rickard, 1986; Vining *et al.*, 1977).

5.6   Farmers, landowners and labourers

The many changes wrought by the New Agricultural Revolution post-1945 have had a dramatic impact upon the farming population and rural society in general. The nature of these changes is best indicated via separate consideration of three groups within the rural community: farmers, landowners and labourers.

5.6.1   Farmers

For Britain, Bell and Newby (1974) classified farmers on the basis of their degree of market orientation and degree of direct involvement in husbandry (Table 5.9). Their simple four-fold classification represents highly distinctive groups of farmers and, perhaps, could be simplified into a clear division between those to whom farming was still seen as a 'way of life' and those who treated farming strictly in business terms. In the former group comes the *family farmer*, a more familiar figure in continental Europe, but there, as elsewhere, being placed under increasing pressure by both suppliers of farm inputs and processors of farm produce.

   Throughout the Developed World the family farmer has been seen as a central element in rural society, occupying a pivotal role in the rural idyll or what has been referred to in the United States as 'agrarianism'. This has also been referred to as the 'agrarian myth', 'pastoralism', the 'Jefferson creed' and 'agricultural fundamentalism'. It emphasises the uniqueness of the farmer in society, viewing farmers as independent, honest and occupying the most satisfying and basic occupation in society (Carlson and McLeod, 1978; Flinn and Johnson, 1974). In Britain it has been the modern day equivalent of the yeoman farmers that have perhaps best represented the agrarian myth, their role changing as owner-occupation has become the norm.

   For at least two centuries in Britain, farming has been dominated by small (in world terms) owner-occupier or tenanted farms run largely by the farmer and family labour. Especially in the north and west, the areas

Table 5.9 Socio-economic classification of farms

	Low market orientation	High market orientation
Administrative involvement in farm operations	Hobby farms	Agribusiness
Manual involvement in farm operations	Family farms	Agribusiness

Source: Bell and Newby (1974)

where depopulation has tended to be greatest, these farmers have been tenants on large estates, though from 1919 onwards owner-occupation has increased greatly so that by the early 1970s three-quarters of all holdings in England and Wales were owner-occupier. The growth of dairy farming and of horticultural production from the late 19th century increased the importance of the family farmers (e.g. Robinson, 1983a), but post-1945 they have diminished in numbers and have also had to reorientate their farming to become better businessmen (see examples in Blythe, 1969). New strategies have been adopted to permit their survival and this has carried the family farmer away from farming as 'a way of life' in the sense that might still be applied to some farmers in the Mezzogiorno or northern Greece for example. The small family farm, unable to benefit from scale economies, has become more marginal and reliant upon various forms of government support for survival. Reductions in this support, as seen in the removal of various subsidies in Western Europe and New Zealand (Supplementary Minimum Prices) (Cloke, 1989c), have brought reduced numbers of family farms and further encouraged the trend towards fewer and larger holdings (e.g. Bouquet, 1982).

Family farmers have been forced to operate more efficiently in order to survive the threats to their existence. This has meant cutting down the input of non-family labour, perhaps relying on contract or casual labour at peak times and joining co-operative groups in conjunction with other family farmers to acquire certain benefits of association and of scale. Such co-operation may give the small farmer access to expertise otherwise absent within the context of the family farm.

It is this expertise, for example in the sphere of financial administration and accounting, which would be more common on farms run by 'active managerials' or 'agribusinessmen'. The latter are at the opposite end of the farming ladder to the small family farmer. Their background is likely to resemble that of managers in industry and, generally, they will be company executives. As such they may have close ties with either the supply or processing side of farming, though, increasingly, non-agricultural organisations such as pension funds and insurance companies have become involved in agribusiness operations (Munton, 1977; 1985).

The agribusinessman is often well educated, equipped with business skills rather than qualifications in farming, and tends to be more mobile, expansionist and outward looking than the family farmer. In the United States agribusinessmen have formed a larger component of the farm population. Here too corporate control of agriculture has been more extensive. One of the most well-known studies of the social effects of this emergence of a higher proportion of large-scale non-family farming operations is Goldschmidt's (1978a; 1978b) work on Arvin and Dinuba, California. This proposed that there was a strong negative relationship between the scale of farm operation and the quality of life in rural communities. The proposal has been corroborated in several other studies (e.g. Heffernan, 1982), with an emphasis upon growing disparities within the rural class structure. However, the nature of such a clear link between farm size and the social and economic indicators of quality of life do seem to vary between different types of farming system (Green, 1985; see also Rosenfeld, 1989; Whatmore *et al.*, 1987a; 1987b).

The 'active managerials', as recognised by Bell and Newby, are an intermediary group between the family farmer and the agribusinessman. One major difference between the active managerial and the family farmer concerns scale of operation. The former operates on a larger holding, with a larger labour force that might include a farm manager. Such farmers may have special expertise in aspects of plant and/or animal husbandry. Concern with profit and innovation is also likely to be a greater concern. However, in Britain the threat of high taxation for the owner-occupier active managerials has limited their expansion.

Potter (1986a) argues that much attention has been devoted to documenting the pattern of countryside change. Only recently has the academic focus shifted to understanding the decision-making processes underlying change. In particular, he argues that the behaviour of farmers, 'the change agents and policy operatives' needs to be better understood. His own work on farmers' investment decisions in land improvement and landscape maintenance highlights the controls of policy, institutions and the farm family, with the farmer solving problems within the limits of these controls. Different investment styles illustrate different relationships between the 'determining controls' and the 'intentioning' problem-solving farmers (Potter, 1983; 1986b).

Numerous studies by Ilbery (e.g. 1978; 1983; 1985) have illustrated fundamental variations between farmers, based primarily upon age, education, experience and access to capital. Some of these studies suggest there are differences in the decision-making of owner-occupiers as opposed to tenants. Yet owning land is not always perceived as the most advantageous strategy. For example, in a comparison of wholly owned, wholly tenanted and mixed tenure farms in the Farm Management Survey sample for England and Wales, Hill and Gasson (1985) found the mixed tenure farms performed noticeably better than 'pure' tenure types with respect to productivity, income and output. Wholly rented farms performed relatively well on small acreages and wholly owned farms on larger acreages. The success of the combined tenure group reflects the gravitation towards this group by the most successful tenant farmers.

5.6.2 Landowners

The problem of taxation, in the form of Inheritance Tax in Britain, formerly Capital Transfer Tax (CTT), has affected that group of farmers referred to by Bell and Newby as the 'gentleman farmers', the direct link with the traditional landowning aristocracy. Newby (1980b) quoted an example of a holding of 400 ha (1,000 acres) valued at £1,975 per ha (£800 per acre) which, if its owner died, would have left the heirs a CTT bill for £290,750 or 29 per cent of the value of the land. Given the tremendous rise in land values from the early 1960s, this placed a great burden upon individual landowners and encouraged institutional ownership. It has accentuated the need for maximum efficiency and profitability, especially for those owning large amounts of high quality farmland. It has also reduced the numbers of 'gentry' owning land. In southern Sweden, for example, Moller (1985) sees this as a long-term process fostered by increased market orientation of agriculture and the change in the organisation of labour. The need to raise capital has often been met by sales of 'marginal' land whilst there has also been the creation of larger tenanted holdings or owner-occupier farms. This has helped to reduce the number of hamlets in southern Sweden.

Traditionally, large landowners have often been more concerned with the maintenance of a distinctive life-style involving the 'gentlemanly' pursuits of hunting, shooting, charitable activity and public affairs. Some aspects of this remain, with the large landowner still being acknowledged as the local squire, a leader of local society and a steward both of the land and of a distinct hierarchy within rural society. Despite a popular view that such landowners were 'above' the need for their estates to yield a profit, few have sufficient wealth to indulge the notion of a concern that does not produce a return. Although many large landowners have often been more concerned with business ventures in other spheres, few have been able to tolerate continued losses from their agricultural estates for any length of time. Hence these estates have been run as commercial enterprises, usually with their day-to-day operations attended to by an agent or manager. The gentleman farmers themselves have tended to focus their attention on activities outside the farming sphere, though, more recently, economic pressures have often prompted more concern with estate management and the acquisition of qualifications in farming, management and business in order to meet the new challenge.

In Britain holdings of less than 20 ha (50 acres) account for less than 4 per cent of farmland, those from 20 to 200 ha (50 to 495 acres) for 48 per cent, and the 6.6 per cent of holdings in excess of 200 ha (495 acres) for 47 per cent. Within this latter group is concealed holdings belonging to the traditional aristocratic landowners. Massey and Catalano's (1977) study shows that a small number of titled families own nearly one-third of the countryside in Great Britain, with just over 200 families each owning at least 200 ha (495 acres). For Scotland, McEwen's (1981) survey shows 87 per cent of all land to be in private hands, with 63 per cent of the entire country privately owned in blocks of at least 400 ha (990 acres). There is a marked contrast between these private landowners, the 13 per cent of Britain that is publicly owned and the 1 per cent in the hands of financial institutions such as large insurance companies and pension funds.

The problems associated with a small number of landowners owning such a large proportion of the countryside have been considered from a variety of standpoints. Shoard (1980; 1987), for example, examines the power that this small group has over decision-making in rural areas. In addition to controlling activities on their own land and restricting access by the general public, there continues to be a heavy involvement by the landed gentry in local affairs. For example, in 1985 26 of the lord lieutenants of the 31 counties of Scotland were landowners, a position enabling them to chair committees advising the Lord Chancellor on the appointment and conduct of magistrates. At national level the landowning aristocracy have automatic membership of the House of Lords whilst other large landowners have access to, and exercise power in, highly influential organisations such as the Country Landowners' Association, the National Farmers' Union and Timber Growers UK. The large landowners have also had a major voice in shaping the Forestry Commission and the Countryside Commission. Thus the 'landowning lobby' has been able to exert an extremely powerful influence upon government decision-making despite the massive decline in importance of the 'farm vote' consequent upon the reduction in the numbers of farmers and the agricultural population in general.

Parallels can be found with the large landowners in other parts of the Developed World, but despite references above to 'gentlemen farmers' and 'landowning aristocracy', it is unrealistic to view large landowners as a homogeneous group. For example, Massey (1977) recognised three distinct types of landowners in Britain on the basis of the relations of landownership and the role of that landownership within society. She described them as:

(a) *Former landed property*: including the landed aristocracy and gentry, the Church of England and the Crown Estates. This accounts for nearly 40 per cent of the land area of Great Britain, and consists predominantly of rural land. In this type of ownership the ownership of land is an integral part of a wider social role.
(b) *Industrial landownership*: the ownership of land as a condition of production. This comprises the majority of industrial capital, but includes owner-occupier farmers too.
(c) *Financial landownership*: unlike (a) this operates completely in

capitalist terms so that land and property represents just another sector for investment. Typical financial landowners are property companies, pension funds and insurance companies (Goodchild and Munton, 1985).

This division implies that conflicts relating to landownership cannot be treated simply as a result of or arising from the operation of 'landed capital'.

Barlow (1986b) argues that the significance of landowners in shaping rural economy and society depends largely on the local nature of ownership relations and political conflict. He points out that property owners as opposed to landowners play a significant role in economic and social development in rural areas as do owners from beyond the traditional rural aristocracy and gentry (see Newby et al., 1978). Landowners have power in three crucial areas:

(a) *Control of local politics.* This enables them to exclude new developments which might help other rural residents and to limit the influence of national political movements.
(b) *Control of the local economy, employment and housing.* This is often seen in the form of low wages paid to estate workers and in the exercise of power over 'tied' housing for farm and forestry workers.
(c) *Control of rural culture.* This is often performed through continuance of traditional rural patterns of paternalism, deference and recognition of a dominant hierarchy extending downwards from the local landowner.

In many parts of Western Europe the theories of Engels, Lenin and Marx, that capitalist development of agriculture would produce a concentration of land in a small group of large landowners, have been incorrect. In some areas it has been large farms that have declined whilst small and medium-sized family farms have increased rapidly (e.g. Tepicht, 1973; Littlejohn, 1977). Yet this represents merely a different facet of the development of the market economy, either 're-aligning' but maintaining a peasant farming sector or turning this peasant sector into prosperous family farming. The latter has characterised large parts of French agriculture whilst Bennett (1986) notes a similar occurrence in the Algarve, Portugal. However, in the case of the Algarve, agriculture has been transformed into a part-time occupation. This has led to fragmentation and sub-division of land, thereby inhibiting modernisation. This change, repeated in other Mediterranean countries, has maintained a modified peasantry whilst increasing productivity and reliance upon 'the external capitalist economic system', though without producing concentrations in ownership.

5.6.3 Farm labourers

In Western Europe academic study has often propagated the conception of the countryside as a reservoir of traditionalism, 'and of the peasantry as an arsenal of pre-modern characteristics' by studies dedicated to the examination of peasant life or *volkskunde* (Evans and Lee, 1986: 9). Traits attributed to the peasant, such as humility, piety, natural wisdom, simplicity and goodness were held to be a product of the peasant's labour and frugal life-style, and were the antithesis of the moral and social degeneration of industrialisation (Gagliardo, 1969; Ziche, 1968; Haushofer, 1978). In many ways these attributes have also been applied to those seen as the successors to the peasants, the farm labourers. These are the 'honest sons of the soil' occupying the lowest rung of rural society and therefore filling many of the social roles previously occupied by peasant groups. In Britain these links with the peasantry may be more tenuous, but, given the rapid fall in the number of farm workers throughout the Developed World, the character of this group in Britain, the country with one of the smallest proportions of its labour force in agriculture, is being approximated rapidly in other countries.

As noted by Newby (1988a), views of the history of farmworkers have generally been of two kinds. One has portrayed the farmworker in an heroic light, struggling against the iniquities of domination by rich landowners. In Britain this image has been fostered by the way in which the Tolpuddle martyrs, transported to Australia for forming a workers' association (Marlow, 1971), secured the farmworker a place in the iconography of British trade union history. This view is rather too cosy and sentimental for it fails to deal with the fact that farmworkers have tended to ignore radical politics and have often been reluctant members of trades unions. The second view simply portrays the agricultural labourer as the partner of the farmer, viewing labour in terms of productivity, supply and demand, policy-related analysis of comparative wage rates, labour market conditions and migration from the land. Neither view offers a proper analysis of economic and social changes affecting the farmworker under agrarian capitalism. Nor do they explain what Newby refers to as the 'invisibility' of farmworkers. Perhaps the most critical aspect of this 'invisibility' has been the way in which their voice has been so muted in the debates shaping the future of the countryside. Consequently, farmworkers' needs for jobs, housing and rural services are usually overlooked in arguments that are all too frequently polarised between the farmers/landowners and the environmental lobby (Armstrong, 1988).

Barlett's (1986) investigation of full-time farmworkers in Georgia revealed some characteristics amongst the workers similar to those portrayed in Newby's (1977) work in Britain. Pay was low and hours long, but the workers expressed strongly positive attitudes towards their work. They often had close personalistic ties with farmowners and shared with them a particular agrarian ideology emphasising respect for farm life. This was true for both white and black workers, both sets of whom had educational levels similar to national norms and had often previously held jobs in the non-farm sector. These were not workers trapped in farm-work or lacking other options. This suggests that some views of agriculture labour as the repository for the least enterprising and lowest achievers in rural society have been misplaced.

Any 'revision' must take account of the presence of conflicting trends within the actual job performed by farmworkers. On the one hand increased specialisation and mechanisation has meant that farmworkers can often require particular skills, e.g. in dairying or operating equipment on arable farms. But, on the other, the reduced number of workers per farm can mean that workers have to be generalists capable of performing a wide range of tasks. Added to this there is the growth of 'intermittent' employment of labour in agriculture, as seasonal or casual workers, which is hinted at in published agricultural censuses. Survey work by Ball (1987) suggests that official figures may underestimate by a factor of five or more, failing to enumerate both directly- and indirectly-recruited (contract) labour. Increasingly this intermittent labour has substituted for regular part-time and full-time workers. Contract labour is usually drawn from beyond the immediate rural area and this has important implications for local rural economies which are being pushed towards an ever-growing dependence on external agencies. The reservoir of local agricultural labour has been replaced by travelling gangs of contract workers, eliminating some of the myths of the farmworker's traditional role and, in some cases, developing their own, e.g. the sheep-shearing gangs in Australia.

Much of the academic work on agricultural labour has focused exclusively upon male labour, ignoring the role of both female hired labour and that of farmers' wives. In effect, this highlights one of the traditional views of women in rural society: as housewife and mother. Other views have emphasised the female role in maintaining both family and community norms whilst, more recently, there has been greater realisation of the extent to which women are still forming a cheap labour pool within farming, including part-time farming (Pfeffer, 1989), and in the manufacturing and service sectors being established in rural areas (Gasson, 1980; 1981; Little, 1986; 1987; Symes and Marsden, 1983). Even so, despite this increased involvement, it is clear that there are high levels of female under-employment in rural areas (Lichter, 1989).

It is also apparent that women have increasingly played a prominent role in rural communities' decision-making (Bokemeier and Tait, 1980). This partially reflects the views that it is at this level rather than at regional/state or national level that women can most easily exercise political power (Almond and Verba, 1963). It is also associated with the growing politicisation of issues traditionally regarded as 'women's issues', e.g. conservation, welfare, child care (e.g. Bozeman et al., 1977), and women's growing ability to solve the problem of lack of leadership in rural communities.

6

Agricultural development and government policy

Agricultural policies in the Developed World have had as their primary objective the production of a high proportion of food requirements from domestic resources in order to reduce external dependence and to reduce the risk of food shortages. They are therefore partly precautionary, but also apply directly to agriculture's fluctuating production by regulating prices and marketing in order to achieve stability. The complexity of the various policies is illustrated in this chapter by special reference to the Common Agricultural Policy (CAP) of the European Economic Community (EEC), though there are many parallels with agricultural policies in other parts of the Developed World.

Self and Storing (1962: 218) recognised both *utilitarian goals* and *equity goals* in agricultural policies (see Figure 6.1). Utilitarian goals refer to the contribution made by agriculture to the national economy, and have been developed to a greater extent than the equity goals. In particular, policies have tended to stress agriculture's contribution to economic growth and economic stability, e.g. farming's role in improving the balance of payments and stabilising domestic food prices at a satisfactory level. In some countries, two other important utilitarian goals have been those of agriculture providing a strategic reserve of food for war or unforeseen circumstances, and of agriculture as a land use to be retained against the tide of advancing urban development. The notions of economic stability are connected with those of the equity goals of social stability in rural areas and suitable remuneration for farmers and farm-workers, but their implementation has not prevented measures which have led to reduced farm labour and which have failed to reduce disparities between the wages of agricultural labourers and industrial workers.

Policy goals have been translated into a variety of measures designed to influence various elements of the farming system and the economic environment in which they operate. Bowler (1979:21) lists four basic types of measure designed to influence:

(a) the growth and development of agriculture, e.g. subsidies on the

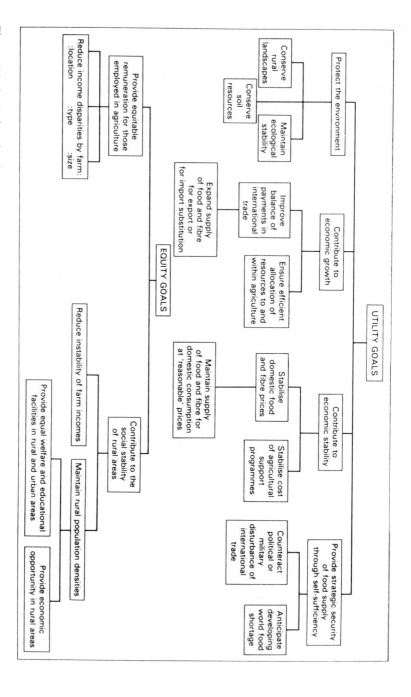

Figure 6.1 Agricultural policy goals (*Source:* Bowler, 1979: 9)

costs of inputs, preferential taxes, preferential interest rates, financial assistance to co-operatives for purchase of inputs, financing assistance for educational extension and research establishments;
(b) the terms and levels of compensation of production, e.g. demand management, supply management, direct payments;
(c) the economic structure of agriculture, e.g. financial assistance for land consolidation schemes, farm amalgamation grants, land reform schemes, retirement grants, financial assistance to co-operatives for production/marketing, inheritance laws, farm wage determination;
(d) the environmental quality of rural areas, e.g. pollution control measures, land zoning ordinances.

There are numerous possible combinations of these different measures, some combinations providing conflicting aims or being insufficient to overcome other forces. In Alaska, for example, numerous schemes of inducement were advanced to encourage pioneering farming settlement: liberal land laws, a subsidised government railroad, and a demonstration group settlement scheme. Yet the counter-attractions of urban settlement helped to limit agricultural settlement on this 'last American frontier' and there has been little growth of the once hoped for yeoman farmer communities (Shortridge, 1976; Anderson, 1961). Rarely have there been situations where the aims of agricultural policy have been clearly defined and have been linked closely to other rural policies. A possible exception, though, is in Prince Edward Island, Canada, where the provincial government's solution to continued decline in the rural population and the threatened demise of family farming has been a 15-year Development Plan. This has espoused a 'small is beautiful' philosophy in an attempt to maintain family farming through controls on land sales and corporate farming (Crabb, 1985; Rawlyk and McDonald, 1979). Other measures have attempted to reverse the drift from the land by encouraging crop diversification, labour intensive commodities, part-time and small-scale farming, e.g. the Small Farm Development Program and the Family Farm Program.

Aspects of the first three of the measures referred to above have been employed on a concerted basis within Britain, particularly under the auspices of the 1947 Agriculture Act which aimed to ensure adequate supplies of food at moderate cost and to make efficient use of domestic agricultural resources. A brief outline of the elements of policy included within this 1947 Act provides one illustration of ways in which the utilitarian and equity goals have been tackled by specific policies.

6.1 Agricultural policy in Britain, 1947–73

In Britain a major factor in the transformation of farming post-1945 has been the influence of government intervention. After very little involvement in agriculture before the 1930s, subsequent direct intervention then through a series of measures brought a commitment to agricultural support yet to be revoked (Holderness, 1985). Four main elements of policy became integral parts of government agricultural support until entry to the European Economic Community (EEC) in the 1970s. These were:

(a) The introduction of *grants* and *subsidies* to operate either as encouragement to farmers to adopt certain favourable practices or as guarantees of minimum prices for produce. For example, under the 1932 Wheat Act, producers of wheat received a subsidy of the difference between average prices and a standard (or guaranteed) price. This *deficiency payment scheme* was introduced in the 1930s for wheat, barley, oats, bacon pigs, sheep and milk for manufacture. It was continued in the 1947 Agriculture Act, setting the tone of post-war policy, and extending the scheme to wool, eggs, sugar beet and fat cattle (Whetham, 1953). Each year the *guaranteed prices* were reviewed for the next harvest or twelve months, and for two decades this Annual Review became a great source of debate between farmers and governments.

(b) The establishment of *producer-controlled marketing boards*, initially under the Agricultural Marketing Acts of 1931 and 1933. At first the boards covered hops, milk, pigs and potatoes, but, post-1945, new ones were established for wool, eggs and tomatoes. Also, the Home Grown Cereals Authority and the Meat and Livestock Commission were created in the 1960s to improve the marketing of home produced cereals and meat.

(c) The greater *control of imports*. In the 1930s tariffs of 10 to 20 per cent were imposed with seasonal rates on horticultural produce and, although restrictions were largely removed post-war, from the early 1960s more negotiations were made with foreign competitors to restrain imports.

(d) A complex *infrastructure of research, advisory and educational services*. This enables research into new methods of farming to be translated into practical schemes for the farmer to adopt, aided by special credit facilities. Close government control and good organisation aided farming efficiency whilst making farming more receptive to new methods. In this there was assistance in improved channels of communication from government to farmers, e.g. the Agricultural District Advisory Service. So the 'Green Revolution' was accommodated very rapidly by British farming, especially in terms of the more scientific application of fertilisers, pesticides and use of irrigation. This led to greatly increased yields while livestock farmers benefited from the first major improvements in breeding techniques since the 19th century. Optimal feeding schemes were introduced for pigs and poultry to produce the 'factory farming' of the 1960s, with animals housed on intensive or battery lines. For beef cattle, barley-based silage feeding was developed and the age at which beasts were slaughtered was reduced. More efficient feeding also raised milk yields from dairy cows.

One major paradox within these agricultural policies, also apparent elsewhere, was that whilst farming consumed large amounts of public finance, especially via price supports, the ownership and marketing of land, which was mainly in private hands, had minimal public controls. In fact, collective public finance supports a selective concentration of private ownership over the means of production. So the state encouraged control of food production by private individuals, trying to maintain both reasonable food prices for consumers and satisfactory incomes for farmers.

Bowler's (1979: 55–66) analysis of the guaranteed price system indicated both how conflicts can arise between various objectives of agricultural policy and how government policy can play a significant role in the spatial patterns of agricultural activity. Although the deficiency payment system provided consumers with food at relatively low prices, the costs to the taxpayers were often unpredictable. The twin aims of the system were to provide farm income support and to orientate production. However, under the inflationary conditions of the 1960s, the latter aim tended to assume a secondary role behind the need to compensate farmers for rising production costs. At times in the early 1960s, deficiency payments for wheat, barley, oats and fat sheep reached 30 per cent of total average returns to farmers, and developed a strong bias towards cereals in terms of the degree of price support. Yet, when analysing the effects of the guaranteed prices scheme, Bowler (1979: 64) argued that 'the nature and pace of technological change in the production of competing products appears to have been a more important influence on agricultural trends than guaranteed prices.'

Evaluations of the policy followed from 1947 until entry to the EEC in 1973 suggest that many of the objectives set out in the 1947 Act were reached (e.g. Bowers, 1985; Winegarten, 1978). Although this system of support may have been a somewhat imperfect system of bolstering farm incomes, they were maintained at acceptable levels during the 1950s and 1960s whilst government intervention meant food prices were relatively stable. Various studies by Bowler (e.g. 1976a; 1976b; 1976c; 1979) suggest there was no clear cut relationship between guaranteed prices and farming trends. Increasingly, in the 1950s, policy was directed towards tying aid to inputs and particular desirable techniques. Initially, such aid had dramatic effects upon the use of fertiliser, the numbers of cattle and acreage of temporary grass. 'The combination of production grants, the work of the National Agricultural Advisory Service and the manipulation of guaranteed prices helped to produce a farming sector in which both change and increased efficiency were generated. Whilst government policy may not have played a leading role in change, it certainly promoted a favourable climate in which new practices could be easily adopted' (Robinson, 1988a: 210).

Winegarten's (1978) study of the evolution of British agricultural policy indicates how successive Annual Reviews altered both individual measures and the general direction of policy, especially in the 1960s when thought was given towards moving in line with the policies of the CAP (Neville-Rolfe, 1973). The nature of these new policies will now be considered with respect to some of its chief impacts on agricultural development.

6.2 The Common Agricultural Policy (CAP) of the European Economic Community (EEC)

6.2.1 The support system

The EEC is the leading importer and the world's second ranking exporter of food and other agricultural products. Its share of world food imports is around 20 per cent compared with that for the United States of 10 per cent, Canada 2 per cent and Australia 0.6 per cent. Traditionally most of the Community's members were net importers of food, but since the enlargement of the Community in 1973 exports have expanded, in annual percentage terms, more rapidly than its imports. Its share of world exports of food and other agricultural producers has risen to 10.5 per cent.

Under Article 39 of the Treaty of Rome, signed in 1957, five objectives of the CAP were stated:

(i)   to increase agricultural production through the rational development of agriculture towards the optimum utilisations of resources;
(ii) to ensure a fair standard of living for farmers;
(iii) to stabilise agricultural markets;
(iv) to guarantee continuity of food supplies to consumers;
(v)   to ensure reasonable food prices for consumers.

In order to assist the alignment of the different methods of farm support then in existence in the six member countries, special clauses were inserted in the Treaty to allow the authorities to establish marketing organisations, monetary restrictions and control of trade both internally and externally. By the mid-1960s basic market regulations had been established for dairy products, beef, vegetable oils and fats, cereals, pigmeat, eggs, poultrymeat, fruit, vegetables and wine. Of these, cereals took pride of place because of their importance in the financial structure of European farming and also through their increasing significance as a source of animal feedstuffs. Initially, the policies were focused upon external protection against cheap imported food, but, gradually, internal common pricing policies were applied so that by the early 1980s only potatoes, wool and agricultural alcohol remained outside the scope of this structure (Fennell, 1979).

The CAP has operated by means of a dual control system. One aspect of this is the application of levies and customs duties at frontiers so that imports from non-member countries cannot be sold within the EEC for less than the desired internal market price. Secondly, the policy involves

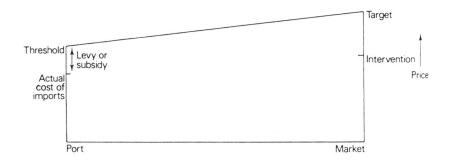

Figure 6.2 Schematic guide to the basic mechanisms for determining the price of agricultural products within the EEC (based on Baudin, 1979)

intervention or the purchase of supplies surplus to the market when prices start to fall below an agreed level.

The cost of supporting the subsidies, levies and interventions is met by the European Guidance and Guarantee Fund (FEOGA), which receives its finance from the member governments. The Guarantee Section of this Fund is responsible for the pricing arrangements of 96 per cent of farm output in the Community, and administers price support for over 70 per cent (Bowler, 1985: 65–70). It accounts for 80 per cent of the total EEC budget. In contrast, the Guidance Section represents less than 3.5 per cent of total CAP expenditure (Clout, 1984: 168). Guidance attempts to improve the structure of production, processing and marketing of agricultural produce. Capital projects are subsidies and financial contributions made to general structural development programmes, regional schemes and policies aimed at improving marketing (Pearce, 1981).

There are three basic elements in the Guarantee Fund (see Figure 6.2): the Target Price, the Intervention Price and the Threshold Price (Bowler, 1985: 70–81; Hill, 1984: 51–9). The *Target Price* is the desired market price set by the Community, or, in effect, the price the farmer should hope to obtain for his produce. However, if the market changes and the Target Price is not reached, the price at market may fall to a level referred to as the *Intervention Price*. This is the price at which the member countries buy in, guaranteeing the farmer a satisfactory return on production. One of the most fundamental arguments about the CAP has arisen over the setting of this Intervention Price. It has been argued that in many cases, and especially with respect to grain and dairy produce, the Intervention Price has been too high, thereby encouraging farmers to overproduce because they are aware that the Community will guarantee them a reasonable return despite the possibility that the produce may not have immediate buyers. The overproduction leading to butter and beef 'mountains' and wine 'lakes' has become both notorious and a major political issue within the Community.

The third component of the Guarantee Fund is the *Threshold Price* which is applied to imports. Imports from outside the Community which are cheaper than the Threshold price, are not allowed into the Community until a levy has been paid to raise their cost to meet the Threshold Price. This prevents domestic producers from being undercut by outside competition, though this type of regulation has been criticised for encouraging inefficiency within the EEC as producers can hide behind this tariff barrier. If producers within the EEC wish to sell their produce on the world market then the difference between the Threshold Price and the world market price forms a subsidy which has to be paid to the producers to compensate them for the comparatively low price they may be receiving. Again, this can tend to protect inefficient and costly production within the EEC from foreign competition. The system is a complex one, though, with the various inputs to the three critical prices being varied frequently, sometimes on a daily basis (see Bowler, 1986b).

The system of common agricultural prices depends on the rates of exchange between the national currencies remaining stable. This was largely the case until 1969, but, subsequently, parities changed affecting all currencies. The European Monetary System cushioned the impact of parity adjustments, but it became necessary to establish a correcting mechanism, *monetary compensation amounts (MCAs)*. These bridge the difference between official parities and the so-called green parities. The divergence between official and green parities arises because currency revaluations, which can undermine the stability of farmers' incomes and food prices, are phased in gradually through the use of special *'green' rates*. Thus 'a country which revalues its currency pays compensatory amounts on exports and charges them on imports; the opposite is the case for a country which has devalued its currency' (Burtin, 1987: 29).

This system has enabled the unity of the market to be maintained, but it has proved expensive, has distorted competition, limited structural adjustment of agriculture and jeopardised optimum allocation of resources. Hence the member states have been urged to align the green rates on the official rates and so eliminate MCAs. Yet, the most effective method of elimination, that of monetary union, seems remote despite the introduction of the European Monetary System in 1979. Indeed, all intra-Community transactions are expressed in terms of a common currency, called *units of account*, and each country has an agreed exchange rate between their national currency and the units of account.

Following the introduction in March 1979 of the *European Monetary System*, whereby all EEC members with the exception of the United Kingdom agreed to align their currencies, the *European Currency Unit (ECU)* was introduced. The ECU could be valued with respect to the pre-existing units of account.

In the United Kingdom, entry to the EEC in 1973 brought fears of increased retail food prices because high support prices for agricultural produce in the EEC had contributed to high food prices. Adoption of this support system was also accompanied by the adoption of import levies preventing cheap imports from entering the United Kingdom market. However, initially, two factors reduced the impact of EEC membership upon food prices. The first was a general rise in world food prices through world shortages of cereals, minimising the need for import levies. And the second was a fall in the international value of sterling. This meant that sterling was over-valued against the EEC units of account. To counter this a 'green pound' was created, representing pounds used for calculating the value of agricultural production and trade between the EEC and the United Kingdom. Whilst the green pound was worth more than the pound in the mid–1970s, consumers were partially shielded from the full effects of both EEC membership and the falling international value of the pound. However, farmers also received less for their produce than they felt entitled to expect. Hence, at the end of the 1970s the new Conservative government devalued the green pound, thereby raising the value of guaranteed prices to United Kingdom farmers and raising consumer prices (Tarrant, 1980a).

The importance of the valuation of the green pound was illustrated by farmers' claims in early 1988 that its strength was costing them about £25 a bullock or £14 a tonne of cereal. This restricted competitiveness and brought about substantial decreases in farm revenue. An estimate of total losses to the British farming industry in 1987 is £800 million, a contributor to farm incomes falling to their second lowest level since 1946 and investment dropping by 30 per cent in two years.

One anomaly produced by the currency variations in Western Europe is reflected in the smuggling of agricultural produce between Eire and Northern Ireland. The green Irish pound has been devalued more rapidly than the green pound, resulting in higher farm prices in Eire and the payment of MCAs for cross-border trade between Eire and the United Kingdom. The land border, which runs between Eire and Northern Ireland, has seen extensive smuggling in pigs and butter. As Tarrant (1980a: 42) remarks, 'in theory, at least, pigs could spend their whole lives in transit between the north (Northern Ireland) and the south (Eire), earning a substantial subsidy on every trip.'

6.2.2   Impacts of the CAP

During the lifetime of the CAP the relative cost of food in the United Kingdom has fallen with respect to that of inputs such as fertilisers, machinery, seeds and feedstuffs (Figure 6.3). To maintain their standard of living farmers rely upon a combination of support from the EEC and increased production. Hence the CAP attempts to meet the need for a decent standard of living for its farmers whilst balancing the demand and cost for consumers. It is this balance that has been much criticised, with a common view expressed that the CAP has favoured farmers at the

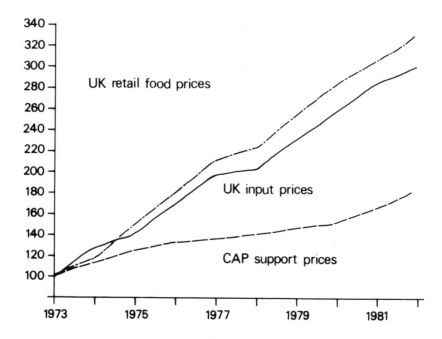

Figure 6.3 Trends in the prices of outputs and inputs in the United Kingdom, 1973–82 (based on Schools Unit, University of Sussex, 1983)

expense of the consumers, i.e. food prices are too high whilst farmers receive too much support from a CAP which is very costly and takes too great a share of the overall budget of the EEC (Body, 1982; Bowler 1986a; 1987; Nicholls, 1978). Unequal treatment of the individual member countries is also claimed as a defect in the policy.

Ironically, with the exception of Eire, it has tended to be the wealthier countries who have received income supplements through the CAP, paid for by the poorer ones. So Denmark, with the highest percentage increases in GNP from CAP payments (Hill, 1984: 93). This reflects the way in which payments from the CAP have been calculated: the net food importing countries transferring income to net food exporters.

Similarly, richer farming areas within states have also benefited at the expense of poorer ones because CAP payments have differentiated between farmers, especially on the basis of type of enterprise. For example, Josling and Hamway (1976) recorded the initial effects of Britain's entry as bringing substantially increased incomes for cereal-growers, smaller increases for sheep farmers and reductions for pig and poultry producers.

The result of favourable subsidies can be seen in the increase in milk production in France by 40 per cent between 1960 and 1975, accounting for nearly half the total increase in milk production in the EEC during that period. This was achieved by rises in yields rather than increases in the number of cows whilst the number of milk producers fell from 1.33 million to 0.67 million. Even so, one-fifth of all dairy herds had less than 10 cows and only 4 per cent had over 50 cows (Dexter, 1977: 216; Nidenberg, 1978).

As illustrated by this example, milk production has been one of several sectors in which CAP has stimulated major increases in output leading to surplus in terms of production exceeding demand. This overproduction has given rise to the infamous butter and skimmed milk 'mountains' and the milk 'lake'. Direct attempts to reduce the surplus milk production were made in two general forms:

(a) The non-marketing of milk premium. For a five-year period farmers give up supply of milk and milk products.
(b) The Dairy Herd Conversion Premium. For farmers producing at least 50,000 kg of milk or milk products per annum or with at least 15 dairy cows, including in-calf heifers, there is a premium to be paid for cessation of supply for four years (Hollingham and Howarth, 1989; Riddell, 1986). Where these controls have involved subsidising the slaughter of dairy cattle to reduce herd sizes (Knowles, 1987), these were not extensive enough to be really effective though (Tarrant, 1980b), with milk and milk products absorbing over 40 per cent of FEOGA's guarantee funds.

In France milk quotas were introduced in 1984 to control surplus production. This reduced output by 2 per cent over a two-year period whilst over 40,000 farmers left dairying through payment of an outgoers grant. Land released from dairying has been converted to cash crops or livestock rearing or fattening (Naylor, 1985; 1986). In Britain the milk quotas reduced the size of the national dairy herd by 15 per cent from 1984 to 1989 and the production of milk by 18 per cent whilst increasing output per producer by 3 per cent. However, buying and selling quotas has become a multi-million pound business, with 42 per cent of dairy farmers in England and Wales involved in lease or transfer of their milk quota in 1988–9. The quotas have had the effect of maintaining the milk price paid to producers and hence made quotas a sought after commodity.

Further limitations on agricultural production are likely in the near future, as presaged by the European Parliament voting to limit agricultural spending in 1988 to £19.25 billions, a large cut. Other proposals, for other aspects of production, relate to redefinition of 'small' for cereal producers as a means of limiting exemptions from proposed price penalties for overproduction. Since its expansion in 1973 the EEC's individual member states have been able to define the size of small-holdings which have therefore ranged considerably, e.g. 40 ha (100 acres) in Denmark, 100 ha (250 acres) in the United Kingdom. An EEC-wide definition of 20 ha (50 acres) has been suggested, which would mean very few British farmers would be eligible for exemption.

These attempts to limit the surpluses have been quite successful in recent yeas, e.g. the number of storehouses of surplus farm produce in Britain has been cut from 618 in 1985 to 145 in 1989, saving the taxpayer an estimated £45 million. Amongst the reductions in this four-year period have been cereals (from six million tonnes to 640,000 tonnes), beef (85,000 to 18,000 tonnes) and butter (204,000 to 12,000 tonnes). The implication is that the Intervention Prices are returning to their intended role of a 'safety net' and not a licence for overproduction.

One of the major influences of the CAP has been upon overall production. For example, during the 1970s cereal output in Britain rose by 24 per cent whilst the sales of milk off farms increased by 33 per cent (Soper, 1979; 1983). In contrast, there was a decline in the output of eggs, bacon and horticultural produce, especially top fruit, in the face of competition from within the EEC. Self-sufficiency in food in Britain has risen from 49 per cent to 60 per cent whilst the figure for self-sufficiency in temperate foodstuffs was 75 per cent in 1986. This increase has been most dramatic for butter, whose self-sufficiency has trebled, and cereals, in which Britain has become a net exporter for the first time since the first half of the 19th century, largely because of higher prices and security of market.

Increased output of certain commodities has been accompanied by a general rise in the costs of production to such an extent that costs have grown more rapidly than returns. The cost of agricultural inputs in Britain rose four-fold between 1970 and 1981, with labour costs alone quintupling. In comparison, producer prices rose only by a factor of 3.4. This disparity has been offset to some extent by the ever increasing subsidies provided by FEOGA, which, within the EEC as a whole, more than doubled its level of price supports from 1977 to 1983 (Bowler, 1985: 70).

It is this level of support that has increasingly been questioned by the rest of society, with the feeling that farmers are 'bankrolled' by the taxpayer. In Britain resentment of the farmers' receipt of grants and subsidies has been associated with the rise in the price of food following entry to the EEC. During the first five years of membership, when the prices of Britain's farm produce were being brought into line with the rest of Europe, food prices doubled, a rise exceeding that of the retail price index as a whole. However, between 1978 and 1982 food prices rose by 45 per cent compared with the rise in the retail price index of 64 per cent (Schools Unit, University of Sussex, 1983: 9–15). Also, without the CAP to protect against the effects of changes in world food prices, consumers would probably be paying 10 to 15 per cent more for food today.

Three critical aspects of change associated with the CAP have been those affecting type-of-farming, the landscape, and the concentration of wealth:

(a) *Type-of-farming*: accompanying the rising productivity has been a series of sectoral biases. For example, British horticulturalists have faced high costs for energy for heating whilst a large proportion of their French and Italian competitors benefit from more sunlight and higher ambient temperatures. Meanwhile, Dutch horticulturalists have benefited from subsidies paid by their national government. Hence, the acreage under horticulture in Britain has fallen since 1973. In contrast, the increase in the acreage of cereals has been accelerated, especially on the larger farms of eastern Britain where the production of cereals for fodder has been accompanied by the growing of oilseed rape as a break crop. Both cereals and oilseed rape have had favourable intervention prices.

An example of the direct impact of the CAP upon land use is the 'yellow revolution' in British farming. This refers to the adoption of the distinctively yellow-flowered oilseed rape (*Brassica napus*) which has diffused rapidly from south to north since the United Kingdom entered the EEC in 1973. Given the deficiency in vegetable oils and protein meal within Western Europe, the CAP established attractive target prices to encourage farmers to produce oilseed rape, as the rapeseed, when processed, gives an edible oil suitable for salad and cooking oils and margarine. A by-product, rapemeal, is a useful protein concentrate in animal feeding-stuffs (Wrathall, 1978: 42). The rise in world commodity prices in the early 1970s also favoured rape-growing as did the introduction of new crop varieties making for production of less acidic animal feed.

The initial diffusion of rape in the early 1970s was due largely to the increased value of rape compared with barley. In addition, rape was a suitable 'entry crop' to grow before wheat, whose price was also rising at this time (Wrathall, 1978: 44). As shown in Figure 6.4 the main area of production soon stretched over a broad belt from the Solent to the Humber. The total acreage of the crop rose more than sixty-fold from 1970 to 1984, with an annual rate of increase of more than 37 per cent. In 1983 for the first time it exceeded the acreage of both sugar beet and potatoes to become the largest non-cereal arable crop in the United Kingdom. Over 11,400 farmers were involved in growing the crop in 1983, producing 750,000 tonnes of rapeseed valued at £200 million (Wrathall and Moore, 1986: 352).

The continuation of favourable support prices has been much to do with the great expansion in acreage whilst oilseed producers have also received subsidies to persuade them to take British rapeseed instead of cheaper oil-yielding crops from abroad. The support measures contributed to a nearly four-fold rise in the average market price per tonne, making the gross margins per unit area 20 per cent above those for grains of feed quality. In 1984 oil from rapeseed accounted for 30 per cent of Britain's oil needs compared with just 2 per cent in 1973 (Wrathall, 1986a; 1988a).

Certain policy measures have recognised the variation in agricultural potential of the EEC and the weakness of reliance upon

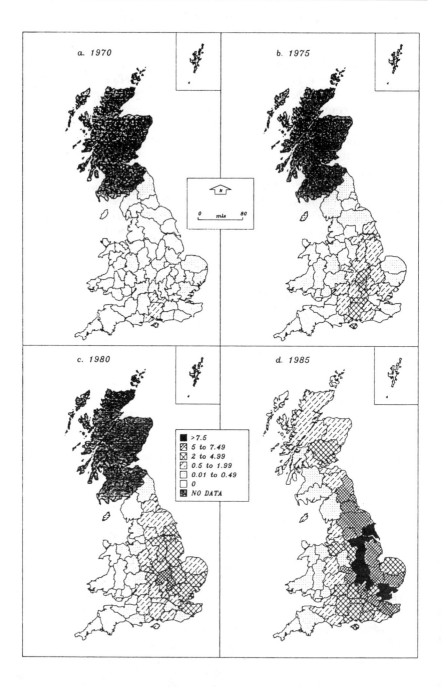

Figure 6.4 Oilseed rape as a percentage of the area under crops and grass in Great Britain. (a) 1970; (b) 1975; (c) 1980; (d) 1985 (*Source:* Robinson, 1988a: 196)

Plate 11 The incentive to maximise output has promoted large-scale removal of hedgerows and dry-stone walls throughout Britain, but especially in eastern arable areas, exemplified here by the Upper Forth Valley, near Stirling

particular types-of-farming, principally livestock production in upland districts. The prime example of a policy aimed at countering some of the imbalances caused by variable natural endowments is the Less Favoured Areas (LFAs) Directive, issued in 1975 (Arkleton Trust, 1982). The designated LFAs were areas in danger of depopulation and where conservation was deemed necessary. Fixed criteria were used to confer LFA status, which for the United Kingdom produced LFAs covering 41 per cent of the surface area (Figure 6.5a). Within these areas farmers are paid Compensatory Allowances (CAs) as compensation for permanent natural handicaps with which they have to contend. In theory this should ensure the continuation of farming in such areas whilst also encouraging farm modernisation, the maintenance of a minimum population level and the conservation of the countryside (Weinschenk and Kemper, 1981). However, the broad extent of the LFAs has meant that the policy has not been sufficiently well focused, and the concentration solely on farming neglects other elements of the rural economy.

(b) *Changing landscapes*: Farming practices supported by the workings of the CAP have altered well-loved components of the rural landscape such as hedgerows, woodland, areas of rough grazing, downs, moors and wetlands (Plate 11). Long-term features have been threatened by the growth of farm mechanisation. In addition, there

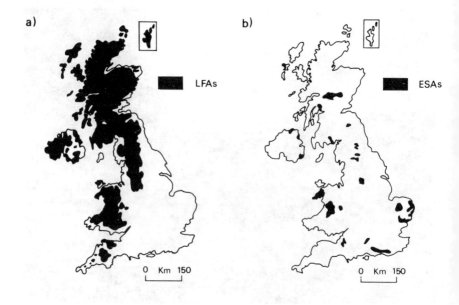

Figure 6.5 (a) The Less Favoured Areas (LFAs) in Great Britain; (b) The Environmentally Sensitive Areas (ESAs) in Great Britain

has been the introduction of new farm buildings of concrete and corrugated iron to house animals or store feed, produce and machinery. This new wave of distinctive farm buildings has been associated, too, with the growth of intensive animal husbandry. The drive towards increased mechanisation and capital investment has been stimulated by the rewards under the CAP for greater production: higher output has meant more income via price guarantees.

Some recognition of the need for controls upon environmentally destructive agricultural practices has been given in the creation of Environmentally Sensitive Areas (ESAs), first established in 1985 (Figure 6.5b). The ESAs are areas recognised as important from a landscape and ecological point of view and in need of special assistance to agriculture if the present environment is to be conserved. In the ESAs financial incentives have been given to farmers maintaining or adopting desirable farming practices. The United Kingdom has been the leader in designating ESAs, but payments are limited in comparison with returns from modern farming methods, so that the driving force of price guarantees seems likely to remain paramount until they are substantially reduced or modified.

(c) The CAP has fostered *accumulation and concentration of capital and wealth*. It has contributed to the pressures upon farmers to increase the size of operation needed to ensure viability (Britton and Hill, 1975; Massey, 1977). Economies of scale have divided farming businesses into the 'haves' and 'have nots' in terms of ability to obtain benefit from these economies. Therefore, it has promoted the growth of agribusiness in which farmers have made heavy investment in plant and machinery, especially on arable farms, producing a highly capital- and energy-intensive industry whilst also fostering the demand for land, as land has remained the essential capital input in farming.

Examples of these three sets of changes are indicated in recent work on agricultural change in Eire since accession to the EEC in 1973. Greater concentration of enterprises has occurred, with an increased divergence between the distributions of different types-of-farming. Both dairy and beef production have assumed a greater significance in the overall pattern of farming, accounting for 70 per cent of total gross output value in 1980 compared with 53 per cent in 1960. Dairy farming has benefited from a guaranteed market offering high and regular incomes plus much government encouragement and aggressive marketing, e.g. the export of Kerrygold butter. Beef production has often been a joint product enterprise with dairying, but has also benefited from expanded markets, improved prices and low labour demands favouring part-time farming (Gillmor, 1987).

For Eire, the accession to the EEC meant a change from dependence upon the low-priced UK market. It brought an increase of 45 per cent in real prices received by farmers between 1971 and 1978 and a growth in capital formation in agriculture. However, price rises have not always been direct stimulants to production increases. Nor have EEC structural policies had very much impact, causing Sheehy (1980) to conclude that the effect of membership upon Irish agriculture was mainly monetary, a rise in farm incomes rather than alteration to types-of-farming or farm structure.

6.3   Improving agricultural structures: agricultural policy and rural poverty

One of the most fundamental divisions within the developed countries has been the extent to which a large proportion of their population has remained on the land. As suggested in Chapter 2, the proportion of the working population engaged in agriculture has tended to remain much higher in the countries of southern Europe, giving them a larger rural population and more widespread and severe rural poverty than found elsewhere in the Developed World (Jones, 1984; King et al., 1983). This poverty has reflected the problems faced by agriculture in southern Europe, which has been hindered by major disparities between large and small holdings, concentrations of land in very few hands, and reliance upon both antiquated production methods and a limited range of products. One particular problem, that of a large number of small,

Table 6.1 Farm-size structure in the EEC, 1970 and 1983

	1–4.9	5–19.9	20–49.9	>50	Average size 1983 (ha)
Belgium	28.0 (34)	45.0 (51)	22.2 (13)	4.8 (2)	15.4
Denmark	2.3 (12)	45.3 (52)	38.9 (31)	13.4 (6)	25.0
France	19.6 (23)	33.8 (43)	31.9 (26)	14.6 (8)	25.4
Greece	72.0 (—)	26.1 (—)	1.6 (—)	0.3 (—)	4.3
Ireland	15.5 (20)	45.0 (52)	30.5 (22)	8.9 (6)	22.6
Italy	68.1 (68)	25.4 (26)	4.5 (4)	2.0 (2)	7.4
Luxembourg	17.9 (21)	24.3 (37)	36.0 (38)	21.8 (4)	27.6
Netherlands	23.9 (33)	47.5 (51)	25.3 (15)	3.4 (1)	15.6
UK	12.4 (19)	27.9 (29)	26.5 (26)	32.1 (27)	68.7
West Germany	31.3 (37)	40.8 (46)	23.3 (15)	4.7 (2)	15.3
The Ten (Nine)	45.9 (43)	31.5 (39)	15.8 (15)	6.7 (5)	15.7 (18.1)

% of farms in each country; 1970 in brackets

Source: Winchester and Ilbery (1988)

uneconomic holdings is indicated in Table 6.1.

Given the extent of the problems that have faced agriculture and rural society within the southern European countries, it is not surprising that many wide-ranging policies have been directed at the problems, at regional, national and supra-national levels. Therefore southern Europe provides numerous illustrations of the way in which agricultural and broader rural policies have been put into practice. The examples considered below provide a series of demonstrations of the range of utilitarian and equity goals addressed by these policies, and especially the concern for reductions in income disparities within the farming sector through reforms affecting farm structure (Clout, 1987a) and the desire to maintain rural population densities. Four specific examples are used: the Mezzogiorno (Italy), Spain, Portugal and France.

6.3.1 The Mezzogiorno

6.3.1.1 THE CAUSES OF RURAL POVERTY

Southern Italy has been one of the 'poor men' of Western Europe for many years as can be demonstrated by reference to a range of different social and economic indicators. Mountjoy's (1966; 1970; 1973) studies attributed this partly to three interlocking factors of geography, history and socio-economic development. In terms of basic geography the area is endowed with poor agricultural resources, a mountainous interior, highly articulated coastline, extremes of heat, little rainfall and, in the mountains, snow cover for four months plus damaging spring frosts. Its location (Figure 6.6) means it is remote from major European markets

Figure 6.6 Italy's Enti di Reforma

as well as the large industrial centres of northern Italy. Historically the area had a tradition of serfdom and subservience dating to Roman times, and by the 19th century it had tended to become a vast ranching complex paralleling the Spanish Meseta in its *latifundi* (large estates), production of sheep, vines, olives and wheat. During the Middle Ages it was separated from wealthier parts of Italy whilst it was governed as the United Kingdom of Southern Italy and Sicily. Internally only Naples and Salerno developed as rich trading cities. And by the time Italy was united under Garibaldi in 1861, very little infrastructure capable of supporting any form of sustained economic growth had been established. The name given to the south, Mezzogiorno or 'mid-day', seems an apt term for a region which had begun a long 'siesta' even before Italian unification. The limiting economic and social factors included great population pressure upon the land and a landlord–tenant situation that was based closely upon a feudal relationship. A small number of large landowners dominated land ownership with their large estates, latifundi, contrasting with the proliferation of very small holdings, minifundi, in other areas. These smallholdings, formed as a result of inheritance amongst all the male offspring of a farmer, were occupied by a peasant class little better off than the large numbers of landless labourers who worked the large estates. Industrial employment was very limited and agriculture yielded low returns despite being protected by tariff barriers. In contrast, northern Italy had physical advantages for farming, e.g. the well-watered fertile plains of the Po. The north also had cheap hydro-electric power (HEP), making its energy costs 40 per cent less than in the south; it had easier access to major European markets; and it had developed an industrial base with major industries established in Milan and Turin. The Mezzogiorno was left behind as a backward area with a big, largely illiterate, unskilled, agricultural labour force, many of whom sought an avenue of escape via emigration to North America. In the first two decades of the 20th century alone 8 million people emigrated.

The receipt of Marshall Aid post–1945 did little to benefit economic growth in southern Italy as its agriculture responded poorly to this type of capital aid. Unemployment rose as demobbed soldiers returned home. There were riots and much illegal squatting on latifundi land. Agricultural employment actually increased by 17 per cent, or half a million new workers, between 1930 and 1950 partly through the discouragement of emigration under Mussolini. In addition to high unemployment and under-employment, other problems are illustrated by the following: 56 per cent of cultivation was performed on slopes of over 15 degrees; half the cultivated land had no drainage systems; excluding the latifundi the average farm size was under 2 ha (5 acres) with much fragmentation, under-capitalisation, use of out-moded methods and insecurity of tenure; nearly half the population were illiterate; half the dwellings had no sanitation or drinking water; the Mezzogiorno had 40 per cent of Italy's population but produced only 20 per cent of its GDP and had just 11 per cent of its industrial population; 35 per cent of privately owned land in Italy belonged to 0.5 per cent of landowners. In the south, 13,000 landowners owned 4.5 million ha (11.1 million acres) whilst almost 4 million smallholders possessed only 1 million ha (2.5 million acres) (Maos, 1981: 382). The mean income per head of population was less than half that in northern Italy and less than one-third of that in the United Kingdom (Graziani, 1978).

One of the characteristic forms of settlement was the agro-town. These had the majority of their employed population working in agriculture, but were also compact and densely populated, often with over 20,000 inhabitants. This form of nucleated settlement has been associated with long distances from place of residence to work, with the workforce, largely farm labourers, walking several miles to the surrounding large estates or their own small minifundia. For Sicily, King and Strachan (1978; 1980) referred to the independence and self-sufficiency of these agro-towns which they saw as part of the perpetuation of feudal or semi-feudal social control. This control is symbolised in the division between the large estates, employing workers on a day-labouring basis, and the extreme fragmentation of the small minifundia.

Ginatempo (1985) views the problem in terms of a system which had created a surplus population. This system had a dominant ruling class, a small middle class, a smaller group of stable workers but a large surplus or marginal strata dependent on public assistance, a family subsistence economy, a marginal private labour market and irregular labour in the public sector. Until recently the marginal strata represented a reservoir of labour for the whole of Italy and, through immigration, North America.

One characteristic of the out-migration was that it helped bring about a higher proportion of women in the agricultural labour force: from 19 per cent in 1931, to 24.6 per cent in 1951 and to 26.6 per cent in 1961 (Barberis, 1968). Whilst a greater willingness in more recent times to record female farm labour clouds the picture, an implication is that women have increasingly entered the labour market to fill the vacuum left by emigrating males (Strachan and King, 1982).

6.3.1.2 POLICIES: THE CASSA PER IL MESSOGIORNO AND THE ENTI DI
        REFORMA

The key date in the planning development in the region is August 10th, 1950 when a government agency, the Cassa per il Mezzogiorno, was formed. Initially, the Cassa was to operate for 15 years, but its life was extended until 1980. Also known as the Southern Fund, the Cassa was to co-ordinate the main lines of development within every sector of the region's economy using government money and any outside capital it could attract. In fact, this total investment proved to be considerable, over £5,000 million being invested between 1950 and 1970. Also in 1950 a *land reform programme* was established to rehabilitate agriculture by redistributing land from large estates into smallholdings by means of *reform agencies*, the Enti di Reforma. In the first 15 years of this programme 700,000 ha (1.73 million acres) were distributed to 100,000 families (King, 1973).

Although the Enti di Reforma operated in several parts of Italy (see Figure 6.6), the most pressing reforms were needed in the Mezzogiorno where latifundi existed alongside very small fragmented holdings. Provision of basic infrastructure was severely limited so this also had to be brought within the Enti's compass. One of the basic aims of the reforms was to redistribute land to the former workers of the large estates and other landless labourers whilst also increasing the size of existing smallholdings. On newly created farms the new owners were to purchase them over a 30-year period. The intention was also to eliminate excessive fragmentation of holdings, creating new consolidated farms in the 4 to 20 ha (10 to 50 acres) range. New farmsteads were to be built within their own farmland, therefore helping to break up the concentration of farms within nucleated villages. On top of this structural reform, also involving some land reclamation, construction of new roads and installation of water supplies and irrigation systems, attempts were made to transform the type of agricultural production. Traditionally, this had been dominated by small-scale peasant subsistence farming and extensive production of wheat, olives and sheep on the latifundi. By extending the use of irrigation it was intended to break the wheat and olives pattern, using tree crops and sugar beet and obtaining greatly increased yields.

Despite the intention for the land reforms to affect 30 per cent of Italy's territory, only 1.2 million ha (3 million acres), or about 3 per cent of the total area, were affected by the law. Less than 770,000 ha (1.9 million acres) were appropriated and distributed to 113,066 farming families. Maos (1981) argues this was insufficient to change radically the socio-economic retardation of the Mezzogiorno. Therefore emigration has remained a solution to continuing rural decline, alleviated only partially by the creation of new employment outside the primary sector. The average area of new holdings created by the reforms varied from 4.5 to 5.5 ha (12 to 13.5 acres) in irrigated areas and from 6 to 10 ha (15 to 25 acres) in the mountains. To established smallholdings the additional area created was just 2.4 ha (6 acres).

In several of the southern land reform areas the strongly nucleated agro-towns have now been complemented by a dispersed pattern of farmsteads located on the individual holdings or semi-concentrated loose groupings of new farmsteads. In effect, these new settlement patterns have often followed the cross-grid parcellation of centuriated lands in Roman times. However, in more mountainous areas such dispersion has proved impractical, and small, new concentrations of farmsteads have been added to the landscape. Maos (1981) gives the examples of La Murgetta in Apulia and La Martella in Matera province. The latter gathered 200 family farms into one settlement whilst a slightly less concentrated settlement scheme, at Borgo Venuzio in the same province, catered for 100 families. In irrigated areas where horticulture has been the predominant type of farming on the newly created smallholdings, absentee farming has developed as the owner's presence on the farm has not been imperative.

Some of the best results of the reforms were obtained in Apulia and Lucania on relatively flat land bordering the Ionian Sea where latifundi had been dominant. Introduction of irrigation and farm restructuring to create 4 to 5 ha (10 to 12.5 acres) holdings brought increased output. On 200,000 ha (495,000 acres) expropriated from 1,500 landowners, 31,000 families were resettled and given various types of training and assistance to help them start farming the new holdings. However, the families resettled in this area represented less than one-third of the number of applicants for land, illustrating the magnitude of the problems faced by the farm resettlement and restructuring schemes. Yet, the greatest difficulties have been in the mountainous interior of southern Italy where irrigation is unfeasible, the agricultural land is often suitable only for grazing and there are very high rates of absentee landownership. Reorganisation in these districts has focused upon increased stocking rates, the extension of fruit crops and also potato production. HEP production and exploitation of tourist potential have also featured in the plans for these areas, but they have tended to be highly resistant to change, with a conservative population depleted of its most innovative sector by long and continued depopulation, as in Basilicata for example.

On the other hand, the Enti in southern Italy have had some successes. Working in conjunction with the Cassa, agricultural infrastructure has been improved and output increased. The chief criticisms have been that the new farm holdings created have been too small to permit the proper introduction of modern farming methods, and that as only 10 per cent of the designated reform areas have ever been expropriated, the spatial extent of the reforms has been too limited.

In the 1950s and 1960s the increase in total agricultural value added in the Mezzogiorno ran ahead of the rest of the country. But this was not true for agricultural value added per worker, reflecting the higher amount of labour retained in farming in the south. In the 1970s there was a decline in the value of farm production in the Mezzogiorno as the development plans continued to fall short of transforming the structural basis of southern agriculture (De Benedictis, 1981).

As an illustration of continuing problems, Weinrod (1979) likened much recent agricultural change in Sardinia to the involutionary process identified by Geertz (1963) in Indonesia, i.e. a kind of change in which a given pattern persists through time yet undergoes continued internal change (see Goldenweiser, 1936). In Sardinia, despite the local 'action programmes' of the 1970s, the agrarian structure has continued to be characterised 'by small plots of land, fractionalisation, and the prevalence of rental and sharecropping arrangements' (Weinrod, 1979: 252). In the mountainous interior, despite the development of a co-operative move-ment and pressures encouraging larger sheep herds, the fundamental agrarian structure and commodity mix remains the same. In contrast, in small portions of the coastal plain there have been structural changes associated with irrigation, new crops (e.g. flowers), processing plants for sugar beet and wine, and tenurial reforms.

One of the initial aims of the Cassa of settling as many as possible of the rural labour force on the land was altered gradually in favour of creating higher incomes and higher productivity, especially through focusing upon the fertile coastal plains at the expense of difficult moun-tain areas. Industrial and infrastructural investments received priority over agricultural schemes despite the latter's greater creation of employ-ment. This changed priority recognised that, in effect, for too long the resources assigned to regional and industrial development have been far too small in comparison with the population involved.

When the Cassa was established, the intention was to devote over three-quarters of its income to agriculture with relatively minor amounts on promoting industry and tourism, and on establishing better roads, services and utilities. It was soon clear, though, that, without a more effective infrastructure, incentives for industry were incapable of laying a basis for industrialisation. Also the initial improvements to communications which focused on the west coast put the east coast and Sicily at a comparative disadvantage and so hindered development there. In fact, the early years of the Cassa tended to benefit northern Italy more than the south as it was northern firms that built the new roads and supplied equipment. For example, in the 1950s the increase in income per capita in the Mezzogiorno was 72 per cent as compared with 82 per cent for Italy as a whole. Thus the south was relatively poorer.

A major policy change was heralded by the 1957 Industrial Areas Law promoting more rapid industrialisation through a process of inducement, stimulation and concentration. Effectively, this proposed to completely transform the existing economic base. It has been successful in establishing capital intensive industry in the region and in developing industries using local mineral resources. However, some of the focus upon the industrialisation of the south has been at the expense of much needed investment in agriculture, and surplus labour has not been sufficiently reduced partly because of the low labour intensity of much of the new industry.

In evaluating the policies of the Cassa and the Enti, it is possible to recognise significant weaknesses. Despite increased agricultural output, the south has continued to lag behind the north, unemployment has remained high and farming remains labour-intensive rather than capital-intensive as in the north. With favourable EEC price guarantees, there has tended to be a renewed emphasis upon those products which have been traditional staples of the region – wine, tomatoes, olives, tobacco and citrus. However, meat and dairy production has remained limited, and wheat production has declined.

For the Enti, it can be argued that they created too many small and uneconomic holdings whilst also producing social problems. The notion of supporting the formerly landless peasants on smallholdings contrasts sharply with the Portuguese reforms which have opted for the creation of large collectives. Also, reform has usually not been as comprehensive as the structural reforms in France. Consequently, some changes have been introduced more recently, with attempts to create larger holdings in areas of minifundia, e.g. in Campania, Sardinia and Sicily, through the institution of *land consolidation schemes* and establishment of co-operatives (Cesarini, 1979; King and Took, 1983). In some cases plot size has been increased ten-fold by such schemes, but it is pertinent to note the limited attention given to farm consolidation in the Mezzogiorno compared with the Spanish experience (see Section 6.3.2).

It can also be argued that the Mezzogiorno has benefited far less from membership of the EEC than northern Italy. The CAP has supported north European agricultural products whilst the Mediterranean crops have had much more limited price support. Regional disparities in agricultural performance have widened and most Italian regions have suffered in comparison with the rest of the EEC (Cesaretti et al., 1980; Tarditi, 1987; Tarditi et al., 1989; Wade, 1980). Structural measures have also not had a great impact despite the introduction of region-specific measures. The long-term funding of the Cassa ended in 1981 and it went into liquidation in 1984. There is no clear indication of the form a successor might take. For the long-term development of the region the decision regarding the nature of the successor is crucial.

Despite the shortcomings of the Cassa's operations, there are significant differences between the Mezzogiorno now and in 1950. There is certainly a higher 'base-line' from which future economic development can occur but there have also been major social and welfare improvements, especially in Apulia, Basilicata and Sardinia (King, 1987a: 189; King and Burton, 1982). The nature of the south's dependence on the north has also changed, so that, according to Graziani (1983), as well as continued decline and subsidisation, there are now 'productive' areas where 'autonomous local development has emerged, based either on the results of the massive investments of the Cassa in industrial estates and irrigation, or on the development of small firms' (King, 1987a: 192). The provincial capitals and coastal areas attracting tourists have been the main generators of this 'productive' development (Slater, 1984).

6.3.2   Spain

In Spain the proportion of the workforce employed in agriculture fell from 42 per cent in 1960 to 19 per cent in 1979. Some of the more industrialised parts of the country, such as Catalonia, had less than 10 per cent employed in agriculture by the 1980s and showed characteristics of innovation and response to changes in demand previously lacking in Spanish farming (Garcia-Ramon, 1985a; 1985b; Benelbas, 1981). Yet, even in Catalonia, changes were largely exclusive of any transformations of the organisation and structure of agriculture. Small family-owned and operated farms have remained the norm here, increasingly run on a part-time basis. Both industry and the growth of tourism have provided new opportunities for off-farm income, and in some provinces, e.g. Tarragona, this income is as much as 40 per cent of that derived from farming. For Spain as a whole, in 1982, 48.9 per cent of farmers had an entrepreneurial activity other than their own farm that brought in more money than their farm, the ratio of non-agricultural activity to farming for this other source of income was 3:1. This situation is a reflection of the continued need for further agricultural reforms even after major national programmes operating for nearly four decades.

Structural reform of Spanish agriculture dates to the Law on Parcel

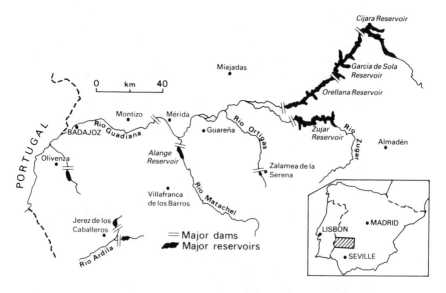

Figure 6.7 Dams and reservoirs in the Badajox Plan (based on Naylon, 1966)

Concentration, approved in 1952, followed a year later by the establishment of the Servicio de Concentracion Parcelaria (SCP) (Naylon, 1959). The need for major reforms had been revealed by the 1945 Servicio de Cadastro which surveyed 28.7 million ha (70.9 million acres) occupied by 19 million different farm holdings with an average of six parcels of land per farmer. Especially in north-western Spain there was a concentration of *minifundia*: small, fragmented properties associated with the system of inheritance of land between all male siblings. For example, the average number of parcels of land per holding in Burgos was 27, the average parcel size being less than 0.4 ha (1 acre). Over 90 per cent of holdings in north-west Spain were less than 1 ha (2.5 acres). In contrast, large parts of Extremadura and the Spanish Meseta were dominated by *latifundia*: very large estates with monoculture and much absentee landlordism. The main intention of the legislation of the early 1950s was to bring about greater efficiency and higher productivity, both through land consolidation in the areas of minifundia and some reorganisation involving large-scale irrigation schemes in areas of latifundia.

The SCP began a campaign to win the support of farmers, as consolidation depended upon villages producing a voluntary petition for land consolidation. Once at least 60 per cent of properties owning at 60 per cent of land requested consolidation, the SCP could then put this into effect by establishing a local commission. The commission's aims were to unite parcels, re-site farmsteads if necessary and to use 'spare' land for settling landless labourers. In conjunction with this, reorganisation of services such as water supplies, electricity and silo construction were organised through the provision of loans by the Instituto Nacional de

Plate 12 Irrigated agriculture near San Antonio, Ibiza, Spain, competes for land with tourist development. Many of the farms on the island are now operated on a part-time basis, the tourist industry providing an additional income

Colonizacion (INC), formed in 1956 to hasten the pace of land consolidation. By 1958 just over 1.2 million ha (3 million acres) had been consolidated, but this was primarily on the *secano*, the dry farming grainlands of the Meseta where minifundia was not present. The requirement for voluntary request of consolidation proved a major stumbling block in areas of minifundia, and the INC also took time to formulate and carry out their village reorganisation schemes (Naylon, 1961). But agricultural change did occur. The 1972 Census of Agriculture in Spain recorded 2.5 million farmers of whom 47.4 per cent had a non-agricultural primary occupation; between 1962 and 1972 319,550 agricultural enterprises ceased operation, but over three-quarters of all holdings remained less than 10 ha (24.5 acres) (Ferrando, 1975).

There was a tendency for schemes to concentrate upon large-scale irrigation projects in which the basis of agriculture could be altered as well as modifying farm structures and settlement patterns. This emphasis upon the need to expand the irrigated area has a long history and is not surprising for a country which has such a large and dry interior where evapotranspiration exceeds precipitation (Plate 12). At least nine-tenths of the cultivated area is subject to low and uncertain rainfall so that irrigation can have a major impact on production. For example, the National Plan of 1902 aimed to add 1.4 million ha (3.5 million acres) to existing irrigated land, which at that time accounted for 6 per cent of the cultivated land but 15 per cent of the value of crop production. However, it was not until 1952, when Spain was granted Marshall Aid from the United States, that major irrigation schemes were put into prac- tice. Between then and 1970 0.5 million ha (1.24 million acres) were irrigated by modern trunk canals, adding to 300,000 ha from older systems. The aim was to bring at least 30 per cent of Spain's river flows into government controlled irrigation schemes, many accompanied by land settlement schemes. The intention was to increase the irrigated area to cover 9 per cent of the nation's territory, representing 22.5 per cent of the cultivable area (Naylon, 1973).

The two major aims of the irrigation projects were the increase of agricultural production both for domestic consumption and export, and the provision of small family plots for landless labourers. In some cases this involved the creation of new villages in the midst of irrigated land, for example in the Palma del Rio and Antequera districts of Andalucia. Yet, such creations have only tended to be successful in conjunction with the emergence of specialist intensive crop production plus the establish- ment of processing plants in the locality. In many cases the irrigation schemes have not substantially altered the structure of rural landowner- ship or brought about redistribution of wealth. Of 1.35 million ha (3.3 million acres) irrigated only 0.23 million ha (0.57 million acres) (17 per cent) were distributed to landless labourers. However, they certainly raised production, created farm employment and laid the basis of future agricultural innovation.

One of the largest of these irrigation and colonisation projects, developed in the 1960s, was the Badajoz Plan, effecting Spain's biggest province (approximately the same size as Belgium). This was part of the Extremadura Meseta with dry farming dominated by latifundia: in 1950 0.59 per cent of agricultural holdings occupied half the area. Mono- culture of wheat was extensive, but also with much sheep-walk, rough grazing and scrub-land. The province had a population of 0.75 million of which half were without guaranteed employment and 30 per cent were illiterate. The Plan, instituted in the early 1950s and running until 1970, attempted to deal with these problems by a combination of hydraulic works, land colonisation, reafforestation, the building of roads, railways, port improvements, the encouragement of new private industry and investment by INI. One of the keys to this ambitious project was the regulation of the area's major river, the Guadiana, involving the construction of the Cijara Dam, the largest in Spain (Figure 6.7). An area equivalent to 9 per cent of the size of Spain was to be irrigated, intended both to increase productivity and also to employ more farm labour. On this irrigated land 10,000 families were to be resettled on individually owned and operated holdings on former latifundia property. These families were granted loans for 25 to 30 years but with special allowances on public works and the cost of buildings. In addition, 49 new settlements were to be created with a variety of services and the establishment of agricultural co-operatives. A new national park was also designated and provision made for 50,000 ha (123,550 acres) of new woodland. In attempts to diversify the local economy, encouragement was to be given to manufacturing industry, although this proved to be mostly primary manufacturing, such as food processing and production of fertilisers, tied closely to agriculture.

One measure of the Plan's success was the creation of 70,000 new jobs in the province from 1955 to 1967. But this can be offset by a net out-migration greater than this and the continuing heavy reliance upon agriculture in the economy. The government recouped much of its outlay through the successes of farming in the newly irrigated areas, but more widespread private initiative in industry was required to provide a broader economic base, and also more needed to be done to alter the structure of farming outside the newly irrigated areas. Much faith had been placed in the effectiveness of the large irrigation schemes as the mechanism for bringing wholesale rural reforms, but Naylon (1966) questions its effectiveness arguing that 'its efficacy as a means of dispensing social equity and economic prosperity is seriously in doubt.'

The regulation of the Rio Guadiana was to facilitate the irrigation of nearly 140,000 ha (346,000 acres) of land, of which 40 per cent were allotted to smallholders and the rest to existing farmers (Naylon, 1967). The newly created smallholdings were from 4 to 5 ha (10 to 12.5 acres), primarily in clusters for 200 to 300 families, though with some dispersed farmsteads so that distances from the farmstead to the fields were largely less than 2.25 miles (3.5 km). The new villages were intended to be self-sufficient in basic services, and some of the larger ones, e.g. Valde-calzada and Palazuclo, have prospered. However, the viability of the smaller communities remains doubtful, especially where private initiative has not stimulated primary manufacturing. In effect, the high expense of the key hydraulic works has not been accompanied by sufficient progress in improved agriculture, housing, education and the rural economy in general (Naylon, 1987: 388–90).

The Plan Badajoz was followed by a programme to stimulate the establishment of industry, initially in the area affected by agricultural transformations, and later extended to cover the entire region. Between 1964–8 and 1974–5, Priority Zones for Industrial Location were promoted, areas for which a series of special grants were available. In 1977 a special quango, SODIEX, was established to oversee and encourage industrial development. Then between 1978 and 1981

Extremadura was designated a Major Area of Industrial Expansion, again with a variety of incentives for new firms locating in the region (Frutos, 1984). These initiatives undoubtedly broadened the economic base in the region, and, under new regional government reorganisation, Extremadura receives compensation for acting as a major power generator for the rest of Spain. However, there remains much rural poverty in the region, the economic base is still in need of greater diversification and average incomes are at least 30 per cent below those in Catalonia.

Agricultural policy in both Franco and post-Franco periods has been confined too much to irrigation and colonisation schemes, without sufficient emphasis on wide-ranging land reforms. In some of the main areas of extensive agriculture the latifundia system has now been dismantled and smallholdings created. For example, in Jaen Province it is probable that the inability to utilise economies of scale in the production of olives has stimulated fragmentation of large holdings (Cabeza, 1984). Similarly, there has been consolidation of plots in minifundia zones (O'Flanagan, 1980; 1982). In Andalucia the regional government introduced land reform in 1984 to create jobs and increase production. These aims were to be met by promotion of intensive crop production on former latifundia expropriated if they failed to meet certain criteria or subject to improvement plans and/or fines until specified production targets are obtained. Not surprisingly, landowner groups have opposed these reforms and limited their implementation (Maas, 1983).

Whilst the shortcomings of government policies can be recognised, it is true that there have been significant changes in the face of Spanish agriculture post–1945, with the growth of commercialisation and mechanisation (Giner and Sevilla, 1984; Naylon, 1987). The pronounced loss of labour from the land has freed land for creation of larger holdings and has contributed to modernisation of many villages (Aceves and Douglass, 1976). Even so, 4 million ha (10 million acres) remain to be consolidated, and there is still no legislation to prevent sub-division of plots under a minimum size (Jones, 1984: 248). The increased amount of irrigated land has contributed significantly to greater output, especially with respect to dairy products and fruit and vegetables. Unfortunately, massive surpluses have been generated for certain Mediterranean crops, e.g. citrus, olive oil, wine and rice whilst self-sufficiency in cereals and meat products remains low.

6.3.3 Portugal

Major agrarian reforms were introduced to Portuguese agriculture after the revolution of 1974 (Porto, 1984). At this time the farm structure was amongst the weakest in the Developed World: over three-quarters of farm holdings with less than 4 ha (10 acres) of land, but the 0.1 per cent with over 500 ha (1,235 acres) accounting for 30 per cent of all farmland. One-fifth of the gross agricultural product was produced by 0.4 per cent of holdings.

The chief reform was the creation of an Agrarian Reform Zone (ZIRA) in southern Portugal. This was pro-collectivist and anti-capitalist in nature, involving land seizures, expropriation and nationalisation. Yet a significant element of the old latifundia system remains today accounting still for 35 per cent of the cultivated area in the ZIRA and employing 25 per cent of the farm labour. Although a similar amount of farmland was converted into new production units, the political turnabout of November 1975 reduced the impetus of the drive towards 'socialist agriculture' and so encouraged some retention of existing agrarian structures (Christodoulou, 1976; De Barros, 1980; King, 1979; Rutledge, 1978).

The newly created collectives were the Collective Production Units (UCPs), combining expropriated farms within the framework of an administrative parish. Their average size was double that of previously existing holdings. The collectives were not the only new form adopted, as Co-operatives de Producao Agricola or Agro-Pecuaria were also introduced, generally coinciding with established farms. Yet the total area covered by reforms represents only 14 per cent of the country's cultivated area and employs only 7 per cent of the rural population. Given that they have also been applied to poorly endowed areas, the limitations of the reforms can be seen to be quite substantial.

The main area in which collective farms were established following the 1974 revolution was Alentejo, formerly dominated by large-scale traditional estates and some smaller and more modern foreign-owned estates. It was a depressed and backward region, largely dependent on non-irrigated extensive agriculture and dominated by latifundia. After the loss of power by the most radical communist groups in 1975, some of the land expropriated from foreigners was returned, but at this time there was a concerted government drive towards collectivisation. For example, at this time the 'Flower of the Alentejo' collective was formed, comprising over 6,500 ha (16,000 acres) and employing 400 people. One million ha (2.5 million acres) of new production units were created, accounting for one-third of the cultivated area of southern Portugal (south of the river Tejo) (Jones, 1984: 244). However, largely because of their technical deficiencies, lack of incentives and insufficient government attention to agriculture, many of the collectives have failed to produce the desired increases in output. Thus the 'Flower of the Alentejo' has had to sell some of its assets to survive: losing 260 labourers, its dairy herd of 130 cows and 3,650 ha (9,000 acres) of land, and employing 40 per cent of its workforce on a part-time basis. Despite increases in the 1970s (Monteiro and Malheiro, 1983), production figures for most agricultural products have decline during the 1980s, increasing the import bill for food and animal feed. This emphasises the continuing extent of the agrarian problems on the full range of holdings in the country, not only in areas still containing a significant number of collectives.

In north-west Portugal, Unwin (1985; 1988) found that capitalist agriculture had taken hold most extensively in the more fertile and prosperous parts of the region. This was leading to an increased marginalisation of the periphery which was being by-passed by the spread of new agricultural techniques and equipment (Ferrao and Jensen-Butler, 1986; Lewis and Williams, 1981; 1985). So many of the inequalities present prior to the 1974 revolution have remained. Even in the areas affected by land reform, change has had relatively limited economic and social impact. A shift in production, e.g. from wheat/ cereals to livestock is one of the more significant changes. Thus production of cereals, fruit and vegetables has fallen, and, though output of most meat products has risen, consumption has outstripped production and raised food import bills (Tsoukalis, 1981). Whether Portugal's entry to the EEC will bring improvements to the agrarian structures and a more balanced pattern of agricultural development remains to be seen.

6.3.4 France

In France agriculture has continued to suffer from structural and social problems. A major retarding factor on the attempts to develop modern, efficient systems of production has been that farmers have been willing to remain on smallholdings incapable of providing a satisfactory subsistence wage, because external sources of income, usually in the form of additional part-time employment, have been available (Clout, 1975; 1987b).

New proposals for reforming land structure were introduced in the Loi d'Orientation Agricole, in 1960, and the Loi Complémentaire, in 1962, creating the Sociétés d'Aménagement Foncier et d'Établissement Rural (SAFERs) and the Fonds d'Action Social pour l'Aménagement des Structures Agricoles (FASASA).

(i) The Loi D'Orientation, 1960: recognised that agriculture should have equal treatment with other sectors of the economy and not merely through price support which itself did not encourage modernisation. It advocated structural changes in farm size and a smaller labour force.

(ii) The Loi Complémentaire, 1962: the Bill implementing the 1960 Act. It introduced the mechanisms for the structural reform of farming.

There are 27 regional SAFERs charged with purchasing agricultural land which comes onto the open market, holding it as a temporary land bank and then utilising it for enlarging existing holdings or settling young farmers on the land in reasonably sized units (e.g. Perry, 1969) (Figure 6.8a). Unfortunately, lack of funds has limited the activities of the SAFERs to only about 13 per cent of all sales of farmland and they are excluded from transfers by lease agreement (Naylor, 1981: 26). Through the FASASA fund, *retirement schemes* are financed to encourage the release of land for possible farm enlargements. There is also finance to retrain under-employed farmers and farm workers, and for the resettlement of farmers from regions with limited farming opportunities to areas where there are more opportunities.

Other attempts to promote efficiency have focused upon the encouragement of co-operation between farmers. For example, in 1964 the Groupement Agricole d'Exploitation en Commun (GAEC), *group farming*, was established to integrate separate farms, create joint management of individual enterprises or partnerships with agricultural workers. The Groupement Foncier Agricole (GFA) reduced tax liability at death for these co-operative ventures.

Under the Loi d'Orientation, the most extensively adopted measure has been the Indemnité Viagère de Départ (IVD), the *retirement pension* (Figure 6.8b). This supplemented the state old age pension and was available to two groups of farmers:

(a) The *basic retirement grant* (IVD Complément de retraite) for full-time farmers aged 65 years or over operating between 1 ha (2.5 acres) and four times the minimum settlement area, varying from 10 to 60 ha (24.5 to 150 acres) depending upon type-of-farming and corresponding to a subsistence unit for a farm family.

(b) The *early retirement premium* (IVD non complément de retraite), available from 1968 in problem areas (the rural renovation zones) for farmers from 60 to 65 years of age. Subsequently, this was extended to farmers aged 55 to 60 years and, from 1974, to the whole of the country. It was combined with a *restructuring grant* (indemnité complémentaire de restructuration) given to retiring farmers crating an amalgamating unit bigger than one-and-a-half times the minimum settlement area. In 1975 the grant was replaced by a *supplementary payment* (Prime d'Apport Structural), independent of the IVD scheme, and given to farmers of any age transferring their land to a SAFER for farm enlargement or installation of suitably qualified young farmers.

The retiring farmers were usually allowed to retain a small plot of land, thereby weakening the effectiveness of the scheme, but land released was added to larger holdings either through SAFERs, direct transfer to an existing holding or installation of a farmer under 45 years of age onto the abandoned holding (Winchester and Ilbery, 1988).

Over half a million farmers have received the IVD pension and one-third of all agricultural land has been transformed, with concentrations in 'problem' areas such as the Midi Pyrenees and Limousin. In these problem areas nearly half of the retirals have been followed by the installation of young farmers, a significant number being French farmers repatriated from Algeria (Toujas-Pinede, 1974). About half the land vacated has been used for farm enlargements (Naylor, 1982: 29).

Naylor's (1976; 1982) assessment of the retirement scheme is that despite the large area affected, change has been slow because of the

Table 6.2 Characteristics of farms in mountain zones in France

Mountain zones	Total number of farms	Mean size of farms (ha)	% land used for permanent pasture	% farms worked on part-time basis
Vosges	7,071	8.8	87.6	37.5
Jura	7,596	28.8	91.3	12.8
Northern Alps	20,548	12.9	89.1	24.6
Southern Alps	15,235	22.9	71.5	14.0
Corsica	5,724	16.5	73.9	15.3
Northern Massif Central	64,126	22.0	86.0	8.3
Southern Massif Central	39,549	25.3	83.8	13.7
Pyrénées	17,131	11.1	84.0	17.3
Total mountain zone	176,980	20.3	83.8	14.5

Source: Winchester and Ilbery (1988)

generally small size of holdings which have become available for amalgamation. Farm retirements have been mainly in the post–65 years age group, the maximum uptake of retirement grants being in 1970. So, although the oldest farmers on the smallest plots have been removed, there is still more untapped potential for creating larger, more efficient units through elimination of small subsistence plots.

Assistance to help young farmers enter farming was introduced to specially designated Mountain Zones in 1974 and, at lower rates, has been available elsewhere since 1977. The Mountain Zones include eight areas (Table 6.2) (Figure 6.8c), delineated on the basis of altitude and subsidiary criteria such as slope, productivity, economic viability and rate of loss of population. Over 175,000 farms are included in the zone, 70 per cent of which rear cattle for meat or milk. One of the forms of aid in the zones is the Indemnité Speciale Montagne (ISM), a subsidy payable to farmers under the age of 65 farming at least 3 ha (7.5 acres). This has been used to meet costs of working in difficult conditions, but rarely to help with structural problems. Therefore it has not contributed greatly to reducing the numbers of farmers in mountain areas or to farm enlargement.

Policies to bring about the consolidation of farm plots date to World War One in areas of northern France where some local exchanges of land were made following the disruption and destruction of the war. But the first major legislation came in 1941 and referred to *remembrement* ('consolidation of lands'). It acknowledged that 14 million ha (34.5 million acres) was in need of reorganisation, approximately 40 per cent of all French farmland. Close to half this area was reorganised during the next 30 years, progress being slow because requests to consolidate land had to come from three-quarters of local landowners before government finance schemes would operate to assist the process. As the land on

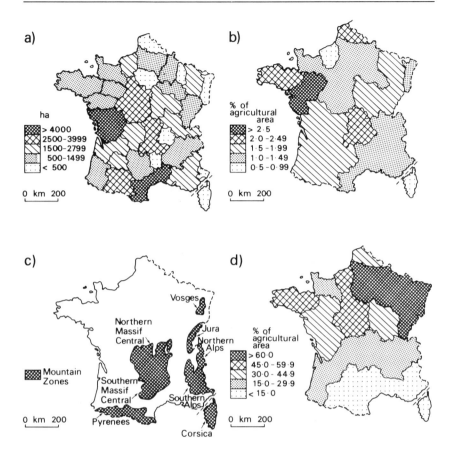

Figure 6.8 Government and agriculture in France: (a) Land acquisition by SAFERs, 1983; (b) Land released as a result of IVD uptake; (c) Designated Mountain Zones; (d) Completed remembrement schemes, 1983 (based on Winchester and Ilbery, 1988)

newly consolidated holdings has to be of equal productive value to the small plots formerly held by each landowner, this has proved a major retarding factor. It has also made complete consolidation into one single block rarely possible.

The largest amount of land reorganised has been in the Paris Basin which already tended to have fairly efficient farm structures. However, in the major problem areas of southern France, the Massif Central and the Alps relatively little has been achieved (Figure 6.8d). This reflects the strong conservatism of these areas, and poverty, as the landowners cannot easily afford the costs of removing hedgerows, earthen banks and the installation of new roads and land drains. This factor of cost is especially important in areas of *bocage* in western and central France where field boundaries are difficult to dismantle. These also tend to be areas where there is a concentration of small, traditionally operated owner-occupier farms highly resistant to change. To overcome this and other problems, finance has been provided to encourage early retirement, to retrain younger farmers in other types of activity and to promote farm amalgamation. For the latter, SAFERs buy land, store it, improve it with better roads and buildings and then releases it to other farmers. A similar body operates in Sweden.

One of the basic underlying problems facing the SAFERs is one which symbolises a fundamental weakness of the CAP. This is the SAFER's role in creating a better, more efficient agrarian structure. However, this may then generate surplus production and conflict with policies to control agricultural output. In effect, the Guidance and Guarantee Roles of the CAP are in conflict, no more so than in the Mediterranean countries with their output of price-supported wine and olive oil. Thus the role of SAFERs in some of the French wine producing areas has been extended to embrace encouragement of diversification and enhanced quality as well as improved agrarian structure, e.g. in Languedoc-Roussillon (Bartoli, 1982). Jones's (1989a; 1989b) work on the SAFER in Languedoc showed that legal and financial constraints plus the influence of pressure groups meant that the SAFER's impact upon structural improvement of agriculture was only piecemeal, e.g. between 1974 and 1982 the SAFER purchased just 17 per cent of land coming onto the market. The SAFER also has no control over the agricultural uses to which the land it sells are put, and so its effectiveness in contributing directly to EEC policy measures for grubbing up vineyards has been limited.

The enlargement of the EEC, through the entry of Greece, Portugal and Spain, places the agricultural measures adopted by these countries as well as Italy and, to a lesser extent, France into a new context. The problem of absorbing the large amount of 'Mediterranean produce' within the CAP without causing sharp fluctuations in price or major surpluses is one that already seems most likely to exacerbate the difficulties of the CAP's agricultural management system. Despite long transition periods prior to full 'harmonisation' of economies there is no promise that EEC membership will significantly benefit the poorer farming regions in the new member countries. In this respect the example of the Mezzogiorno is not an encouraging one. However, in recognition of the low levels of economic development and infrastructure in the Mediterranean, the EEC produced a 'package' aimed at improving the situation. Coming in 1978, soon after the application for Community membership by Spain and Portugal, this recognised the likely need in the expanded EEC for widespread improvement measures.

The proposals of the late 1970s have been translated into the Integrated Mediterranean Programmes (IMPs), geared to channelling funds to stimulate development in the poorer parts of Greece, Italy and France which are likely to lose farm income with the entry of Spain and Portugal to the EEC. The intention is to spend c. US$7 billion by the

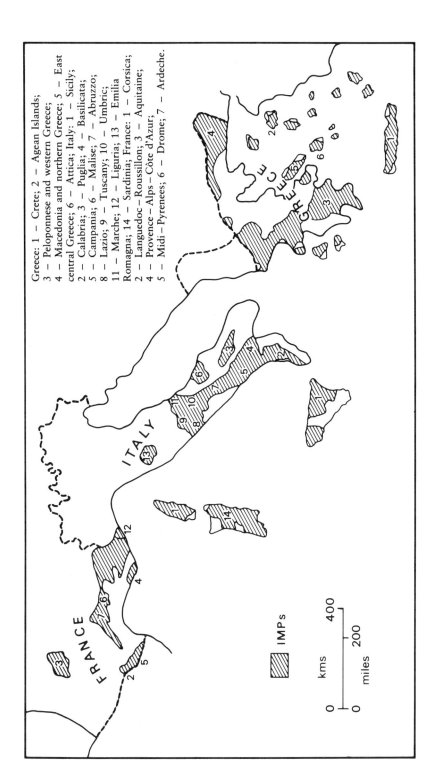

Greece: 1 – Crete; 2 – Agean Islands; 3 – Peloponnese and western Greece; 4 – Macedonia and northern Greece; 5 – East central Greece; 6 – Attica; Italy: 1 – Sicily; 2 – Calabria; 3 – Puglia; 4 – Basilicata; 5 – Campania; 6 – Malise; 7 – Abruzzo; 8 – Lazio; 9 – Tuscany; 10 – Umbric; 11 – Marche; 12 – Liguria; 13 – Emilia Romagna; 14 – Sardinia; France: 1 – Corsica; 2 – Languedoc–Roussillon; 3 – Aquitaine; 4 – Provence–Alps–Côte d'Azur; 5 – Midi–Pyrenees; 6 – Drome; 7 – Ardeche.

Figure 6.9 Programme areas for the Integrated Mediterranean Plan

mid-1990s in designated regions (Figure 6.9). Half of the scheduled grants will go to Greece, with loans to Italy and France, but there will also be heavy price support for those products for which the Iberian countries will provide strong competition to the French, Italian and Greek producers, e.g. wine, olive oil, peaches and tomatoes (Hinton, 1986). So despite some additional structural reforms, the CAP's overriding emphasis on price guarantees seems likely to be seen too in the IMPs.

6.4 Prospects for new policies

In West Germany, although agriculture accounts for less than 2 per cent of GDP and employs about 5 per cent of the labour force, with only 3 per cent full-time, public subsidies to farming are in excess of DM20 billion per annum. This is equivalent to 70 per cent of the country's agricultural gross output and helps maintain West Germany's food prices at 50 per cent above world levels. Examining this type of farm support, the Centre for International Economics (1988) argues that reduced support and fewer trade barriers would actually generate more farm employment in the Developed World not less. For example, the Centre estimates that, applied to the EEC as a whole, removal of farm trade barriers would create three million new jobs, boost manufacturing output by more than 1 per cent and manufacturing exports to the rest of the world by 5 per cent. Similarly, if the USA cut its trade barriers, the US trade deficit could be reduced by over US$40 billion and substantial numbers of jobs could be created in other countries. The critical stumbling blocks to such developments are the fall in land prices that would result and the diminished returns to agriculture and its satellite industries. Consumers and other sectors of the economy would benefit, but the all powerful farming interests would have to be defeated first to permit farm policy reforms.

In the United States government spending on agriculture rose nearly nine-fold from US$3 billion in 1979 to US$26 billion in 1987. The complaint that farmers were capitalising their profits whilst socialising their losses was voiced, reflecting the extent to which protectionist policies were felt to be supporting the farm sector. The value of farm exports has been caught by the cost of subsidies: in 1987 it cost one tax dollar to earn a dollar in farm export income. Meanwhile, subsidies have contributed to rising food prices plus overproduction, especially of grain. Even so, with greatly increased competition, notably from the EEC in the world grain market, the USA has lost nearly half its share of the world agricultural market. With stockpiles of unsold grain increasing, policies rewarding farmers for taking land out of production have effectively removed 25 million ha (62 million acres) from the arable area. The paradox is that despite the subsidies and the high productivity of much American farming, many farmers are unable to maintain profitable production and are going out of business. As in the EEC it has been the largest farms, especially those operated as corporate ventures, that have benefited most from subsidisation whilst smaller farmers have been unable to develop the same economies of scale and have suffered from the price-cost squeeze. It is these smaller farmers from the grain producing areas of the Mid-West who have protested in the streets of Washington DC, but lower rather than higher subsidies are likely to result in the future when attempts to reduce the Budget deficit may pick agriculture as a prime target for cuts given the high levels of farm support maintained at present.

As indicated above for southern Europe, existing policies tend to be caught in the trap of maintaining large producers because of their scale-economy efficiency whilst also seeking to maintain smallholders in order to support 'traditional' rural life. In some cases it is almost as if it has not been fully appreciated by policy-makers that direct income aids and commodity support benefit different groups of farmers. Direct income aids can be targeted at the poorest farmers thereby assisting small farmers and those in the less well endowed agricultural regions whilst this is less so for commodity support which favours larger farmers. Yet, despite the introduction of LFAs and ESAs in the EEC and comparable measures elsewhere, there seems little desire to modify substantially many of the existing farm support systems.

7

Forestry

7.1 Timber production and wood-using industries

Forestry is the second most extensive user of rural land after agriculture and is one of its chief competitors for the use of marginal land, especially in upland areas. However, for most developed countries the story of indigenous woodland has been one of decline in competition with other land uses, notably agriculture, or substitution of species as commercial forestry has focused upon particular species, especially *coniferous softwoods*. Loss of woodland has been especially marked in the Mediterranean and the British Isles where destruction has occurred over 4,000 years, reducing the forest cover in the latter to just 5 per cent of the land surface in 1900. Reafforestation has increased this proportion to 9 per cent, but this is still well below the overall proportions for Western Europe (21 per cent) and North America (c. 40 per cent).

About one-third of Europe is classified as forest or other wooded land, with the Nordic countries having the highest proportion of land covered by forests and other woods (53 per cent). The Nordic countries have 38 per cent of Europe's exploitable forest, 29 per cent of growing stock and 31 per cent of the increment; conifers dominate, representing 84 per cent of growing stock.

The forests of the Developed World include two of the three main forest types which are related directly to climatic zones. These two are the temperate zone forests and the forests associated with cold high latitude and alpine climates (Figure 7.1). It is the latter group, in the cold climate boreal belt extending through Alaska, northern Canada, Scandinavia and Siberia, that dominates production in the Developed World (Plate 13). As the *boreal forest* is composed mainly of coniferous trees, production is largely of softwoods (91 per cent of Canada's production for example), and the main producers are the pines, spruces, hemlocks, larches, firs and Douglas firs). The main species composition of the boreal forests are pine and spruce, whilst the sub-alpine forests, which occur in the Alps and Pyrenees for example, have pine, spruce, larch and fir. Much of the evergreen mixed forests that were once widespread

Plate 13 Logging of spruce and mineral extraction has promoted settlement along the John Hart Highway in the vicinity of Chetwynd, British Columbia

throughout the Mediterranean are now confined to relics or degraded forms. The original woodland vegetation now appears mainly as scattered clumps of helm oak, cork oak and pine. Broadleafed deciduous trees predominate in the temperate zone of much of Western Europe, but are also important in North America and Australasia (Table 7.1).

In the primary sector the United States and Canada are the major producers, accounting for nearly two-thirds of the Developed World's roundwood, but the smaller amounts in Finland and Sweden are extremely important within these countries' economies. The same pattern is followed in the manufacturing sector, though, as shown in Table 7.1, the United States dominates in the production of paper and Finland in that of newsprint.

Other measures of the importance of forestry in Scandinavia and North America are the industry's contribution to employment and exports. In Canada, in 1982, 7.4 per cent of the labour force relied upon the forests for their livelihood whilst exports of forest products were Can$12.1 billion or 14 per cent of exports by value. Forest products' net earnings contributed more to the positive balance of trade than any other major commodity group (Northcott, 1981). Forest products also occupy a very important position in the economies of the Nordic countries. In Sweden one-fifth of total export earnings comes from forest products, and the pulp and paper industry represents 6 per cent of the total industrial sector's output and employment. In Finland the paper industry

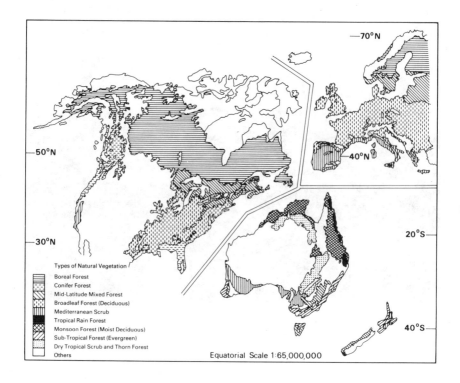

Figure 7.1 Distribution of types-of-forest in the Developed World

is even more important, with 15 per cent of the workforce employed in the production of pulp, paper and newsprint. Forest products account for 21 per cent of Finnish GDP and 36 per cent of the export earnings (John, 1984: 171–2).

The United States is the world's major producer of industrial wood, providing one-quarter of the world's production. About three-quarters of this is coniferous wood. It is also a major producer of most solidwood and fibre products, but it is still a net wood importer because of the very large domestic market (Sedjo and Radcliffe, 1981). In recent years there has been a steady increase in production in the southern USA so that nearly half the timber harvested comes from this region. Despite continuing fears of timber shortages in the south (e.g. Brooks, 1985), there is a substantial scheme of tree planting and plantation establishment. In contrast, production in the north-western states has fallen as old timber has not been fully replaced.

Table 7.1 Timber production and wood-using industries, 1985

	Roundwood million cu m	%	Sawnwood 000 cu m	%	Other	
Australia	16	2.2	3,314	1.8		
Austria	14	1.9	6,782	3.7		
Belgium	2.7	0.4	685	0.4		
Canada	161	21.9	41,929	22.6	wp	19,295
Denmark	2	0.3				
Eire			143	0.1		
Finland	47	6.4	10,275	5.5	wp	7,440
					np	1,556
					p	5,923
France	30	4.1	10,113	5.4		
Greece	2	0.3				
Iceland						
Italy	8	1.1	2,717	1.5		
Luxembourg			26	0.0		
Netherlands	1	0.1	337	0.2		
New Zealand	9	1.2	1,902	1.0		
Norway	8	1.1	2,463	1.3		
Portugal	8	1.1	2,230	1.2		
Spain	11	1.5	2,050	1.1		
Sweden	53	7.2	11,302	6.1	wp	8,699
					p	6,182
Switzerland	4	0.5	1,746	0.9		
United Kingdom	4	0.5	1,721	0.9	p	3,791
United States	322	43.7	75,339	40.6	wp	45,835
					p	59,131
West Germany	33	4.5	10,604	5.7		
Totals	735.7		185,678			

p = paper; wp = wood pulp; np = newsprint (000 tonnes)

Source: United Nations, Forest Production Yearbooks

7.2 The world's forest industries

7.2.1 Canada

In Canada 47.8 per cent (4.4 million sq km) of the national land area is forestland, representing 7 per cent of the world's forests. Most of this is administered by provincial (67 per cent) and federal (27 per cent) governments. The latter's jurisdiction applies to the Yukon and North-West Territories as well as the federal Crown lands of the Indian reserves, military bases and national parks. Each province has its own forestry agency which deals with a wide range of forestry matters, including subsidising and influencing private forest management through taxation policies and protective regulations.

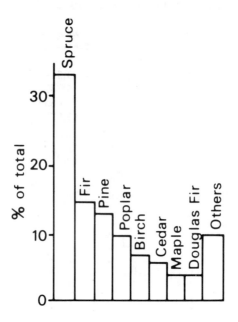

Figure 7.2 Tree species in British Columbia (%)

Total forest land in Canada amounts to c. 440×10^6 ha (c. 1087×10^6 acres) of which 340×10^6 ha (840×10^6 acres) has been inventoried, including 260×10^6 ha (642×10^6 acres) of the productive acreage. However, a significant amount of this 'productive' forest is of low quality and understocked. Therefore a unit of forest land in Canada supports much less wood than the same unit supports in Europe or the United States. There are also large areas of forest that are inaccessible, e.g. 14 per cent of forest land in eastern Canada is suitable only for local harvests as it occurs in scattered patches or has low yields or other limiting factors (Bickerstaff et al., 1981: 22 and 36). The largest producing province is British Columbia which has 63×10^6 ha (156×10^6 acres) of forests, or 68 per cent of the land area, and over 80 per cent of this forested area is classed as productive. Large-scale commercial operations are in progress on over 70 per cent of the productive forest, making up the 'primary supply area'. The remaining forest land tends to be either at higher elevations or in the north of the province where there are problems of accessibility. Favourable climatic conditions contribute to high productivity, the principal species being western hemlock, western red cedar, Sitka spruce and Douglas fir (see Figure 7.2). British Columbia accounts for two-thirds of Canada's softwood lumber and nearly half the wood-pulp. The province has also accounted for major increases in paper and board production since the late 1960s (USDA Forest Service, 1982: 93).

Despite the huge area under timber, Canada has moved steadily towards a situation of serious wood shortage. The depletion of reserves has not been accompanied by sufficient forest renewal whilst other deficiencies are related to losses through fire and disease, the establishment of parks and wilderness preserves, and environmental constraints placed upon timber harvesting. As over half the country's hardwood forests consist of polar species unattractive for commercial exploitation, those areas currently supporting commercial species are coming under greater pressure for improved management and increasing productivity. Some estimates suggest gains of up to 100 per cent in volume of wood produced are possible through application of intensive forest management and improved protection against fire and pests. Such improvements and a programme of timber-stand improvement may bring attainment of the target set for 2000 AD by the Canadian Council of Resource and Environment Ministers, of a 65 per cent increase on the 1982 harvest level. This will entail intensive sustained renewal and less attention paid to the demands of wilderness preserves and parks, urban expansion, highways and reservoirs (Stanton, 1976). A renewal programme could also expand employment in the industry, perhaps by as many as 100,000 additional jobs (Government of Canada, 1981).

Forest products constitute 12.5 per cent of all manufactured goods in Canada, and are especially important in British Columbia (45 per cent) and New Brunswick (30 per cent). However, it is in British Columbia that forestry has its greatest effect upon economy (Plate 13), representing 6.7 per cent of the province's employment, 11.7 per cent of its expenditure on capitalisation and repairs, 53 per cent of its exports through Canadian ports and 42 per cent of the value of the province's manufacturing shipments. Indeed 'there are few jurisdictions in the Developed World that depend so heavily on forest resources. Moreover the economic development of British Columbia has been to a large extent a history of forest resource development' (Pearse, 1980: 1). So a drop in output could have severe economic consequences. Such a drop may be caused by lower timber volumes from second-growth stands, reductions in the forested area and increasing competition in traditional markets which could reduce unit value and total value of exports of provincial forest products (Percy, 1986).

The ability of the forest base in British Columbia to permit expansion of the forest products industry is in doubt as the province may have reached the limits of its exploitable timber supply (see Figure 7.3). In addition, the strong competition with the United States may cut returns from timber exports. However, the recent free trade agreement between the two countries may benefit British Columbian timber exporters in an unrestricted market. If this is the case there is likely to be an increased demand for logging in areas currently claimed by Indian groups or proposed for conservation. Such areas have already been the subject of great debate and political controversy, symbolising different attitudes to the province's resources and the wider conflict between economic and non-economic interests (e.g. Sewell, 1988).

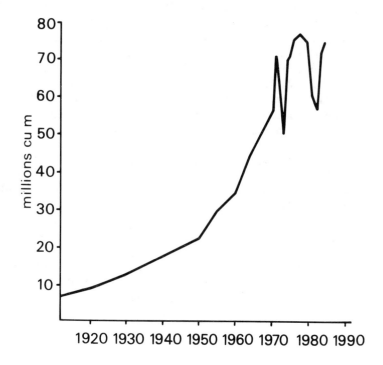

Figure 7.3 The timber crop in British Columbia, 1920–84

7.2.2   The Nordic Countries

Forestry has been extremely important to the economies of the Nordic
Countries, playing a significant role in both the growth of manufacturing
industry and also international trade. Yet there are major differences
between the constituent countries in terms of how their forests have been
managed and in how national forest products industries have evolved.
For example, in Sweden one-third of the private forest area is owned by
joint stock companies, compared with a much smaller role for such
companies in the other Nordic Countries (Table 7.2). This accounts for
some of the largest privately owned forests in Scandinavia being located
in southern Sweden compared with the predominance of smaller forest
plots in Denmark and Norway where 90 per cent of forestry is combined
with agriculture. In Finland half the forestry is essentially farm forestry
compared with 40 per cent in Sweden. Other key differences can be
illustrated by a separate consideration of the two main Nordic timber
producers, Finland and Sweden (Table 7.3).

Finland's forests have been the resource base for a dramatic improvement in incomes and living standards post–1945. Timber has been the
'green gold' forming the basis for two-thirds of the country's exports.
The transformation dependent upon timber has effectively turned Finland

Table 7.2 Ownership of production forests in the Nordic Countries

Owner groups	000 ha			
	Denmark	Finland	Norway	Sweden
Public				
State forests	150	4755	647	4425
Other public	59	844	404	1788
Private				
Companies	59	1812	301	5630
Other private	225	12653	5308	11797
All productive forests	493	20065	6660	23639

Source: Yearbook of Nordic Statistics, 1985, p. 112

Table 7.3 Estimated annual removal of timber in Sweden and Finland, 1950–84

	millions cu m				
	Coniferous Timber	Pulpwood	Fuelwood	Other	Total
Finland					
1950	9.5	9.1	13.8	6.4	38.8
1960	15.5	15.6	12.3	3.2	46.6
1970	14.3	19.8	7.7	3.3	45.1
1975	9.4	14.1	6.2	2.0	31.7
1980	20.7	19.9	3.5	3.0	47.1
1984	16.2	18.2	3.2	2.8	40.4
Sweden					
1950	12.5	10.0	9.0	2.0	33.5
1960	15.8	20.8	4.8	1.7	43.1
1970	23.0	32.8	2.1	0.9	58.8
1975	21.1	33.0	1.2	0.9	56.2
1980	22.2	22.0	2.0	0.9	47.1
1984	24.3	23.4	2.9	0.9	51.5

Source: Yearbook of Nordic Statistics 1985, p. 114

in recent times from domination by a poor subsistence economy into a modern manufacturing and trading nation. The roots of this dramatic alteration appear in the early years of this century in the gradual growth of Finland as a supplier of the products of the wood-working industry to European markets. This has been developed and improved upon consistently through both technical and managerial expertise so that for more than two decades Finland has been producing over one-third of all Europe's output of cellulose, newsprint, printing and writing papers (Fullerton and Williams, 1972: 233–4).

The basis of this commercial forestry and forest-products' boom has been the development of saw-milling from the 1880s when steam-powered mills were introduced. Exports of sawn goods still account for 15 per cent by value of forest product exports. There are over 600 exporting sawmills and numerous others producing for local consumption, so that most communities have access to at least one. This prevalence of local, small-scale wood-using is indicated by the high proportion of land under productive forest: 52.9 per cent, with a further 17.6 per cent not producing. The dominant species are Norway Spruce and Scots Pine whose technical properties make them highly suitable for cellulose manufacturing and paper-making. It has been the development of cellulose manufacturing and wood pulp production, over and above the output from sawmills, that has brought increased wealth post–1945. It is the pulp and paper sector (including mechanical wood pulp, cellulose, paper, board, converted paper and board products, and fibreboard) which accounts for three-quarters of Finnish forest industries' exports by value. It is also this sector which has developed gradually by exploiting the European market to great effect from the 1950s following vertical integration within the timber industry, to form large mills of a high technical standard and ensuring high productivity at internationally competitive prices.

Britain, itself greatly deficient in timber, has been a major recipient of Finnish exports, and currently accounts for half of Finland's pulp exports as well as large quantities of plywood and newsprint. The processing of pulp has altered significantly in favour of tapping lucrative export markets, such as the British, by means of combining pulp and milling operations for production of paper and board. Adoption of economies of scale has enabled output of up to 0.5 million tonnes per year from single complexes with commercial quantities of by-products such as alcohol, oil resins and turpentines produced. This expansion has been stimulated by the rising European demand for newsprint, and this now accounts for over half of Finland's paper output. In addition, the growth in popularity of well-packaged merchandise of all kinds has brought increased output of card and board for various forms of packaging, plus packaging made from recycled waste from sawmills. Other products manufactured from waste include softboard for insulation, hardboard and enamelled, oil-tempered and melamine surface-treated boards.

The other major product from Finnish timber processors has had to compete with similar goods from the rest of Scandinavia and West Germany. This is furniture manufacture, in Finland's case making use of the softwood birches which account for nearly 20 per cent of the national forest. Birch has also been used for plywood and pre-fabricated houses, exploiting the Russian market. Increasingly, emphasis has been placed upon high quality production for higher returns, and with great improvements in the production and marketing of finished products.

Much of the country's forests are in private hands, with a significant number of forest owners combining farming with forestry. Indeed, one-

Table 7.4 The United Paper Mills Ltd operations in Finland

Turnover	60% paper and paperboard 25% converted paper and board products 15% metal and chemical products, sawn timber and shipping
Exports	75% of turnover
Employment	9,000 people in nine locations – the 'Profit Centres' (see Figure 7.4)
The Profit Centres	include sawmills, mechanical and chemical pulp mills, engineering works, chemical factories, printing works, forestry centres, shipping bases.
Kaipola	Has 1,000 employees producing 350,000 tonnes of paper per annum from a mechanical pulp mill and two paper mills. Fibre raw materials are supplied from company mills at Tervassari and Jamssankoski. Talc is supplied from a company plant at Sotkamo. Wood is supplied from a large procurement area and transport by water, road and rail. Consumes 2,000 cu m of wood daily, obtaining four-fifths power through the national grid and the remainder from the mill's own power plant. Was formerly a newsprint mill, but now specialises in lightweight paper grades. New pulp mill constructed in 1977 with a capacity of 100,000 tonnes per annum. 92% of the paper mill's output is exported.

Source: John (1984)

third of farm income is derived from farm-based forestry and often a farm woodlot is the single greatest source of farm income. However, the pattern of fragmented ownership and small forest plots has limited necessary forest renewal programmes. In the 1960s one solution to the need for more new young timber was to drain and plant peat bog, using machinery specially developed in Finland. This and other programmes are intended to increase the annual cut of Finnish timber from just under 50 million cu m in 1970 to 100 million cu m in 2040. Even so, industrial demand for wood may exceed supply and careful control will be needed.

The post-war expansion of pulp and paper industry relied upon mechanical pulp production using water power from rivers such as the Kumi. However, the mechanical pulp mills were partly displaced from the 1960s by chemical pulp mills linking pulp-making and paper-making in large integrated concerns which also produce packaging materials and chemicals. John (1984: 173) gives a good example of these large integrated complexes, the United Paper Mills Ltd, who symbolise the scale and range of operations that have transformed the forest products

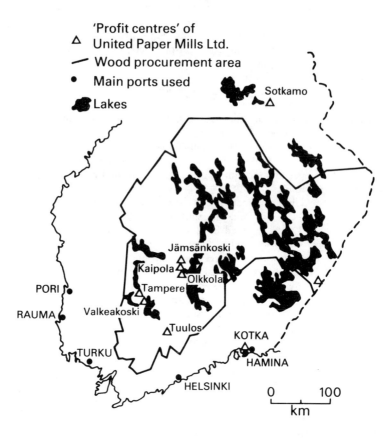

'Profit centres' of
△ United Paper Mills Ltd.
⌒ Wood procurement area
• Main ports used
⬛ Lakes

Figure 7.4 The 'profit centres' of the United Paper Mills group in southern Finland (based on John, 1984: 174)

industry in Finland (Table 7.4). Its vertically and horizontally integrated operations have enabled it to reduce costs and thereby compete in international markets (Figure 7.4).

Sweden has a similar volume of timber production to Finland, but with a higher output of wood products. Of Sweden's exports the timber and wood-working industries account for 25 per cent, representing 15 per cent of the world trade in timber. Sweden has a long history of exporting wood products and has some advantages over Finland in favourable cutting and transportation conditions. Similar trends to those described for Finland have characterised Swedish timber production and processing, especially in terms of the development of large units. For example, the majority of production comes from eight firms, including two co-operatives and a state-owned group. Sweden has specialised more in profitable furniture-making and also pulp manufacture, Sweden being the world's largest exporter. Private ownership of forests is dominant, but there are conservation laws and extensive reafforestation schemes to ensure forest renewal. Some of the largest areas of timber are now owned by joint stock companies which have developed vertically-integrated enterprises combining timber production and processing industries, e.g. in Dalarna and Varmland in the Swedish Midlands. Here and in other parts of the country there is a close link with hydro-electric power (HEP) generation to support integrated plant operations, e.g. at Skoghall, near Harlstad on Lake Vaner, HEP from the Klar watershed supports saw-mills, chemical, pulp and paper plants and the production of electro-chemicals (Fullerton and Williams, 1972: 212).

Twelve large companies dominate Sweden's 5.6 million ha (13.8 million acres) of company forest; Svenska Cellulose AB have nearly one-third of this. Several of the large companies, Uddeholms AB for example, have a range of industrial interests extending beyond forestry but using some of is products, e.g. in the chemical industry, and using HEP for various industrial processes as well as pulp and paper production (John, 1984: 139). As in Finland, the largest share of the forest products industries are concentrated in the south of the country, where the most productive forests are located. Lake-shore and riverside locations are important as they provide the abundant water required by the pulp and paper industry.

Generally, in the Nordic Countries, the rate of felling has remained approximately at levels reached in the 1960s, but the greater application of mechanisation has greatly reduced labour demands. For example, in 1960 a single worker could deal with 1 cu m of timber per day compared with the present volume of 10 to 15 cu m. In Sweden this has had the effect of cutting the number of full-time workers from 250,000 in 1960, to 25,000 in 1985 (Oscarsson and Oberg, 1987: 324). This has fuelled rural depopulation despite the persistence of farm forestry which has enabled smallholdings in many parts of Scandinavia to remain viable when an income from farming or forestry alone would have been inadequate.

7.3   Economic forestry versus conservation: the greening of the south

The major land use conflict within forest land itself arises between the forest industry's desire to exploit timber for profit and society's desire to protect and maintain the forest for grazing, hunting, recreation and protection of watersheds. For example, in the United States, as part of the National Forest System, 66 million ha (163 million acres) or nearly 20 per cent of forestlands have been set aside for protection. An addi-tional 32 million ha (79 million acres) are protected as national parks (IUCN, 1985). Europe has over 1.9 million ha (4.7 million acres) of closed forests within national parks whilst the total area of wooded land managed for protective functions is 33 million ha (81.5 million acres) or 23 per cent of all closed forests. In addition to this protection, forest renewal has also been extensive in some cases, e.g. 1.8 million ha (4.45 million acres) is reforested annually in the United States, 1.1 million ha (2.7 million acres) in Europe as a whole, and 0.72 million ha (1.8 million acres) in Canada. Whilst the conflicts between economic forestry and the need for preservation are found throughout the Developed World, the nature of these conflicts is perhaps best illustrated in the United States, especially in its southern states.

In the United States, deforestation has continued at a rapid rate in some of the most densely forested parts of the country, e.g. there is a loss of 24,000 ha (59,300 acres) of forest per annum in Washington and Oregon states. Indeed, tree-cutting is at an all-time high in the 156 designated national forests: 2.7 billion board feet in 1987, of which almost half came from ancient stands in Washington and Oregon. This partly reflects the problem faced by the US Forest Service which has obtained increased funding from the lumber industry for allowing more timber to be cut and more forest roads to be built through wilderness areas, and yet still at a price which fails to cover the cost of managing the forests. Proposals made in early 1989 to raise more revenue from commercial development could double both the number of trees cut and the network of roads in the six Appalachian forests. Conservationists argue that this is tipping the balance, away from the Forest Service's commitment to preserve wildlife habitat and valuable ecosystems, more towards its promotion of lumbering and recreation. This highlights the inherent conflict of interests in the remit of the Forest Service: between management for commercial development and management for conservation.

For the southern United States, Clark (1984) paints a vivid picture of the transformation of landscape during the 20th century, stressing the move from destructive to constructive phases in woodland management. The former was triggered by the removal of the restrictions of the Southern Homestead Act in 1876, after which lumbermen and speculators rushed in to acquire millions of hectares of cheap virgin forest. Thus the forest, which remained largely intact at the end of the Civil War, was rapidly exploited, meeting a rapacious drive for timber by the wood-using industries. Scant concern was shown for conservation and renewal of forest resources, and hundreds of sawmills brought large-scale removal of timber, exposure of fragile soils and mass denudation. This 'sawmill era' can be contrasted with the greater emphasis upon forest renewal that emerged in the 1930s with the development of scientific forestry, establishment of forest schools and the advance of chemical research into the use of wood pulp. Together these and new attitudes, promoted by greater awareness of the severity of previous destruction, helped to develop major reclamation programmes as seen in the work of the Civilian Conservation Corps and the Tennessee Valley Authority. To

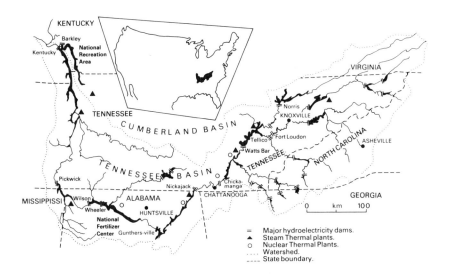

Figure 7.5 The Tennessee Valley Authority (based on Bradshaw, 1984)

these efforts at forest conservation and renewal were added post–1945 changes within the wood-using industries which were concerned not solely with exploitation of timber but more with the maintenance of the forests as a renewable resource. It is this new attitude to the southern forests that has promoted reafforestation and brought about what Clark calls 'the greening of the South'.

The Tennessee Valley Authority (TVA) was established in 1933 at the same time as several other New Deal agencies, e.g. the Federal Soils Survey and the Civilian Conservation Corps. The TVA and the other agencies addressed the question of regional economic development at a time of large-scale abandonment, deforestation, erosion and economic depression. The law creating the agency gave it broad powers of direction and control over land within the Tennessee River system and extending across territory within seven states (Figure 7.5). The TVA's chief areas of control were to be applied to flood control, the generation of electricity, the maintenance of open navigational channels and general conservation of resources including the 'proper use' of marginal lands and selection of the best methods of reforestation. These individual areas were to be dealt with in terms of assuring economic and social well-being for the inhabitants of the river basin (Clark, 1984: 87).

At its inception, one of the TVA's prime roles was seen as the generation of electrical power by utilising the streamflow of the Tennessee River system. Hence, major activities in the region in the 1930s tended to be linked to construction of dams and lakes for HEP production. Despite this the increased demands for electrical energy, made during World War Two by new industries that had already moved into the

Valley, promoted increased coal mining and steam-generated electricity production. Subsequently, the TVA began purchasing coal from strip mines and was charged unsuccessfully in 1972 with violating the 1969 Natural Environmental Policy Act through its promotion of environmental destruction. That year the TVA purchased nearly 30 million tonnes of coal, only half of which came from deep mines and with 12,500 ha (31,000 acres) of land disturbed by mining. By this time energy production had become the chief TVA programme, with nuclear-thermal electricity plants being operated from the late 1960s (Bradshaw, 1984). Subsequently, environmental concern has limited the expansion of nuclear power, though power generation retains its prime role within the Authority.

Although the TVA was originally conceived as having a significant regional planning function, today it acts more as 'a kind of hybrid industrial development corporation' (Knox, 1988: 139). It is a vast enterprise, employing 45,000 people, spending $2,200 million annually and supplying electricity to over six million people, a far cry from some of the collectivist afforestation programmes and agricultural improvement schemes of its early years. Nevertheless, the TVA has played a major role in the 'greening' of the South, bringing the proportion of the Tennessee Valley under forest cover to 60 per cent by adding at least 1.5 million ha (3.7 million acres) of forest land during its first fifty years of operation, whilst actually losing 70,000 ha (173,000 acres) of forest to the creation of new lakes and streamside areas plus new urban and industrial areas on former forest land. In the 1970s the volume of growing timber was increasing at 3.33 per cent per annum as both forest regeneration and productivity were advanced.

Yet, despite renewal and new types of forest management, the fourteen southern states of the USA lost 3 million ha (7.4 million acres) of forest in the 1970s. This loss highlights the comparatively limited extent of government ownership and direct control of woodlands (around 19 per cent of the South's forests), and the importance of private corporate owners. Lack of government control and the numerous competitors for land mean the South may lose over 80,000 ha (197,700 acres) of forest lands annually in the last two decades of the century (e.g. Wallach, 1981).

Post–1945 one of the most important attempts at 'greening', in the sense of preservation of forest and wilderness, was the establishment of a National Wilderness Preservation System in 1964. This was intended to prevent wholesale destruction of certain parts of the natural environment, and, together with national parks and various miscellaneous lands designated by Congress, accounted for 162 million ha (400 million acres). However, in the 1980s federal government policy has stressed individual and private enterprise in regulation of environmental standards rather than government controls and limitations which were symbolised in the Carter administration's Alaska Land Bill which created over 42 million ha (104 million acres) of national parks, conservation areas and wildlife refuges. Indeed, in 1983 over 300,000 ha (740,000 acres) in Alaska were removed from protected status as wilderness areas, and larger public areas elsewhere in the country were proposed for sale. This 'frontier-style capitalism', in which the interests of the individual may over-ride the need for long-term environmental considerations, is likely to be a recurring feature of development in the USA given the traditional concern with the rights of the individual (Pope, 1984).

7.4 Quasi-governmental control of forestry: the British experience

The individual countries of the Developed World have taken widely differing attitudes to their national forests, though many have attempted to develop programmes of conservation and reafforestation via the control of a substantial amount of the national forest by a specially established agency. The situation facing such agencies has also varied from management of substantial timber reserves and the presence of large commercial lumber concerns as in Canada, to management of a small amount of forest in need of much reforestation and wholesale re-establishment of sylviculture. Britain is perhaps the best example of the latter, with a long history of deforestation (James, 1981), and a small timber industry. It is also a good illustration of the way in which a government agency has been invested with powers to oversee forestry development (Ryle, 1969), but with certain conflicting aims and critical conflicts with other government agencies, in this case most notably with the Ministry of Agriculture, Fisheries and Food (MAFF). These conflicts, and the ways in which changing policy has affected the extent and composition of British forestry, provide an effective contrast with the examples of forestry in other countries quoted above.

7.4.1 The Forestry Commission and forestry policy

The 1919 Forestry Act brought into being the Forestry Commission whose initial task was to make good the timber depletion brought about during World War One. Subsequent Acts, principally those in 1947 and 1981, and the Forest Policy White Paper of 1972, have diversified the Commission's role, but the basic functions remain those of *forest authority* and *forest enterprise*. As a forest authority its remit is to:

(a) advance knowledge and understanding of forestry;
(b) carry out research and combat disease;
(c) ensure the best use of timber resources and promote an efficient timber industry;
(d) promote training and education;
(e) administer controls and schemes to the private sector in pursuance of sound forestry;

(f) pursue good use of land and integration with agriculture.

As a forestry enterprise its remit is to:

(a) plant, manage and market products;
(b) stimulate the local economy;
(c) protect and enhance the environment;
(d) promote recreation where appropriate;
(e) integrate forestry and agriculture.

Of these two functions that of forest enterprise has attracted most attention, through the expansion of the area under forestry. The concept of establishing a strategic reserve of timber was met only gradually between the wars, attempting to replace over 160,000 ha (395,350 acres) of woodland felled during the German blockade of Britain. Between 1919 and 1939 the Forestry Commission planted 149,000 ha (368,175 acres) whilst 81,000 ha (200,150 acres) were planted by private landowners. This recovered the woodland lost during World War One. Then, during World War Two, a considerable acreage was felled or destroyed. Whilst this renewed loss of timber stocks was still occurring in 1943, a White Paper on Post-War Forest Policy urged an increase in the acreage under forestry, both by replanting in existing woodland areas and by planting on land which had not previously been used for forestry. These suggestions were followed by the 1945 Forestry Act under which the Forestry Commission was reconstituted. It was brought under the jurisdiction of the MAFF, with the Commissioners' earlier power, to acquire land compulsorily, transferred to the Minister of Agriculture. The Minister could also purchase land for afforestation and could manage it himself or delegate its management to the Commission which would prepare management plans. In practice the powers of compulsory purchase have been used rarely, partly through reluctance to use them and partly as there has generally been enough land available on the open market.

Whilst the 1945 Act had applied to forestry on publicly owned land, the 1947 Forestry Act dealt with private land, on which the intention was also to extend the area under timber. Under this Act, landowners could enter into Forestry Dedication Covenants through which they could receive grant aid to develop woodlands. Tax concessions were instituted under the Act to further encourage private planting (Watkins, 1984). In effect, fiscal concessions were awarded in return for a management plan.

Further schemes to assist private landowners with forestry projects were introduced in the 1967 Forestry Act. In addition to up-dating Dedication Agreements, two new planting schemes came into being. Approved Woodland Schemes were instituted for landowners not wishing to part with rights of management in the long term. In accordance with an agreed plan, planting grants were made available for planting, but with no financial help to follow for later management. Small Woods Planting Grants (SWPGs) were organised on similar lines, with grants available for small areas of land suitable for tree planting. The 1967 Act also introduced a new advisory body in the form of the Home Grown Timber Advisory Committee which, in addition to aiding the Forestry Commission, was charged with promoting forestry interests and helping to liaise between the Commission and private interests.

Mather (1978) recognised a series of phases of forest development from 1945 to the mid-1970s. Until 1957 the private sector met its target plantings under the stimulus of the Dedication Scheme. Meanwhile, the Forestry Commission, reluctant to use its power of compulsory purchase and so short of land, did not meet its targets. Indeed by 1955 the rate of afforestation was falling because of the strong competition for land in the uplands between forestry and hill farming. MAFF were concerned that agricultural land was not sold for forestry and this helped to limit afforestation – to only 71 per cent of the target for 1945–55. In eastern Scotland and north-eastern England the dual use of land as grouse moor and sheep grazing limited release of land for forestry whilst in other parts of England and Wales the holding of rough grazings in common was also an obstacle. This encouraged a focus upon south-west Scotland where the sale of small hill farms provided rough grazings suitable for afforestation.

This focus upon Scotland was intensified in the 1960s as both an increased Forestry Commission planting rate and a more conscious regional policy were followed (Figures 7.6 and 7.7). By the late 1960s a 'flight' from hill farming was under way, stimulating the greater release onto the market of upland suitable for coniferous afforestation. Once again southern Scotland was a chief source of land for forestry. However, in the early 1970s more support for hill farming reduced the supply of land for forestry. Reduced targets were set for the state sector (Figure 7.8) and reduced grants for the private sector reduced the area of planting by private owners (Figure 7.9). The focus of planting shifted northwards to the Highlands as some of the larger estates there were sold.

A major review of forest policy was undertaken in the early 1970s. A government White Paper was published in 1972, drawing heavily on a study in the same year by the Treasury which took the form of an inter-departmental cost-benefit study of forestry. A comparison was made with timber production in other temperate areas of the Northern Hemisphere, and, whilst this was largely favourable, it was shown that timber production in the uplands, where Forestry Commission plantings were concentrated, was not economic. However, one of the major findings of the White Paper was that government aid to the Forestry Commission should continue on social grounds. It was stated that without such assistance the demise of forestry in certain cases would greatly reduce the viability of small rural communities. The economics of forestry showed a lower rate of return on investment than from hill farming, but forestry created more jobs for a commensurate capitalisation. To further the amount of employment in forestry and increase the rate of return, further expansion

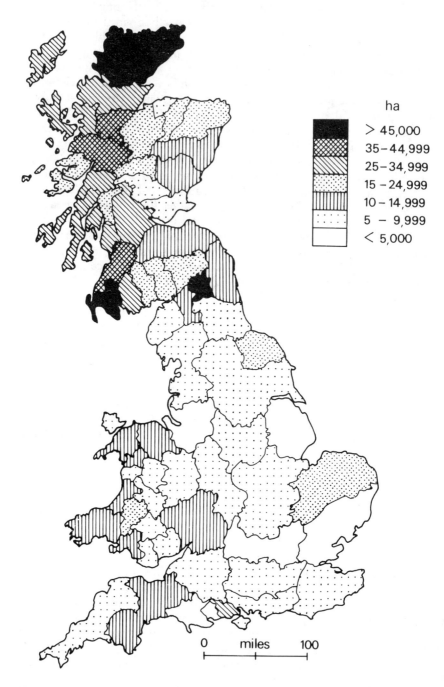

Figure 7.6 Area of forest by Forestry Commission conservancy, Great Britain, 1987 (*Data source:* Forestry Commission, *67th annual report*)

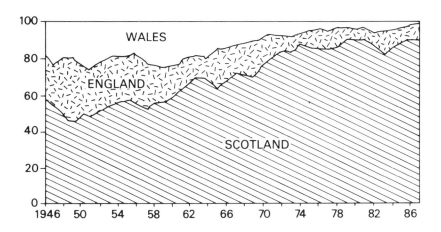

Figure 7.7 State afforestation in England, Scotland and Wales (%), 1946–87 (*Data source:* Forestry Commission, *Annual Reports*)

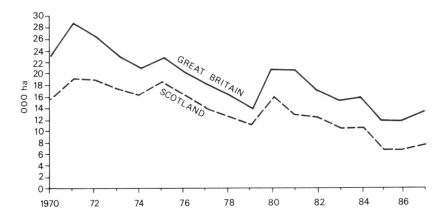

Figure 7.8 Forestry Commission plantings in Scotland and Great Britain (000 ha), 1970–87 (*Data source:* Forestry Commission, *Annual Reports*)

of recreation in forests was suggested. While the notion of a strategic reserve had now been replaced by considerations of import substitution, the potential for reducing imports of timber and timber-based imports was noted. The role of private forestry as a contributor to this was also recognised as were the difficulties faced by private timber producers by the length of time between planting and harvesting. The White Paper proposed that government aid for private forestry should be continued but with a simplification of grant aid.

The 1972 Review saw a more flexible Forestry Commission in which

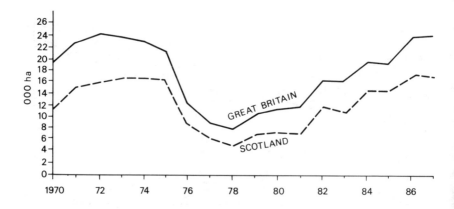

Figure 7.9 Timber planting in Scotland and Great Britain by private owners for which grants were paid (000 ha), 1970–87 (*Data source:* Forestry Commission, *Annual Reports*)

conservation and recreation would play a more important role. Indeed, this has been one of the features of the Commission's work in the 1970s and 1980s. The Commission has diversified its role into amenity, conservation and access, and more employment.

One result of the 1972 Review was the abandonment of the Dedication Scheme which was felt to be too complex and impractical. However, grant aid for private forestry was subsequently reintroduced and then revised, with the establishment in 1981 of the Forestry Grant Scheme (FGS) which replaced both Dedication and the SWPG. Under the FGS, differential subsidies are paid for planting and replanting if the primary aim is production of timber. Grants are available for a five-year Plan of Operation, higher grants being awarded for broadleaves.

The change in philosophy towards forestry brought about by the election of the Conservative government in 1979 was illustrated by the 1981 Forestry Act. In this, powers were given to sell land under forestry, though with due regard to maintaining national forestry resources. The intention was to recoup £40 million between 1981 and 1984 through sales, but, in practice, the level of sales has been much less – partly because of the concern expressed by a number of interested parties. The new government attitude to forestry has been indicative of the view that the more varied functions of the Forestry Commission in the 1970s and 1980s do not require more land to be brought under timber whilst, at the same time, the economics of home-produced timber have been questioned. Attention has also been drawn to the monotony of the large coniferous plantations in several areas, the general lack of access to upland areas consequent upon new plantings and the need to reappraise the viability of *farm forestry* (Plate 14).

In most parts of the Developed World there has tended to be an

Plate 14 Dunloinn Forest and Loch Cluanie in the Scottish Highlands where the Forestry Commission carried out extensive afforestation in the 1970s

'apartheid' between farming and forestry in terms of government planning (though see Reid and Wilson, 1986). This was certainly true of the Forestry Commission in the United Kingdom where 'forestry has been sponsored by the state and agriculture has been subsidised by the state for the advantage of each' (Donaldson, 1969). Yet, this clear separation was not the case on most private landed estates, and the notion of effective combination of farming and forestry interests has been fostered by recent legislation aiming to encourage farm forestry. This has promoted the development of forestry on better land, where trees will mature more quickly, be less prone to wind damage, and produce better quality timber, notably hardwoods (Macpherson, 1988). However, such farm forestry schemes must overcome the problem of the long length of time before crops are ready for harvest: twenty years before the first thinnings and between 45 and 65 years before the main harvest. New incentive schemes for farmers have had to be introduced to overcome what is a crucial economic deterrent.

'Standard' grants have existed for some time in the United Kingdom to assist farmers who wish to plant trees (Blunden and Curry, 1988: 65–70). For example, the Forestry Grant Scheme and the Broadleaved Woodland Scheme, superseded in April 1988, were available for areas as small as 0.25 ha (0.6 acres), but were not intended to provide an annual income for farmers during the long pre-harvest period. Rather, they often supported the establishment of shelter belts and amenity planting.

However, following the 1987 consultation document 'the Farm Wood-lands Scheme', new ideas have first been aired and then subsequently incorporated in a new Woodland Grant Scheme as part of the Farm Land and Rural Development Bill.

The new scheme is intended to promote planting of trees on 12,000 ha (29,650 acres) of farmland in an initial three-year period whilst providing annual payments to farmers. The highest rates of £190 per ha (£77 per acres) per annum are to be paid to farmers on better quality land, thereby encouraging planting on arable land. Cheaper rates will be paid on poorer agricultural areas, with areas classed as rough grazing only qualifying for a £30 per ha (£12–15 per acre) grant. The payments will operate over twenty years for conifers and over thirty years for purely broadleaved plantings. However, these payments seem unlikely to prove attractive to farmers, especially tenants.

The Forestry Act of 1981 empowered the Forestry Commission to dispose of plantations and planting land as part of government wishes to reduce the Commission's dependence on public funds and to make provision for private investment in the Commission's assets. The revised target set for disposal of assets was around £100 million between 1981 and 1989 (Mather and Murray, 1986: 109). In the first three years of disposals over 25,000 ha (61,775 acres) were sold, representing 2.7 per cent of the Forestry Commission's estate of forest land. In Scotland the majority of land was purchased by farmers and other private investors. Less than one-sixth of the area sold went to institutions, but these sales accounted for over one-third of total receipts. This institutional invest-ment represented large units in the best timber-growing areas, e.g. Argyll. The majority of sales, though, have been of small blocks of land at some distance from a major concentration of Forestry Commission land. Such sales have become more common as pressure to meet financial targets has been reduced (Mather and Murray, 1988).

Despite various suggestions aimed at promotion of broadleaved deciduous species (e.g. Watkins, 1986) new grants introduced in 1988 increased grants for conifers at a greater rate than for deciduous trees. The grants, designed to replace tax incentives for private forestry which had been removed in the 1988 Budget, aimed to reach 33,000 ha (81,550 acres) of commercial planting per annum. This will ensure further large-scale planting of conifers in the uplands of Scotland and Wales despite criticism by government-sponsored bodies such as the Countryside Commission and the Nature Conservancy Council.

Shoard (1987) cites three ways in which afforestation in the uplands has had negative rather than positive effects:

(a) by causing a decline in the number of species of wildlife, especially rare and distinctive ones;
(b) by reducing employment, primarily from land well stocked with sheep;

(c) by destroying the character of the landscape and thereby reducing its tourist potential.

Grove (1983) is critical of the effects which the planting of coniferous trees in the uplands has on pre-existing moorland plant species, water quality and bird communities. In Britain large-scale planting of conifers has been fostered with government assistance whilst possibly environmentally beneficial planting of broadleaved species and retention of ancient woodland has been neglected or threatened by government policies (see Ramblers Association, 1980).

Johnson and Price (1987) investigated the link between the growth of afforestation and depopulation in Snowdonia National Park, Wales. No clear link was established despite the fact that since 1951 a marked increase in labour productivity in forestry has led to reduced employment and the growth of short-term contract work. The development of forestry in this area had been encouraged by the notion that it would be a more labour-intensive land use than agriculture. However, planting has often been carried out in areas of severe depopulation and, because of the hiatus in job provision between planting and first harvesting, forestry has been unable to provide an immediate stable economic base to maintain population.

A more wide-ranging examination of the link between forestry and employment in rural Britain concluded that afforestation involves neither the loss nor gain of jobs at the local level in the large majority of afforestation schemes (Mather and Murray, 1987). However, afforestation does result in a net increase in labour input per unit area compared with hill sheep farming, at least in the short term. This increase has tended to be supplied by mobile squads of workers and therefore may not have particular benefits for communities in the vicinity of the newly afforested area.

The most recent initiative to increase the amount of forested land in Britain shifts the emphasis away from the uplands. It has come from the Countryside Commission who, in July 1989, announced a plan to establish twelve forests on the fringes of industrial areas in England and Wales. The land intended for these forests is either derelict, poor quality scrub land or agricultural land within designated green belts. Each block of new forest will cover between 10,000 and 16,000 ha (25,000 and 40,000 acres) and will cost c. £25 million. They would add 10 per cent to the existing forested area and would cater specifically for recreation and environmental education. Described as 'community forests', the first three announced are in south Staffordshire, the Hornchurch area in east London, and between Gateshead and Sunderland, in Tyne and Wear. The Forestry Commission will provide advice on forest management whilst funding will be from central and local government, private companies and individuals.

7.4.2   Recent forestry problems in Scotland

The chief beneficiary of agricultural land loss in Scotland has been forestry. In the mid–1980s the rate of transfer was 26,000 ha (62,250 acres) per annum, a rate expected to exceed 30,000 ha (74,150 acres) in the 1990s when future upland afforestation in Britain will be concentrated solely in Scotland. In the 1980s it was hill land of poor agricultural value that was primarily the subject of afforestation, though this may change as more incentives are given for forestry on better land.

Most recent afforestation in Scotland has been concentrated in the private sector which has also grown through transfers from the Forestry Commission following policies of the partial disposal of the Commission's holdings. In 1984 private sector ownership of Scotland's 929,000 ha (2.3 million acres) of forest accounted for 44 per cent of the area. However, over half of this private sector area was owned by personal or corporate 'investment' owners rather than the traditional landed estates who dominated this sector in the 1960s. The tax benefits of forest ownership have obviously stimulated this significant change (Mather, 1987).

Since 1979, under the Conservative government, the aim has been to increase the role played by private forestry. Thus, in the 1980s, over three-quarters of the annual afforestation in Scotland has been carried out by this sector, encouraged by a combination of planting grants and tax relief (Forestry Commission, 1987). Whilst private forestry has increased, there has been an accompanying change in the nature of ownership within this private sector, with a growth of non-traditional personal investors, i.e. from outside the long-established estate owners. Typically, these newcomers have been high-rate taxpayers, not residing locally and with no related land enterprises. The tax relief on forestry has been a major incentive for such owners. Meanwhile, corporate ownership of forests has also risen, though more often involving purchase of established plantations as opposed to afforestation. This growth, of what Mather (1988) terms *investment forestry*, has been paralleled by the expansion of forest management companies, vigorously promoting forestry as a form of investment, and acquiring, planting and managing forest land on behalf of their clients.

Mather's (1978; 1979; 1985) examinations of the different types of forestry ownership in Scotland have revealed significant variations in terms of their influence upon afforestation and the physical attributes of new forests. For example, most schemes carried out by traditional estates were smaller, lay on better land and had a higher broadleaved content in comparison with investment forestry. Within the investment sub-sector, corporate management generally involved larger acreages than its 'personal' counterpart. The different considerations of a range of different types of investors in forestry appear to have been translated into variations in both afforestation and type of forest (Campbell, 1984). In particular, the role of the forest manager has become highly significant

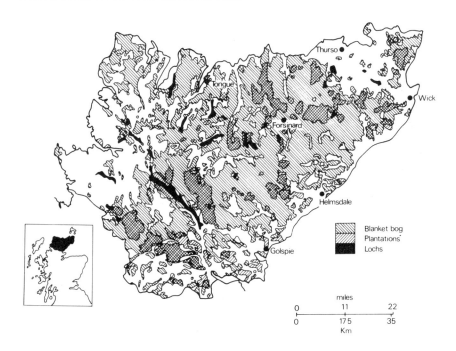

Figure 7.10 The Flow Country, northern Scotland

in cases where ownership and management are separated.

The spread of coniferous plantations in Britain through the direct intervention of the Forestry Commission has been extended in the 1980s by the surge in the amount of afforestation resulting from the private initiatives. These have reflected a taxation system which has made forestry a highly attractive investment. However, some of this afforestation has promoted increased conflict between forestry interests on the one hand and conservationists on the other. This conflict has arisen largely because of the changing location of new plantings, for example, away from areas in the Southern Uplands, formerly dominated by rough grazing for sheep. The new focus has become areas in northern Scotland, relatively little affected by afforestation in the 1960s and early 1970s, but where many feel that significant ecological damage will be produced by afforestation. This concern reflects the differences between afforestation of rough grazings in southern Scotland and afforestation of ecologically diverse and sensitive land in the north of the country.

The greatest concern has been raised over the Flow Country of Caithness and Sutherland (Nature Conservancy, 1988) (Figure 7.10). This area contains large expanses of peat bog, small lakes and marshland, and has been recognised by international conservation bodies as an ecosystem as distinctive and significant in world terms as parts of the Brazilian rain forest or Africa's Serengeti. As the home of rare plants

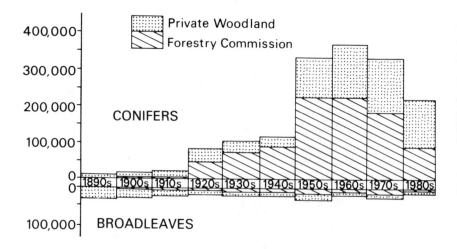

Figure 7.11 Area of forest in Great Britain, 1890s–1980s (*Data Source:* Gilg, 1978 and Forestry Commission, *Annual Reports*)

and wading birds, it has been regarded as one of the few extensive 'wilderness' areas in Western Europe, though until recently it has been subject to little protective legislation. This lack of protection and the existence of tax relief for forestry made it a target for investors looking for the large blocks of afforestable land which have become both more scarce and more costly in the rest of Britain. Hence there was a direct clash between the interests of new private landowners, purchasing land for forestry, and conservationists; a clash heightened by the development of new machinery enabling trees to be planted in areas formerly too wet and marshy to be deep-ploughed for tree-planting. The newly planted trees then soak up water, reducing its availability for wetland plants and birds.

An additional concern associated with this expansion of forestry in environmentally sensitive areas has been the role of the Forestry Commission in giving grants to individual owners, not only in the Flow Country but also in 'sensitive' areas in England, e.g. on Ashtead Fell in the Lake District where planting was opposed by the local National Park Authority. Grants and favourable taxation legislation undoubtedly encouraged the spread of private forestry (Figure 7.11), though the Budget of 1988 largely removed the tax relief on forestry. In place of this relief, specific grants were introduced for both conifers and broadleaved trees, in an attempt to attain a rate of commercial planting of 33,000 ha (81,550 acres) per annum. However, the role of the Forestry Commission in affecting the location of this afforestation remains crucial as it is one of the arbitrators regarding the environmental impact of afforestation (Stewart, 1985). For the Flow Country it has been argued that the available grants and the former taxation policy not only pushed forestry into 'sensitive' areas but also these areas were not capable of giving good yields of timber.

Although parts of the Flow Country are protected under four international treaties, there is relatively little specific designation for conservation purposes. Thus it is not classified as one of Scotland's National Scenic Areas, and designation of a large area around Forsinard as a Site of Special Scientific Interest (SSSI) has not been finalised, partly because of the limited availability of state funds to compensate landowners in the area for not planting conifers.

In January 1989 the conflict between forestry interests and conservationists was extended by a report from a group established by the Highland Regional Council to arbitrate between forestry and conservation interests. The group's suggestion that a further 39,000 ha (96,370 acres) in the Flow Country should be planted over a twenty year period was supported by the Highlands and Islands Development Board, the Regional Council, the Nature Conservancy Council, the Countryside Commission for Scotland and the Forestry Commission. In contrast, the Royal Society for the Preservation of Birds and some other conservation groups protested strongly against the likely threats to wildlife and visual destruction of the landscape.

The report divides the Flow Country into four categories of land in terms of their suitability for forestry, and gives an overall target of 100,000 ha (247,000 acres) of forest as desirable to sustain an annual production commensurate with a viable forestry industry. This would then require the additional afforestation, supplementing an existing 61,000 ha (150,730 acres) already planted, to be focused on two of the designated categories of land: 'possible for forestry' and 'preferable for forestry'. This would involve planting an additional one-sixth of the area of Caithness and Sutherland, but would preserve an area of western Caithness and central Sutherland as a conservation area. Nevertheless, the proposed area for afforestation includes land regarded as having potential for designation as an SSSI, and excludes some further land included in the plantable 'reservoir' of land recommended by the Forestry Commission.

Underlying this continued expansion of planting recommended by the Forestry Commission is the prospect of privatising the Commission. Despite some sales of land in the 1980s, the Commission still owns 230,000 ha (568,330 acres) in Scotland, employs 8,000 people and received £73 million in state subsidy in 1987. Yet a recent National Audit Office report revealed that the Commission was not attaining its minimum Treasury target of a 3 per cent return on its assets, partly through 'uneconomic' planting in low yielding areas. In addition, much of its employment was clerical and concentrated in Edinburgh rather than in rural areas. There is a possibility that developments akin to those followed in British Columbia will be adopted. There the Provincial Government sold shares in the British Columbia Natural Resources

Corporation in a privatisation exercise which was well subscribed. It would seem that the future development of large-scale forestry enterprises in Britain will be tied more closely than before to the resolution of the conflict between economic motives in the forest industry and the conservation lobby. This might concentrate forestry in the high yield areas thereby giving greater protection to environmentally sensitive districts such as the Flow Country, provided that government subsidy doesn't tilt the balance in favour of 'economic' interests. Future forestry policy seems likely to be a continuing balancing act between the multiple objectives set out in the Countryside Commission's (1987c) recent review: to produce a national supply of timber as a raw material and as a source of energy; to offer an alternative to agricultural use of land; to contribute to rural employment either in timber industries or through associated recreation developments; to create attractive sites for public enjoyment; and to enhance the natural beauty of the countryside and create wildlife habitats.

8

Rural industrialisation

8.1 Advantages and disadvantages of rural locations for industry

Ever since the development of factory organisation and the coming of the railways, rural areas have lost their traditional craft industries as well as suffering from a decline in their local services. In many countries this long-term demise has been met in recent decades by attempts to reverse the trend through government-encouraged introduction of new manufacturing and tertiary activities. In Britain the advantages of such policies were pointed out in the 1942 report of the Scott Committee (see Table 8.1), but awareness of problems associated with rural locations for industry has also been apparent. For example, disadvantages include the loss of productive farmland, dislocation of agricultural activities, the introduction of noxious fumes and effluents, disfiguration of landscape, removal of labour from traditional rural work, and the possibilities of 'undesirable' social changes associated with a new industrial workforce.

Despite these problems, many countries have sought to encourage industry to locate in rural areas, especially in regions of long and continuing depopulation (Bryant, 1980). Two examples in the United Kingdom are the policies of the Highlands and Islands Development Board in Scotland and the Development Board for Rural Wales. In the United States there have been substantial efforts to channel industry into 'depressed' rural areas on the assumption that this will increase economic opportunity and improve the quality of life (Hausler, 1974). Three key pieces of legislation in the 1960s followed this reasoning: the 1964 Economic Opportunity Act, the 1965 Public Works and Economic Development Act, and the 1965 Appalachian Regional Development Act. These were followed by the 1972 Rural Development Act. Local communities have often used similar arguments about economic advancement and improved living standards to promote industrial development (e.g. Garrison, 1970). The overall result was a growth in the non-metropolitan industrial workforce in the United States of 2.3 million between 1954 and 1978. About two-thirds of this increase came in the late 1960s (Lonsdale, 1985: 164–5).

Table 8.1 Advantages of industrialisation according to the Scott Committee (1942)

1. Country life revived – influx of personnel
2. Source of employment for wives and daughters of rural workers
3. Provide alternative employment for rural school leavers, i.e. halt depopulation
4. Industries could provide seasonal jobs for farm workers
5. Factories associated with improved physical amenities and social standards – electricity, gas, piped water; education, recreation, higher rates

Out-migration from the cities has been most common for firms seeking low cost, unskilled labour, and so labour-intensive industries, such as clothing, electronics, fabricated metals and textiles, have been involved. Such movement has also featured as a strategy adopted by large corporations to maintain profit margins whilst faced with declining productivity and overseas competition (Peet, 1983). Frequently it has taken the form of the location of branch plants in rural areas, concerned with standardised, routine, production processes (e.g. Malecki, 1986). In regional terms in the United States this set of changes has favoured the 'Sun Belt' states of the South, South-West and West (Beyers, 1979). However, in the mid-1980s the two states with the highest proportion of their workforce in manufacturing were North and South Carolina where low wage furniture factories, textile mills and high technology branch plants were important. Also, the fall in oil prices in the mid-1980s has reduced some of the Sun Belt's comparative advantage and may assist restructuring in older established industrial areas.

Two particular situations in which the establishment of industrial processes in rural areas has been common have been industries associated with agriculture, especially the processing of agricultural produce, and where the location of mineral deposits or energy supplies dictates the location of industry. However, even for industry in any of these situations, or where government encourages industry, requirements for success are suitable available land plus planning permission, appropriate utilities, services, transport and housing for workers. Generally, the type of manufacturing industry established has been small- to medium-scale, with a small workforce and few bulky inputs unless these are produced locally.

Advantages of rural location may include lower land prices, and hence more available space at a lower cost than in major cities, a lack of congestion permitting quicker transport of goods, perceived higher quality of life in rural areas, and lower wage rates (Keeble and Gould, 1985). In a summary of these advantages, Summers et al. (1974: 2–11) referred to four reasons why industry may prefer rural locations:

(a) the lower cost of water and land;
(b) tax exemptions and other inducements offered by local communities;

(c) the growing difficulty of attracting workers to plants in the centre of metropolitan areas;

(d) the belief that there is more of a 'work ethic' amongst the residents of small towns and rural areas.

They suggested that other contributory factors may be the growth of a labour pool in rural areas caused by the shedding of labour from agriculture and craft industry; the improvement in communications enabling greater ease of shipment of goods to and from new rural sites; the positive attractions for workers to live and work in a rural environment; and the development of technology permitting and encouraging the dispersal of both industrial production and the tertiary sector, e.g. advanced telecommunications, computer links.

The four reasons referred to above can be compared with those given by Lonsdale (1985), also referring to the United States:

(a) Substantial geographic differences in hourly wage rates. Rates have continued to be lower in rural areas, and greatest for semi-skilled jobs.

(b) Regional distinctions in labour unionism. Weaker union organisations in some of the more rural states in the South and the Plains may be an attraction to decentralising industries.

(c) The improved image of Southern labour. The long held view of Southern labour, both urban and rural, as unsuitable for factory work has been broken down.

(d) Advances in heating and air conditioning treatment. These have encouraged companies to consider a location in the largely non-metropolitan South.

(e) Excellent highway and trucking facilities. Major improvements in the 1960s have given greater geographic flexibility in plant location.

(f) Widespread automobile ownership. This has greatly extended labour-sheds and permitted deconcentration of industrial location.

(g) Aggressive recruiting of new industry. Many of the less industrialised states have practised major recruiting campaigns, offering low cost labour-training programmes, tax concessions and cheap rates.

(h) Limited legal barriers to establishing new plants or closing old ones.

(i) The deteriorating business and social environment in the larger cities. This 'push' factor may have been the most important and reflects the numerous negative aspects of urban life.

By the late 1970s the effect of these various push and pull factors was to bring the share of America's factory employment in non-metropolitan areas to almost a par with its share of overall population. Subsequently, general business recession had a stronger impact on non-metropolitan areas, partly reflecting the greater presence there of industries susceptible to cyclical fluctuations in the economy, e.g. electrical equipment, transportation equipment, farm equipment and wood products. Another factor was the greater prevalence of branch plants in rural areas, these being more prone to closure or reductions in labour force than their 'host' factories (Ericson, 1976; Lonsdale, 1981).

A fundamental 'fact' of the ruralisation of industry is that whilst capital has relocated and restructured, labour has remained relatively static. New directions of capital flow have produced *locational footlooseness* and technical divisions of labour resulting in *branch plant developments*, often in the rural periphery. In many cases this periphery has proved attractive because it has a captive labour force provided from the labour shedding nature of other activities in rural areas (Cloke, 1985). For example, De Smidt (1987) attributed the decentralisation of manufacturing in the Netherlands in the 1960s to differential growth among the divisions of major corporations, generating establishment of branch plants and subsidiaries in new locations. Subsequently, economic downturn reduced this process, prompting mergers and acquisitions which acted in favour of fewer and larger plants.

Standardisation of production processes has permitted the extensions of operations from large manufacturing centres to smaller settlements as a form of cost savings for parent firms. Such branch plant development has been influenced by the role of transport systems, available sites, labour supplies and labour competition, though, in rural Kentucky, Cromley and Leinbach (1981) found that community infrastructure rather than labour supply alone was more important in accounting for employment levels. In this particular area, external factors tended to account for short-term changes in employment.

8.2 Restructuring, decentralisation and the location of industry in rural areas

Keeble's (1980; Keeble *et al.*, 1983) examination of the decline in manufacturing industry in the United Kingdom in the early 1970s revealed that whilst traditional areas of manufacturing activity had generally lost over 10 per cent of their manufacturing employment from 1971 to 1976, growth was recorded in traditionally non-industrialised areas located peripherally to the central urban-industrial axis in England and to the Central Lowlands of Scotland (see Figure 8.1). The two leading 'growth counties', both in rate and volume terms, were the Highlands of Scotland and rural Northumberland.

Keeble's attempts to explain this in terms of the operation of a series of factors in '*the restructuring hypothesis*' were not supported. This hypothesis included factors such as the impact of government regional policy incentives, residential preferences (generally favouring non-metropolitan locations in southern England), and availability of female labour. However, other work, also in the United Kingdom, has lent support to those aspects of restructuring which have stressed the importance of government agencies, availability of space in rural areas and the role of the female labour force (Dicken, 1982; Fothergill and Gudgin,

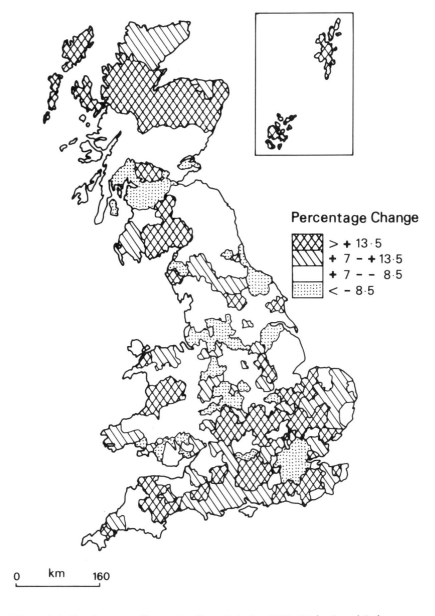

Figure 8.1 Employment Change in Great Britain, 1971–81 by Local Labour Market Area (based on Champion *et al.*, 1987: 64)

1979; Hodge and Monk, 1987).

The increasing capital intensity of manufacturing production has increased demand for floorspace, which is often at a premium in urban locations (Fothergill *et al.*, 1985). This has not only been an important factor in the general out-movement of firms from urban to rural areas but also in specific 'targeted' movements encouraged by government activity. For example, in both Scotland and Wales, agencies promoting industrial development in rural areas have created significant direct and indirect employment (Howes and Law, 1972; Hodge and Whitby, 1979).

A recent survey for mid-Wales, where various development agencies have operated since 1957 (Thomas, 1966; Development Board for Rural Wales, 1984), showed that the creation of jobs in industry had been maintained in the 1970s and early 1980s in a ratio of 3:1 direct:indirect employment in new factories surveyed in five towns in Powys and Dyfed (Thomas and Drudy, 1987). Over half the direct employment was female, suggesting the factories have helped to increase female participation in the labour force and have provided alternatives to the other main sectors of female employment: tourism and retailing. However, the character of the employment created was limited, with a small proportion of skilled workers and little training offered. This deficiency is a crucial weakness as it is the continued lack of professional, intermediate or skilled occupations that is fostering out-migration of young people (Thomas, 1986). The creation of a science park (at Aberystwyth) and a technology and research park (at Newtown) may help redress the balance.

References above to the importance of a pool of female labour which can be tapped in rural areas reflect the fact that one characteristic of employment in rural areas has been lower female activity rates than in towns and suburbs. The presence of a higher proportion of elderly women in the rural population or in 'less active' age bands within the working age group is an important determinant. However, there also seems to be a strong likelihood that low rates reflect lack of opportunity for women both to secure and reach suitable employment (e.g. Moseley and Darby, 1978). Bokemeier and Tickamyer's (1985) investigation of the participation in the labour force by non-metropolitan women found that education rather than age or family status was the chief determinant of female work experience. Hence, despite rather limited employment opportunities within their study areas in Kentucky, increased education tended to lead to a better work situation.

Cooke and Pires (1985) have argued that non-metropolitan industrialisation cannot be adequately explained in terms of the 'standard' emphasis placed upon environmental attractiveness of rural areas and the shortage of space for *in situ* expansion in existing urban locations. They examined '*productive decentralisation*', or the process in which modern industry has located or relocated in areas distant from large urban concentrations (e.g. Bagnasco, 1981), and found that some recipient locations were not especially environmentally attractive. Using three European regions, South Wales (United Kingdom), Emilia-Romagna (Italy) and Aveiro (Portugal), they showed that decentralisation has often been to sizeable urban as well as rural locations. In each case, metal-based and engineering industries had been prime decentralisers, changing their location so that economies of scale could be achieved in small productive units. This and the search for lower production costs, social characteristics of the workforce and product-market factors were the prime influences upon relocation. Only in the Portuguese example was rural industrialisation a consequence of productive decentralisation. In this case the new industry was either centred upon the processing of primary materials, making use of a pool of female labour, or had developed through the increased demand from the internal market for cheap, metal products. Local capital and small businesses were dominant in this decentralisation, aided by low production costs and market demand protected by tariff barriers. Therefore removal of these tariffs with entry to the EEC could have removed a major element in the decentralisation process in Portugal.

There have been high rates of enterprise formation and growth of employment in rural areas of central Portugal in the 1980s. This has been true even in more inaccessible areas with a poorer industrial base. Lewis and Williams (1987) regard this as an example of productive decentralisation associated largely with decisions by small- and medium-sized firms. However, their survey found differences between the Portuguese case and that in northern Italy where different legislation has stimulated a thriving small business sector but limited decentralisation. In contrast, the recession of the early 1980s led to absolute and relative declines in manufacturing industry in both Canada and the United States. The decline was greatest in the US North-West and Mid-West and in the Atlantic Provinces, Quebec and British Columbia in Canada. However, the type of industrial specialisation in the non-metropolitan areas remains unchanged: textiles, clothing, furniture, paper and wood products, non-electrical machinery and food products. These have tended to be industries with cyclical instability, slow growth and/or low wages (Kale and Lonsdale, 1987). Their lack of growth in the mid-1980s may be contributing to a decline or even reversal in the 'population turnaround'.

The type of industry traditionally located in rural areas has been highly specialised, with heavy reliance upon the processing of primary produce and minerals. A retention of this specialisation has tended to distinguish those rural areas where population growth and a more broadly based economy, including service industry, have not developed from those where it has. There is evidence for this in Aitchison's (1984) work using coefficients of specialisation and diversification to illustrate the spatial differentiation of employment in rural France. His indices, subjected to a factor analysis, revealed various types of rural employment structure. Diversification was associated with the industrial départements of eastern and north-eastern France whilst specialisation characterised those dominated by primary activity, e.g. Gers, Mayenne and Cantel. A contrast was revealed between east (diversified) and west (specialised) in terms of breadth of job opportunities.

Specialisation has often reached its peak in conjunction with the *extractive industries*, the exploitation of mineral resources, both metallic and non-metallic, continuing to be an important economic activity in some rural areas. The presence of commercially viable deposits of minerals has given rise to settlement in near-deserted and inhospitable parts of the Developed World, e.g. the Pilbara iron ore field in Western Australia, as well as promoting settlement in areas which otherwise would have been dominated solely by agriculture or forestry, e.g. the potash and oil deposits of the Canadian Prairies. In some cases, notably coal mining, the extraction of the mineral resource has been followed by the large-scale growth of industrial activity attracted to the proximity of the mines by factors such as the high cost of transporting the mineral. In such cases, the extraction process has become part of a non-rural environment, with the tentacles of urbanism capturing this activity by means of large-scale urban settlements as well as the development of a distinctive society. The 'closeness' of many mining communities has led to frequent references to 'mining villages', but the question of whether these can be treated as definitively 'rural' is one seldom considered, even if such communities often approach some of the ideals of gemeinschaft. However, small, traditional and remote mining settlements, for example those in the Appalachian coalfields, certainly retain what many would regard as rural social characteristics and present vastly different social and economic problems to the 'town-in-the-wilderness' style of development typified by Mt Isa, Queensland, or Kalgoorlie, Western Australia, where both community development and life-styles represent off-shoots of distant metropoli.

Many of the different types of economic and social development associated with the extraction of mineral resources can be related closely to the nature of the resource. The classification in Table 8.2 makes the distinction between metallic and non-metallic, which represents a rough divide between those capable of attracting urban-industrial incursions into the countryside (metallic) and those tending to be more ubiquitous and generating less 'concentrated' forms of development (non-metallic).

Much of the recent academic and public concern with mineral extraction in rural areas has been related to environmental impacts of mining operations, with little attention given to economic and social implications (e.g. Dean, 1985). The same has tended to be true for water resources, with a focus upon environmental concerns and competition for water (e.g. Park, 1982). Power supply based upon water power is only one of the many demands made upon water resources within rural areas. For example, Porter (1978) also identifies water supply for domestic, industrial and agricultural purposes, flood damage reduction, effluent disposal, navigation, amenity and recreation. Several of these uses are conflicting, although some are complementary (Figure 8.2). The multifunctional nature of water resources causes special management problems

Table 8.2 Classification of mineral resources

(a) *Metallic mineral resources*
1. Abundant metals; such as iron, aluminium, manganese, titanium, and magnesium
2. Scarce metals; such as copper, lead, zinc, tin, tungsten, gold, silver, uranium and mercury

(b) *Non-metallic mineral resources*
1. Minerals for chemical, fertiliser and special uses; such as sodium chloride, phosphates, nitrates
2. Building materials; such as cement, sand, gravel, gypsum, asbestos
3. Fossil fuels; such as coal, petroleum, natural gas and oil shale
4. Water; lakes rivers and ground waters

Source: Skinner (1976: 10)

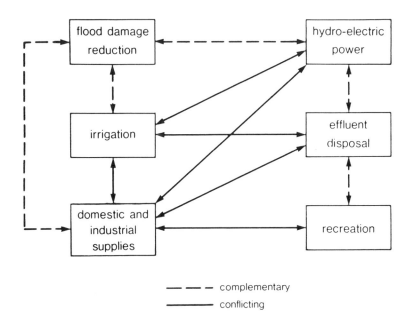

Figure 8.2 Complementary and conflicting uses of water resources (*Source:* Cloke and Park, 1985: 73)

whilst some uses have also impinged upon other rural activities, promoting conflict, e.g. creation of new reservoirs in upland areas.

In New Zealand the feasibility of using the Clutha and Kawarau Rivers as a major source of electrical power has been considered since the early 1940s. Various development schemes have been proposed to tap an energy source equivalent to nearly 2 million tonnes of coal. Under the

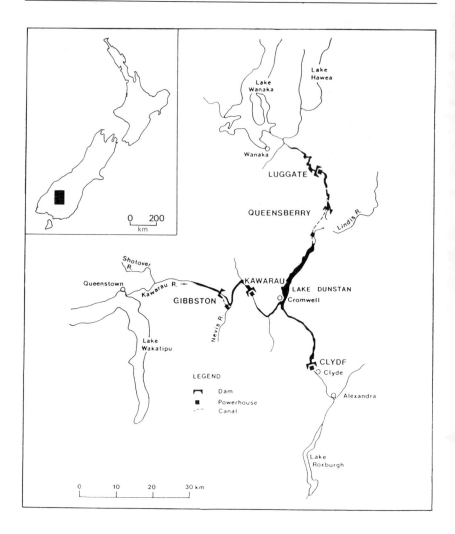

Figure 8.3 Dams and power installations in the Clutha Valley Development,
South Island, New Zealand (based on New Zealand Ministry of Works maps)

Clutha Valley Development Plan of the 1980s, over NZ$1,200 million
will be spent on damming the Clutha at Clyde, Luggate and Queensberry
to create 900MW generating capacity (Figure 8.3). The dams will create
Lake Dunstan, covering 10.2 sq miles (26 sq km), and flooding over
2,300 ha (930 acres) of productive land, some of it very high quality and
supporting production of apricots and other fruit. Part of the town of
Cromwell will be flooded, but the Plan includes relocation of residences
and businesses affected. In addition to the hydro-electric power
generated, new recreational opportunities will be produced for yachting,
rowing, power boating, swimming, fishing and water-skiing. The other

gain will be in terms of the increased potential for irrigation. This part of central Otago has relied upon irrigation for over 100 years in order to maintain fruit and vegetable production, and feasibility studies indicate that several existing schemes in the Upper Clutha and Fraser rivers can be upgraded.

Some general points are made by Radetski (1982) about the character of new extractive ventures. He characterised these as often producing very limited employment opportunities, especially for local inhabitants, and also restricted business linkages which therefore restricted any local multiplier effect. In many cases, government policy towards mineral extraction actually prevents additional industrial development in the vicinity of any mining or other extractive industry on environmental grounds (Bradbeer, 1987). Such concerns can be contrasted with attitudes to establishing processing plants for agricultural products in rural areas, though the New Zealand example considered below may be deemed a special case because of the importance of food processing within the country's economy.

8.3 Food processing: the New Zealand example

Food processing may be considered as an intermediary stage between the farmer and the consumer. In reality it fulfils two roles. It may involve either processing primary produce for forwarding to the consumer or it may send processed output to a manufacturing stage (Scott, 1970). For example, wheat may be processed to produce flour which may then be sold to consumers, or it may be manufactured into bread for sale to consumers. Both the processing and manufacturing stages may generate wastes and by-products which can be returned to farms, for example whey from cheese factories, or which will provide consumer goods.

In New Zealand food processing has played a major role in sustaining the nation's economy. Around 80 per cent of farm output, mainly pastoral products, is exported, mostly after processing in which produce is converted into less perishable and more easily consumable and acceptable commodities. The main items processed are milk, meat, wool, and fruit and vegetables. Food processing is the largest manufacturing industry in New Zealand in terms of value of output, accounting for NZ$7 billion of sales in 1983–4. Until the 1980s it had also been the largest employer in the manufacturing industries, with a labour force of over 75,000 in 1984. It also accounts for 5.6 per cent of national employment and one-quarter of all employment in manufacturing. This position of pre-eminence has been retained for many decades, despite the development of a more diverse manufacturing base and major changes within various sectors of the food processing industry, the three largest of which are the manufacture of dairy products, meat, and fruit and vegetable processing. During the past decade the opportunities for processing primary produce have become more varied and complex, as

Plate 15 The trend towards large-scale processing plants with integrated operations is exemplified by this dairy processing complex. Like many of the other large plants in New Zealand it is located in the heart of a major dairying area, in this case the Manawatu (near Palmerston North)

illustrated by the nature of changes in long-established freezing works where packaged meat, leather goods and animal-based chemicals have been added to the range of products. These changes have included locational shifts, with both greater concentrations in the major urban areas and also new greenfield developments of large-scale processing plants (Robinson, 1988c).

New Zealand enjoys an advantage over its competitors with regard to on-farm production costs of *dairy produce*. Farm collection and manufacturing costs are also lower in New Zealand because of lower wages and scale economies in the dairy industry. A slight advantage in export costs is maintained through centralisation of marketing and the handling of large volumes of exports. Exports of dairy produce account for around 20 per cent of total exports by value.

Dairy processors have been faced with diminished profit margins for their products for which they have compensated by increasing output and rationalising their operations (e.g. Ward, 1975). Rationalisation has involved the closure of branch plants, the concentration of production, larger units for benefits of scale economies (Plate 15), and the maximum utilisation of milk solids through complex inter-factory transfers of milk, cream, skim-milk, buttermilk and whey (Clark, 1979: 53). The new large processing plants have generally been located on greenfield sites at locations in rural areas with good transport links for collection of milk supplies and shipment of processed dairy produce. Many butter factories were established pre-1939 when farm-separated cream was transferred direct from farm to factory. Today, milk is collected from the farm by tanker, cream is separated at the processing factory and then transferred to a butter factory. Smaller supplies of cream are also supplied by cheese factories, contributing to a highly complex spatial pattern of transfers (Clark, 1979: 57–61). The change to collection of milk rather than cream brought a decline in the number of suppliers, with small producers unable to meet costs of upgrading their milking sheds or meet the requirements imposed by the collection of milk in bulk by tanker which commenced in the 1950s. Contraction has also been encouraged by a rising demand for a range of dairy products obtained from milk, so that large integrated processing units are able to switch production from one product to another according to market requirements whilst also benefiting from scale economies. Rationalisation has often been extensive, for example in Taranaki the number of dairy factories was reduced from 44 in 1969 to six in 1983 (Willis, 1984: 8) as large complexes such as the Kiwi Co-operative at Hawera were established.

The changes in New Zealand's dairy processing industry are illustrated by the Hawera plant, 145 miles (c. 230 km) north-west of Wellington. Established in 1973 as the country's largest single dairy manufacturing unit, in 1985 it received 3.6 million litres of milk by tanker from 1,500 suppliers, representing over 10,000 tonnes of product. This product can be varied according to season and demand, reflecting the product diversification that has characterised dairying in the 1970s and 1980s. The Hawera plant represents an expansion of production in rural areas and the full utilisation of economies of scale. However, there has also been a development of smaller processing plants in close proximity to consumers in urban areas, producing milk, cheese and ice cream.

The majority of livestock reared in New Zealand are destined for *freezing works* which are located throughout the country, performing the operations of killing, dressing and freezing livestock carcasses. There is a range of different types of organisation owning and running the works, but all have sought to diversify output, thereby stimulating upgrading of existing plants, some closures and also opening of new plants. As with dairy production, several of the larger new plants are on greenfield sites outside major centres, using modern technology and a small labour force to process meat obtained from a large catchment area (Sulzberger, 1980). In effect, they have copied the strategy used by American meat companies in the United States in the 1960s.

Good examples of this have occurred in the rural hinterland of Wellington. Whilst plants in Wellington itself have closed, new ones have opened, e.g. at Eltham, Kaiti, Oringi, Takapau and Whakatu (Figure 8.4). Smaller establishments in the meat industry, e.g. non-exporting abbatoirs, producers of ham, bacon and small goods, and game packers have developed a greater concentration on urban locations. Similarly, for

Figure 8.4 Major meat processors in New Zealand, 1986 (*Source:* Robinson, 1988c)

the *fruit and vegetable processors*, the tendency amongst the smaller firms has been to focus their activities close to the major markets provided high transport costs of bulky produce from farms to factory have not proved limiting. However, this has not been the case for wine production, for which location close to vineyards has been the norm, e.g. the Montana wineries in Marlborough.

This New Zealand example illustrates the attractions of rural sites for new large-scale developments, especially for plants able to draw upon produce from a wide surrounding primary production area. However, the pull of the market and port facilities remain great and can draw even 'natural' rural industries such as food processing into urban areas unless there are significant gains from rural location, such as those provided by government inducements.

8.4   Policies promoting rural industrialisation

There are three principal reasons for encouraging new industry in rural areas: to absorb labour shed by agriculture; to stem out-migration by providing more local employment; and to help maintain market or service centre functions. Three types of strategy have been used to promote the establishment of new industry in rural areas: improvement to locations as sites for industrial development; reduction of relative costs to industrial activities in rural locations; and augmentation and speeding up of information flows.

Examples of these types of basic policy aims can be seen in Eire, where government policy has been a key factor in influencing the growth of industry in rural areas in the west of the country (Breathnach, 1985). Policy measures, allied with the restructuring practices of multi-national investment (O'Farrell, 1980; Perrons, 1981) and the changing perception of rural districts and small towns, have greatly accelerated employment in the west (Table 8.3). Some of the most influential policies have been those of the Industrial Development Authority (IDA), established in 1949 and given statutory responsibility for the promotion of manufacturing. The IDA's differential operation of grants in favour of the west, grant allocations to rural areas plus a site acquisition and advance factory building programme, have provided significant stimuli to manufacturing, e.g. between 1972 and 1981 approximately half of the 600,00 sq m of advance factory floorspace in Eire was sited at 75 locations in the eleven western counties (Brunt, 1988: 110) (Figure 8.5). The IDA's Small Industry Programme, established in 1967, has also contributed to the dispersal of indigenous industry to rural areas (O'Farrell, 1984), the small businesses tapping the previously under-utilised pool of female labour in the rural areas plus males displaced from the agricultural workforce and part-time farmers (e.g. Lucey and Kaldor, 1969).

In Eire in the 1970s, attempts to reduce long-continued rural out-migration were articulated by way of a policy for industrialisation of

Table 8.3 Industrial employment in the west of Ireland, 1926–81 (%)

	1926	1961	1971	1981
Employment in manufacturing	4.8	7.9	12.9	19.2
Share of national manufacturing	19	13	16	21
Growth in Irish manufacturing		62.8	19.6	9.1
Growth in manufacturing in the west		8.9	44.5	44.9
Geowth in the rest of Ireland		75.6	15.8	2.5

Source: after Breathnach (1985)

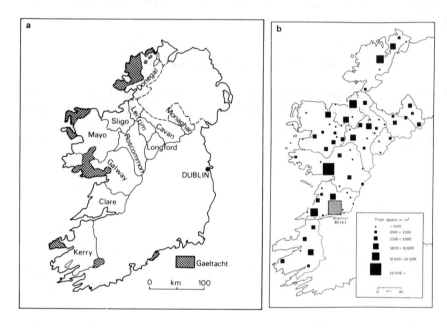

Figure 8.5 (a) The Irish Gaeltacht; (b) Advance factories in western Eire (based on Breathnach, 1985)

small towns, largely between 3,000 and 16,000 inhabitants, which would act as growth points (Industrial Development Authority, 1972). This strategy was intended to create alternative employment for those leaving the agricultural labour force. It is questionable whether such a dispersal policy of rural industrialisation is capable of limiting depopulation beyond a very limited area. Yet some success can be claimed for the policy in its creation of new manufacturing employment in rural areas, even if spread effects of the policy have been limited. Another critical weakness of the Eire policy can be traced to the type of jobs created. If the new employment is anything other than in the form of manual labour, it would be unsuitable for most surplus rural labour without considerable investment in education and training programmes. However, one effect of rural industrialisation policies in Galway has been for members of the farm population to adopt urban-based employment. Of 344 farms surveyed in Cawley's (1979) study, 68 (20 per cent) were classed as part-time, the head of household having another full-time or part-time urban-based job.

One of the major defects of schemes to promote industrial decentralisation has been the retention of financing and control within traditional industrial ares. Some employment growth in peripheral areas may be generated through establishment of simple production units, usually branch plants, but major benefits often fail to accrue to the periphery (e.g. Strachan, 1988). Belil and Clos (1985) argue that this situation is only overcome effectively if good transport links are maintained between industrial centres and the periphery, citing the El Vendrell-Valls-Montblanc axis south-west of Barcelona.

Knox (1988: 126) highlights one significant factor in the ruralisation of industry in the United States as being differential rates of local taxation. These have favoured suburban and non-metropolitan locations and have been reinforced by the availability of relatively low cost labour, inexpensive supplies of easily developed land and low levels of unionisation (Haren and Holling, 1979; Kale and Lonsdale, 1979). The great advances in transport and telecommunications systems have also reduced the need for city centre location and have enabled manufacturing, wholesaling and tertiary operations to take advantage of opportunities presented in non-metropolitan areas (Estall, 1983).

Commenting on federal programmes designed to increase the flow of industrial capital into rural areas, e.g. under the 1965 Economic Development Act and the Rural Development Act of 1972, Tweeten and Brinkman (1976) concluded that in many rural communities the incentives used to attract firms exceeded the benefits. The types of firms attracted mainly employed women, but grew only slowly and paid low wages.

In Britain a recent report from the Countryside Commission's Policy Review Panel (1987) has stressed the need for the creation of a more diversified rural economy. This sentiment is also echoed in the recent report from the Department of the Environment (DoE) (1987), *Rural enterprise and development*. These documents urge not only diversification within agriculture and the need for conservation of valued countryside but also more emphasis upon a variety of non-primary activities. To date, attempts to generate this variety and to encourage rural industrialisation have been led by the Development Commission and its agency, The Council for Small Industries in Rural Areas (CoSIRA). These organisations were recently merged and their regional organisation strengthened.

The Development Commission, established in 1909, has operated to encourage new industry in rural areas in England through the building of *advance factories*, constructed without specific occupiers in mind and

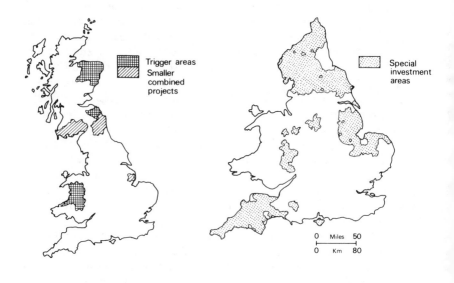

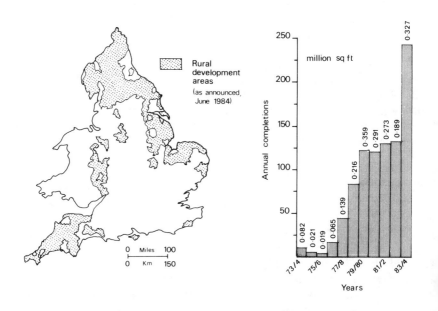

Figure 8.6 (a) Trigger Areas and Smaller Combined Areas; (b) Special Investment Areas (SIAs); (c) Rural Development Areas (RDAs); (d) Advance factory completions, total numbers and million of square feet, 1973/4–1983/4 (*Data source:* Chisholm, 1985)

then let out to suitable tenants. Similar policies have also been followed in Scotland and Wales. From the mid-1960s the focus was upon Trigger Areas, areas experiencing high rates of depopulation but still thought to have good potential for recovery and future growth (Figure 8.6) (Tricker and Martin, 1985). Other measures employed by the Development Commission include collaboration with local authorities in provision of factory premises, and contributing part of the cost of converting redundant buildings. Part of this programme is now restricted to specially designed Rural Development Areas (RDAs) (formerly Special Investment Areas) (Figure 8.6) (Chisholm, 1985; Tricker and Martin, 1984).

The Development Commission's target was to create about 1,500 jobs in Britain between 1975 and 1985, which they succeeded in meeting, whilst also helping to retain population in small communities and therefore contributing to population retention or even growth in rural areas. The success in meeting the target has also been related to the Commission's sponsorship of training schemes, introducing 'key' workers to rural areas through provision of housing guarantees, and compilation of local skills registers. One other set of measures operated under the auspices of regional policy has been the attempt to reduce the relative costs of locating in rural areas. These measures have included grants for new buildings, removal costs, rent free periods in factories and tax concessions.

CoSIRA has focused upon support for small businesses, including retail and other concerns in addition to manufacturing enterprises. Its activities have been confined largely to settlements with less than 10,000 population, dealing with business units of not more than twenty skilled workers. Its activities include the giving of advice, loans and training. Chisholm (1985: 289), quoting data for 1983-4, refers to advice given to 15,000 small firms employing 72,700 people and 309 loans in the year at a value of £4.6 million.

For the future, *Action for rural enterprise* refers to the need of more provision of new premises, aided by more liberal planning controls, and recommends a focus upon particular industries. These include high technology manufacturing, food processing, timber products, service industries, retailing (including pick-your-own schemes), transport, fish farming, 'alternative' crops and livestock, organic farming and tourism and leisure. Tourism has been given special priority in the RDAs and, increasingly, is being seen as a provider of jobs and income in the countryside.

8.5 Tourism and recreation in rural areas

8.5.1 Types of rural recreation

Two major changes in society post-1945 have contributed to a dramatic growth in a new economic activity. The changes are the rise in the ownership of private cars and the reduction in the number of hours per week spent in paid employment. And the new economic activity has been one catering for the increase in people's discretionary time (Dower, 1965). This time has been spent in a myriad of different ways, but it is possible to distinguish a clear rural dimension within the range of 'non-work' time. For example, Simmons' (1975) chief definition of *rural recreation* is that it involves activities in a 'natural' setting, so excluding 'nature-divorced' pursuits such as football and cricket. It has incorporated a wide range of recreational activities as well as *tourism* which has been closely associated with extensions of annual paid holidays. Both tourism and recreation have become major factors in local and national economies, e.g. over 1.5 million jobs in Britain depend in one way or another on tourism.

The move from the sixty hour working week, common in the Developed World at the end of the nineteenth century, to a forty hour week and below has increased dramatically the amount of an individual's time that can be regarded as leisure time. For example, Clawson (1963: 5) classed 34 per cent of the time of US citizens as leisure in 1950 (time after work, sleep, housekeeping and personal care), with a prediction that this would rise to 38 per cent by 2000. Of this large amount of leisure time, it has been weekends and special holidays that have been most important for rural recreation, though annual holiday leisure has often been directed to the countryside. Given the increasing leisure time budget, the other factor contributing to the growth in rural recreation has been growing affluence and with it the rise in personal mobility. Affluence has brought more disposable income and, in turn, this has meant an increase in the rate of expenditure on leisure activities (see Coppock and Duffield, 1975).

Various studies have shown that, increasingly, people's leisure time is being used in a space-extensive way: a move from passive recreation to participation. Growth has been fastest in informal pursuits taking the form of day or half-day trips in the countryside. With the rise in the ownership of private cars the urban population has 'discovered' the recreational potential of both the countryside on its doorstep and also more remote and less occupied areas.

This participation in recreation in the countryside has been strongly influenced by social class: the affluent and better educated have the highest rates of participation whilst women have lower rates than men (Birch, 1979). However, there is evidence that a greater proportion of the working classes may have a preference for spending their leisure time in the town rather than in the countryside (Sidaway and Duffield, 1984).

Recreation directed at the countryside evaluates and utilises a series of resources. In terms of evaluation, intrinsic qualities of scenery and landscape are often cited as the chief attraction underlying a long walk, a canoe trip or a drive through the countryside. The aesthetic value of the countryside can therefore by regarded as a resource and a stimulator of the use of leisure time in such a fashion as to derive benefit from that resource. The variation in the perception of rural aesthetics varies from country to country (e.g. Simmons, 1975: 55–67), but a recurrent theme is that time spent in particular rural environments can satisfy a certain desire not met in other contexts. In Britain this may involve long walks over open land whilst in North America it is more likely to be spent completely away from human habitation.

In the early and mid-1960s, at the beginning of the surge of academic interest in the growth of tourism and recreational activities, several categorisations of recreation were formulated (ORRRC, 1962; Clawson and Knetsch, 1966). For example, Clawson and Knetsch divided recreational areas into three types: user-oriented, intermediate and resource-orientated. It has tended to be the latter two that have predominated in the countryside. The prime example of a resource-oriented recreational area is a national park where certain physical resources in the form of distinctive scenery and associations of plants and wildlife, make the area attractive to visitors. Such areas contrast with user-oriented areas, which are usually located in urban and suburban environments and include playgrounds, swimming pools and soccer pitches. These facilities are usually within a short travelling distance of a large range of potential customers whereas the resource-oriented recreational area is more likely to be quite remote from the major centres of population. Intermediate areas tend to be in the immediate hinterland of an urban area and offer diverse recreational opportunities, such as overnight camping, picnicking, fishing, hiking and downhill skiing. A typical example would be a regional or state park located about half a day's drive from home (Nelson and Butler, 1974: 291).

Since this three-fold description was produced there has been an increased blurring of the spatial separation between the three classes. In particular, the growth in the numbers of people visiting resource-oriented sites, such as Yellowstone and Mammoth Cave National Parks in the United States or Banff National Park in Canada, has moved the focus of the intensive use of recreational areas away from the city and its immediate hinterland. Recognising this, Nelson (1970) suggested that a more useful distinction with regard to recreation should be made between facilities-oriented and non-facilities-oriented types of recreation. Included in the latter category are activities such as backpacking, hiking, photography and canoeing. These use relatively little equipment and the effects of technology are relatively limited. Facilities-oriented recreation includes skiing, snowmobiling and activities requiring the provision of services such as hotels, ski-lifts, garages and parking spaces, all of which involve some environmental cost.

More recently, Hockin et al. (1978) sub-divided land-based recreational activities into:

(a) overnight activities, e.g. camping and caravanning;
(b) activities involving shooting, e.g. archery and gun sports;
(c) activities involving a significant element of organised competition,

e.g. golf, skiing, motorcross and orienteering;
(d) activities involving little or organised competition, e.g. angling, informal cycling, horse-riding, rambling, picnicking and wildlife study.

8.5.2   Policies promoting tourism and recreation in the countryside

Until relatively recently outdoor recreation in the countryside was largely unaffected by public control or by public provision. This was because recreational demand remained free to respond to prevailing market forces, and public authorities had rarely acquired and managed land specifically for the purposes of outdoor recreation. So recreation was generally seen as a part of agriculture, forestry and water management, until specific recreational planning and management began to appear from the mid-1960s. Since then both demand for recreational provision in a rural setting and the planning of such provision have increased as recreation has assumed a higher status as a user of land in its own right and also, especially when combined with tourism, as an economic activity bringing income and employment to rural areas.

In terms of promoting recreational activity in the countryside there has tended to be a distinction recognised between facilities-oriented recreation in countryside close to urban centres, e.g. picnic sites, sign-posted walks, natural trails, and recreation in 'deeper' countryside where the presence of wilderness and scenic beauty provide opportunities for different forms of activity, e.g. hill-walking, rock climbing, pony trekking. For the latter type of area much of the planning activity has focused upon designating areas for protection from 'damaging' forms of economic development (see Chapter 10). In contrast, for the rural-urban fringe, planning for provision of facilities has often taken greater precedence.

In the United Kingdom a government White Paper, 'Sport and Recreation', supported the view that more facilities for recreation should be developed in the urban fringe (DoE, 1975), whilst the Countryside Review Committee (1977) reported that green belts should have special priority for meeting leisure demand. However, Harrison's (1981) survey of a number of informal recreation sites in London's Green Belt revealed that such arguments were in need of revision. She concluded that 'the recreational role of sites in the Green Belt is a peculiar blend of urban park and countryside wilderness. Sites acted as urban parks for the residents of suburbia but did not act as substitutes for the deeper countryside or as recreational outlets for the residents of inner London, and many sites were little used' (Harrison, 1981: 114). Hence, the idea of the Green Belt intercepting demand by deflecting urban residents from visiting 'more sensitive' countryside sites was open to question, although the Green belt was providing a distinctive role, especially for car-borne middle-class residents of suburbs.

Ferguson and Munton's (1979) survey revealed that London's Green Belt contained nearly 500 informal recreation sites, covering over 31,000 ha (76,600 acres) or 5.7 per cent of the surface area of the Green Belt. Of these sites over three-quarters covered less than 100 ha (250 acres), but with variation from 1 ha (2.5 acres) to over 500 ha (1,235 acres) (e.g. Epping Forest). Local government authorities were the major managing body (Table 8.4), followed by the National Trust. Common land represented 17 per cent of the sites by number and area. The distribution of sites was most uneven, with Surrey dominating the pattern and Berkshire and Kent having notable deficiencies. No clear policies for eliminating this inequality could be identified.

In Britain it seems that the growth in rural leisure participation has stabilised as people demand more 'sporty' types of pursuit (Blunden and Curry, 1988: 21) (Table 8.5). However, the need for more attention to be given to the planning of rural recreation has been recognised in the Countryside Commission's (1987a) consultation paper, *Enjoying the countryside*, and the Commission's (1987b; 1987c) policy initiatives, *Policies for enjoying the countryside* and *Recreation 2000*. These advocate a greater promotional role for the Commission, special attention to be given to specific groups in society, e.g. the handicapped, improved recreational transport facilities (see Groome and Tarrant, 1985), and an increased access to the countryside for the general public. These views also highlight the economic potential inherent in expanding recreational activities in the countryside, particularly in conjunction with tourism. The general theme behind the reports is that policy must move further in the direction of encouraging recreational activity rather than merely tolerating it. However, this theme is in conflict with some of the needs for conserving the countryside and its realisation is made harder by the large number of agencies involved in producing policy for rural recreation in Britain.

A question that has received more attention in the 1980s has been that of how countryside recreation can be allocated to potential consumers. Increasingly, in the market-oriented economies of the Developed World, pricing of resources has become a commonly used allocative mechanism. This has been the case most especially in North America where charging for countryside recreation has served four roles:

(a) raising of revenue;
(b) allocating demand between recreation and non-recreation expenditure;
(c) allocating demand between alternative recreation facilities;
(d) achieving certain objectives within the provision of countryside recreation:
  (i) differential pricing to divert users from one facility to another;
  (ii) resource protection;
  (iii) variable pricing to smooth uneven patterns of use.

Numerous arguments have been advanced against such pricing policies (e.g. McCallum and Adams, 1980). In North America there is now a substantial literature on the economics of countryside recreation and the

Table 8.4 Number and area (ha) of informal recreation sites in London's Green Belt, by county and managing body

		County Councils	Local Government Authorities District/Borough/London Borough Councils	Corporation the City of London	National Trust	Common land managed by other bodies including commons' conservators	Other managing bodies	All
Bucks	No.	14	4	1	8	11	2	40
	Area	451	57	202	836	263	421	2230
Herts	No.	3	19	0	4	17	8	51
	Area	96	1280	0	678	606	655	3315
Essex	No.	4	15	1	1	7	0	28
	Area	480	508	2023	425	94	0	3530
Kent	No.	0	20	0	8	11	3	42
	Area	0	886	0	113	354	140	1493
Surrey	No.	49	85	0	41	35	13	223
	Area	3657	4063	0	2809	3723	1295	15,547
Berks	No.	0	10	0	7	1	1	19
	Area	0	143	0	168	221	479	1011
Greater London	No.	2	76	7	1	2	0	88
	Area	620	2596	602	36	77	0	3931
All*	No.	72 (14.7)	229 (46.6)	9 (1.8)	70 (14.3)	84 (17.1)	27 (5.5)	491 (100.0)
	Area	5304 (17.1)	9533 (30.7)	2827 (9.1)	5065 (16.3)	5338 (17.2)	2990 (9.6)	31,057 (100.0)

*Figures in brackets are percentages of total

Source: Ferguson and Munton (1979)

Table 8.5 Recreation participation patterns in England and Wales

Activity	Percentage of people		
	1977	1980	1984
Watched sport	5	9	10
Took part in sport	7	9	25
Long walks	19	17	23
Visit sea coast	17	13	20
Visit historic houses	18	10	25
Drives/outings	41	26	41

Source: Blunden and Curry (1988) based on surveys by the Countryside Commission during four-week periods in the summer months

role of charges to users in meeting costs incurred (e.g. Clawson and Knetsch, 1966; Cicchetti and Smith, 1973; Krutilla and Fisher, 1975). However, there is little agreement on the role that charging for recreation should play in provision, and charges on users are still only one of several ways in which outdoor recreation is funded. Thus in the United States other revenue comes from general taxes, specific bond issues, specific taxes and grants-in-aid.

8.5.3 Tourism in rural areas

Whilst rural recreation has grown dramatically post-1945 a more recent related development has been the way in which longer-term recreation, or tourism, has become a significant and even dominant factor in the rural economy. In many parts of the Alps the origins of tourism were the Victorian alpinist holidays, focusing upon mountaineering and hill-walking. Later came summer tourism for middle-class urban families and, more recently, winter sports have attracted large numbers to particular resorts, e.g. Cervina and Courmayur in Italy's Val d'Aosta.

Especially in scenic areas such as national parks, tourism's impact on the local economy has often made it the mainstay (e.g. TRRU, 1981). Measurement of its impact is notoriously difficult, given the part-time and seasonal character of providing services for tourists, but estimates for diverse regions suggest £10,000 worth of tourist expenditure generates 3.5 to 4 direct and 0.5 indirect jobs (e.g. Archer, 1974; Dower, 1980; Smith and Wylde, 1977). These multipliers can vary considerably though, depending upon the nature of tourism, e.g. hotel accommodation versus self-catering, and small-scale, labour intensive facilities offer the best return per unit of tourist expenditure.

The extent of the spread of benefits from the growth of tourism in rural areas has often been questioned (e.g. McLaughlin, 1986a). Tourism tends to be one of the lowest paid service sectors, with especially low pay for females (e.g. Low Pay Unit, 1984). It may also provide a lucrative

Plate 16 Tourism has transformed many Alpine valleys in Austria and Switzerland, attracted by opportunities for skiing in winter and spectacular scenery throughout the year

source of rent income which would be difficult to match from a conventional tenancy. So tourism can contribute to a reduction in the already limited private rented housing stock in rural areas (McLaughlin, 1985). In addition, the notion of conserving the environment in order to encourage tourism may have important consequences for the selective distribution of any resultant benefits (Wenger, 1980).

Kariel and Kariel's (1982) work on the socio-cultural impacts of tourism in the Austrian Alps revealed significant interrelated effects upon economic infrastructure, landscape and way of life. They portrayed the diffusion of the socio-cultural changes in the form of a logistic curve (Figure 8.7), the last stages of which were represented by major changes in the nature of the host communities. Although some of these charges were quite definitely detrimental, e.g. increased marriage breakdown, the overwhelming attitude of long-time residents to tourism was favourable. In particular, income derived from tourism was viewed as more easily obtained and more profitable than that from agriculture. Tourism was also seen as a means of retaining young people and thereby maintaining a viable community (see also Greenwood, 1972; Diem, 1980). However, the need for local control over development was stressed in order to maintain some of the traditional characteristics of settlements and landscapes (Plate 16).

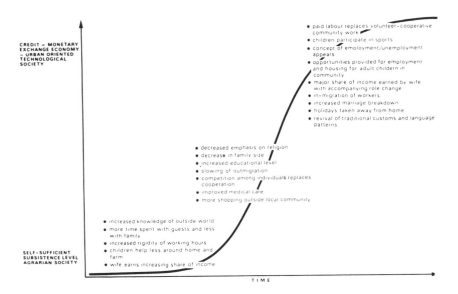

Figure 8.7 Diffusion of socio-cultural changes resulting from tourism (*Source: Kariel and Kariel, 1982*)

Policies implemented by the Highlands and Islands Development Board in Scotland have emphasised tourism as a generator of employment and income (HIDB, 1967: 3). Specific aims stated by the Board have been to attract more tourists, particularly foreign tourists; to extend the holiday seasons; to develop accommodation and catering facilities and improve the quality of these facilities; and to improve and develop leisure facilities generally, and particularly in water sports, winter sports and all-weather centres (HIDB, 1983: 57).

In 1983 the Board spent £645,000 on 37 tourism schemes in the Northern and Western Isles. Principally this money helped to develop the bed and breakfast and self-catering sectors in the Isles. These are the main types of accommodation offered (Table 8.6), and it has been argued that they have limited influence upon employment in rural areas. However, according to the Scottish Leisure Survey of 1981, the first comprehensive survey of holidaymakers in Scotland since 1973, the amount of direct spending by visitors during the summer of 1981 added nearly £4 million to the economy of the Western Isles as a result of 34,000 visitors (Scottish Tourist Board, 1981). The flow has created seasonal employment, though this seasonality is a weakness of many tourism ventures as it so often adds to seasonal unemployment in winter. Nevertheless, in the Western Isles, employment in hotels, bars and catering establishments provides over 500 jobs, or over 5 per cent of all employment.

The ratio of the number of visitors to residents in the Isles is one to one, but for the whole HIDB area the ratio is five to one (Duffield and

Table 8.6 Tourist accommodation in the Orkneys, Shetlands, Western Isles and HIDB region, 1984

Type of accommodation	Ork.	Shet.	W. Is.	HIDB region
Hotels – number	19	20	26	786
bedspaces	550	805	812	27768
Guest houses – number	7	13	12	322
bedspaces	65	108	114	4323
Bed & breakfast – number	51	30	117	1483
bedspaces	268	112	658	9112
Self-catering – number	62	29	45	1548
bedspaces	554	323	367	28948
Caravans for hire – operators	28	4	23	384
bedspaces	264	24	308	14164
Youth hostels – number	10	4	2	57
bedspaces	370	95	68	3079
Caravan/camping – sites	2	0	2	1355
tent pitches	60	0	40	7355
Total bedspaces	2071	1467	2327	87394

Source: HIDB, Supply of Tourist Accommodation in the Highlands and Islands, 1984

Long, 1981: 411), illustrating the significance of tourism in the region's economy. Although there may be questions about the weakness of the multiplier effect associated with tourism, it does provide a useful additional income for numerous rural households in summer, and even year-round in some of the skiing areas. Further measures to improve facilities, to increase the availability of information on tourism, and to diversify the range of accommodation provided indicate that tourist-related employment may increase slightly in the late 1980s and early 1990s in the Highlands and Islands.

Evaluations of tourist attractions in rural Scotland suggest that exploitation of latent potential has proved difficult because easily identified 'tourist regions' lack specific key attractions which in their own right can attract large numbers of visitors. Thus Heeley (1988) cites Loch Lomond as a good example. There are 48 kms (30 miles) of loch side with impressive scenic, wildlife, historic and cultural resources; the West Highland Way runs alongside it; the area has been designated as a Regional park; three million people live within 90 minutes driving time; over two million people visit the area per annum. However, visitors do not stop in the area for any length of time and therefore economic spin-offs are small. The lack of 'honey-pot' sites here and in many other Scottish rural tourist areas is a major structural weakness which is being highlighted in the Scottish Tourist Board's annual Visitor Attractions Survey (VAS).

The VAS also reveals the nature of attractions located in rural areas (see Table 8.7). The majority are historic castles and stately homes set

Table 8.7 Visits to attractions located in rural areas of Scotland, 1987

	('000)
Culzean Country Park/Castle	380
Cairngorm Whisky Centre	329
Balloch Country Park/Castle	300
Loch Ness Centre	NA
Blair Drummond Safari Park	142
Blair Castle	135
Inverewe Gardens	128
Glenfiddich Distillery	127
Oban Sea Life Centre	NA
Glenturret Distillery	119
Glencoe Visitor Centre	116
Baxters of Speyside Visitor Centre	110
Culloden Visitor Centre	109
Scone Palace	100
Aden Country Park	100
David Marshall Lodge	100
Urquhart Castle	99
Land 'o Burns Centre	95
Crathes Castle and Gardens	89
Cawdor Castle	89
Burns Cottage	NA
New Lanark	80
Iona Abbey	80
Balmoral Castle	NA
Brodick Castle	72
Glamis Castle	72
Abbotsford House	70
Killiecrankie Visitor Centre	65
Kelburn Country Centre	NA
Falkland Palace	63
Loch Garten Nature Reserve	62
Glenlivet Distillery	60
Hopetoun House	59
Creetown Gem Rock Museum and Gallery	54
Bannockburn Centre	52
Queen's View Centre	51

Source: VAS

in their own grounds, but these exhibit a varying degree of commercialisation, some with limited ability to generate income. Hence, it is possible for the Highlands and Islands Development Board to argue that Scotland is not making the most of its rural attractions.

In their attempts to attract more tourists Scotland's rural areas have had to compete not only with major urban attractions within the country, but also with holiday resorts abroad, able to offer better weather and a range of 'exotic' attractions. Especially in the Mediterranean countries these attractions have helped to boost some rural economies whilst also developing both existing and new urban centres. Spain has been the best example of this, with substantial tourist development in the Balearics, the Costa del Sol, the Costa Blanca and the Costa Brava. In Spain revenue from tourism accounts for 10 per cent of the GDP, tourism employs 10 per cent of the active workforce and in 1988 earned £10 billion revenue, covering the national trade deficit. Unfortunately, the consequence of the complete transformation of many coastal areas has been the wholesale destruction of some rural areas. The change from small fishing village to major modern resort, experienced by Torremolinos and many other resorts, has effectively moved tourism beyond the rural sphere. In its wake it has increased rural depopulation and introduced two significant trends in agriculture: intensive production for the hotel trade, e.g. in the rural hinterland of the Costa del Sol, and part-time farming in which farmers participate in the tourist trade through seasonal work in hotels and restaurants or by developing farm tourism. However, the latter has also been promoted throughout the Developed World. For example, in upland Britain at least 20 per cent of farmers have an income from farm-based tourism which may account for just over 10 per cent of net income, though primarily as a small supplement (Davies, 1983; Evans and Ilbery, 1989; Frater, 1983; Maude and Van Rest, 1985). In parts of Western Europe, though, farm-based tourism has had a greater impact. In France there is the increasingly popular 'Gites' scheme of farmhouse bed and breakfast whilst in the Alps the contribution of tourism to farmers' incomes has been substantial (Kariel and Kariel, 1982; Vincent, 1980).

8.5.4 The environmental impacts of rural recreation

Awareness and concern has grown over the environmental impact of recreational activity. In part the growing severity of this impact reflects the concentrated form of rural recreation with distinctive foci upon a few 'honey-pot' sites. A large element of recreation takes the form of passive pursuits such as picnicking and scenic viewing, numerous surveys showing that over half the visitors to particular sites of attraction never leave their cars or the vicinity of the car park in order to indulge these pursuits. However, this concentrated use of such sites can promote environmental damage especially on nearby public paths and walkways. Further afield, damage to paths can be severe if they are heavily used, e.g. destruction to parts of the Pennine Way in the Yorkshire Dales. In such upland districts, though, a more common form of damage caused by recreationists is to farm walls, fences and possibly also to flora and fauna through disturbance. Liddle and Scorgie (1980) suggest that if wildlife and habitats are disturbed, fauna tend to decline more rapidly than flora. This can often be observed at sites specially designated for their wildlife interest.

Major conflicts between recreation and agriculture occur through the impact of recreationists upon farmland and animals. Especially close to urban areas, problems arise over trespassing, dumping of rubbish, general pollution, and harassment of stock. These problems have grown partly because recreational use of agricultural land is *ad hoc*, unplanned and generally unwanted (Shoard, 1976). Rights of way across agricultural land are often resented by farmers embittered by damage caused by a small minority of visitors. Yet, in marginal upland areas, this damage can be critical for farmers whose net income is likely to be well below that of the average industrial worker (Phillips and Roberts, 1973).

Conflicts between farmers/landowners and the general public using the countryside for recreation often occur over rights-of-way. In particular, access to public footpaths and bridleways has provoked numerous legal battles throughout the Developed World. In England and Wales 16 million people per annum make use of a network of paths that covers 120,000 square miles (192,000 km). There is a legal framework permitting such paths to be created, diverted or extinguished, but, though 1,800 orders of this kind are made each year, they only affect 0.25 per cent of the network. To rationalise such attractions and as a means of regulating problems of access, the Countryside Commission (1988) has proposed a series of new measures:

(a) give farmers powers to make order to change the line of paths on their land, rather than having to persuade local authorities to act. Farmers would bear the cost and argue their case at an inquiry against any objectors;
(b) allow farmers to divert paths temporarily, e.g. away from fields used for lambing or running bulls;
(c) give ramblers powers to bring about changes, through proposals to local authorities and appeals to central government;
(d) give local authorities stronger enforcement powers, enabling them to impose immediate fines on farmers blocking paths.

These suggestions recognise that existing arrangements for altering rights-of-way are cumbersome, costly and time-consuming. However, their condemnation by the Ramblers' Association and their acceptance by the Country Landowners Association demonstrates the gulf that exists in views on access to the countryside, though these vary tremendously from country to country (Penn, 1988). For example, contrast the more restrictive legislation in England with that in Norway where, under the Outdoor Recreation Act, rights of access are allowed on all land except that under cultivation or within 150 m of a private residence.

The desire for increased access to the countryside and conflicts between recreation, the need to conserve flora and fauna, and various economic interests provide a complex background to the planning legislation of the past four decades. Meeting the needs of recreationists whilst protecting the landscape is just one problem; one which cannot be resolved in isolation from other concerns. It is these multi-faceted issues that are now considered in more detail in Part Three, the conflicts between recreation and conservation being discussed further in Chapter 10.

PART THREE
RURAL PLANNING

9

Planning for development

9.1   The limits to rural planning

Countryside planning has tended to be urban planning writ small and the statutory planning process has been ill-suited for dealing with the resource-based planning which is characteristic of much rural activity (Coppock, 1986: 296)

White (1986: 414-15) defines rural planning as 'the deliberate creation and management of schemes aimed at fulfilling specified goals in the rural environment.' Whilst these goals may be designated at a variety of scales and with several objectives set for various aspects of rural activities, the goals are often in conflict. A typical conflict, for example, is between the needs of nature conservation on the one hand and provision for recreation on the other. Resolution of such conflict by planning has often proved difficult because, in most national planning systems, responsibility for rural planning is split between several different organisations. Thus bodies undertaking 'integrated' rural planning, e.g. Italy's Cassa per il Mezzogiorno and Scotland's Highlands and Islands Development Board, have been few.

Most rural planning has been negative in the sense that it has focused primarily upon control and preservation, e.g. the implementation of green belt policies, land use zoning schemes, and the preservation of areas of ecological or scientific interest. There are critical weaknesses in this role. For example, Gans (1967) points out that planning which focuses upon building controls and land use regulations maintains an inertia in social and economic relationships and distances people from their environment and from one another, i.e. planners plan for the *status quo* by creating and sustaining different settings for different social classes (Glass, 1959; Pahl, 1977).

Scott and Roweis (1977: 1098) have pointed out the 'emptiness' of planning as simply 'a goal-orientated process that seeks to achieve specified desired objectives subject to given constraints.' They argue that planning in the Developed World is one activity of the capitalist state and therefore seeks a form of management whereby the capitalist accumulation process can be maintained, but with resultant class conflict being ameliorated. Knox and Cullen (1981) describe 'the whole apparatus of the town planning movement' as part of the internal survival mechanisms of industrialised capitalism. They view planners not only as managers but as reinforcers of particular values: a consequence of the planning hybrid in which it is both dedicated to humanistic reform and management of land and services according to the imperatives of a particular mode of production. Fincher (1983) goes further than this and recognises such a thing as 'socialist planning' within capitalist states, when referring to the planning performed by socialist municipal governments or local government institutions. These views see the nature of the state, or local state, as the key to understanding planning policy: .

Unless the form, function and mechanisms of the state are fully appreciated, research into policy-making and planning will be dogged by inherent but largely untested assumptions concerning why policies are made, and on whose behalf they are implemented (Cloke and Little, 1987: 343).

One view of planning and the regulation of the land market is that it attempts to organise space in the interests of business and commerce (i.e. capital) through the provision of collective infrastructure and other resources which individual businesses require but are unwilling or unable to provide themselves. In addition, regulation sets state institutions against private landowners by such mechanisms as the compulsory purchase of land for redevelopment or land zoning schemes (Lipietz, 1980). These measures can be viewed as a movement away from the development of economically 'efficient' land use patterns, and can be seen perhaps to be most pronounced in this respect in areas where building development is restricted, e.g. green belts, national parks (Plate 17). State encouragement of land development yielding high returns for capital investment is also restricted by the general limitation in the extent to which planning can direct investment. So even in countries with a well-developed planning infrastructure, such as Britain, 'the planning system is essentially a system of negative planning, for when the local and strategic plans have been established, it is still the private sector which decides whether and where to develop' (Saunders, 1986: 254).

Masser (1980) adds to this three specific limitations in the planning process itself;

(a) limitations in planning/spatial theory and methodology;
(b) uncertainties concerning the procedures and institutional frameworks that govern the planning process;
(c) limitations relating to the goals of planning with respect to social and economic change.

Despite these limitations, planners can play an active role in social change, introducing measures which can have major impacts upon

Plate 17 The desirability of the Lake District as a location for summer holidays
has extended to the market for second homes. The shortage of suitable
properties for both local residents and the 'second homers' has led to a call for a
restriction on second home ownership in the area

altering the direction of prevailing trends. A fundamental question which
arises in the assumption of this 'active role' is what 'guiding image' about
the desired future situation does the planner follow and hence what
ideology is invoked to determine this image? In effect, planning takes an
expression of an underlying ideology and introduces specific measures
with a fixed aim or aims in mind which are 'tempered by the reality of
experiences, limitations of available knowledge, technical feasibility, and
even expediency' (Commins, 1978: 80; see also Ambrose, 1986).
      To these 'limits' to planning can be added the lack of correspondence
between policy and its actual implementation. Increasingly, those
interested in rural change have recognised a need to focus upon the
investigation and understanding of planning and policy-making in rural

localities. Within this broad area of study the way in which policy has been implemented has been viewed as important, in recognition of a significant division between stated policies of rural planning agencies and 'what actually happens on the ground' (Cloke, 1986; 245), or the 'policy-implementation gap' (Blacksell and Gilg, 1981; Cloke and Hanrahan, 1984; Lewis and Wallace, 1984).

Whilst planning agencies have often experienced numerous problems in attempting to put theories into practice, Hebbert (1982) notes that for many regional planning agencies lack of continuity has been a major factor in 'poor performance'. Conflicts between federal and state governments, and national versus local governments have often severely limited the effectiveness or brought about the downfall of planning agencies. Conflicts between agricultural policy and other planning policies can be seen in many areas. With respect to policy in the uplands this has occurred where capital grants to farmers have led to labour shedding through the encouragement of the adoption of labour saving technology. Yet the loss of labour has been increased by higher grants in the very areas where other planning measures are attempting to maintain population levels (Slee, 1981).

It must be recognised that the formulation of planning policy for rural areas is itself associated directly with class, gender and spatial divisions as represented in the form of elites and pressure groups who are involved closely with the politics of the planning process. This 'compromising' of rural planning is described by Cloke (1987) who details various forms of relationships existing between policy formulation and the agencies of government. His basic argument is that these relationships tend to reduce the effectiveness of planning activity from dealing with particular rural problems. For example, McLaughlin (1983; 1985) suggests that the state's response to rural deprivation has been to deflect public attention away from underlying causes to focus upon issues of rural service provision which do not threaten the disturbance of existing socio-economic divisions in the countryside. This can be seen as planning being 'hijacked' for purposes of maintaining the status quo or as planning being essentially concerned with resource allocation rather than fundamental redistribution (Bowers and Cheshire, 1983). There is the implication, though, that the nature of rural planning implementation depends on local politics and the extent of central state intervention. For Britain, Cloke (1987) concludes that the constraints upon planning exerted by its role as part of wider state activity leave it largely impotent to tackle social and economic imbalances in the countryside. Indeed, Little (1987) suggests that if social divisions are 'a problem' then planning policy and action have contributed directly to that problem.

The three most common considerations on which rural planning has been based have been social equity, the desire to overcome urban congestion and the need to eliminate undesirable social consequences of structural change in agriculture. However, planning aimed specifically at rural areas has been of limited extent, prevailing influences on this situation being the views that rural, and specifically agricultural, land is virtually unlimited and that urban problems demand almost exclusive attention. Rural land has often only been considered in terms of its potential as a site for future urban development or as a supplier of goods and services for urban consumption. In Britain it was only when major conversions of land from agricultural uses were documented widely in the 1930s and 1940s that greater concern arose for more effective management of rural resources (Cocklin et al., 1987: 323). This concern has been magnified during the last three decades by growing public interest in the quality of the environment and conservation issues. Increasingly, it has been recognised that rural land cannot meet all the demands made upon it without causing environmental disruption. The need for reconciliation of the divergent claims of these demands has provoked extensive debate upon the appropriate mechanisms and criteria to use. Within this general context the issues that have provoked most interest have been:

(a) agriculture to urban land conversions;
(b) environmental degradation, e.g. soil erosion, pollution, depletion of groundwater reserves;
(c) provision of land for recreational purposes;
(d) conflicts between agriculture and a range of other land uses – forestry, mineral extraction, and recreation;
(e) commercial highly capitalised agriculture versus maintenance of a desirable landscape and environmental quality.

One of the principal ways in which these issues have been dealt with by planners has been through measures that have attempted to regulate competition between competing land uses.

The need to regulate land use competition has been met by government action in all countries of the Developed World, so that government itself has played an increasingly important role in influencing patterns of land use. There are four particular types of government intervention:

(a) regulating the economy. This provides a framework and system within which agriculture, recreation, forestry, housing, industry and transport function;
(b) creation of specific policies affecting the individual land uses;
(c) direct government control or quasi-government control of individual types of land use;
(d) creation of policies with respect to competition between land uses, e.g. the creation of national parks is an example of this although it represents planning primarily for recreation and conservation. This is the government's regulatory function.

When regulating competition, government can take various different stances:

(a) It can stop competition by retaining the present balance of land uses.
(b) It can enable a limited amount of competition by allowing one use

to expand in one area but not in another.
(c) It can allow a free-for-all, though this is generally elusive as, through
the comprehensive nature of most planning machinery, nearly all
changes in land use are subject to some form of planning control.

In practice there are usually some elements of all three stances practised within individual countries, depending upon the scale of competition considered. The retention of the existing balance of urban and rural
land uses is achieved in many cases by limiting the expansion of the
urban area. This is referred to as *development control* and it has a
spatial component: by stopping development in one place it is effectively
channelled elsewhere, e.g. the 'leapfrogging' of development over
London's Green Belt.

Various systems of development control have been employed in the
constituent countries of the Developed World (e.g. Davies, 1988a;
1988b; Edwards, 1988; Hooper, 1988; Punter, 1988). In recognising the
magnitude of the task in trying to summarise the range of individual
national systems, in this chapter and the other three in this part of the
book, special attention will be given to the planning framework and
legislation applying to Britain.

9.2 The planning system in Britain

9.2.1 *The foundations of the planning system*

Whilst legislation enacted post-1945 has played a major role in determining wide-ranging changes within the countryside, the antecedents of
many changes can be traced to earlier planning activity. Legislation has
been the product of a long process involving the evolution of various
organisations and gradually changing attitudes to issues affecting the
countryside. In particular, three crucial periods within the development
of 20th-century politics can be seen as crystallising attitudes and ideas
and enshrining them within government policy.

Initially, Lloyd George's reforming Liberal administration laid the
groundwork for town planning schemes and helped to create a Development Commission. However, the direct influence of these pales alongside
that of the radical Labour government of the late 1940s. Following a
series of highly influential reports during World War Two, legislation
was introduced that shaped the future of various aspects of rural
economy and society. The 1947 Agriculture Act and the 1947 Town and
Country Planning Act were at the forefront of attempts to extend the
control of both local and central government. The 1949 National Parks
and Access to the Countryside Act and the creation of the National Parks
Commission also extended the influence of the state. After three decades
of this more rigorous planning, a third period is now under way in the
privatisation schemes of the Conservative government which have

affected forestry, housing and transport. Concern with conflict of interests in the countryside has also been the subject of attention, notably in the controversial 1981 Wildlife and Countryside Act.

Four war-time reports crucially influenced post-war planning legislation and especially that of the 1947 Town and Country Planning Act which set out the basic philosophy of the British system of planning. They were the Barlow Report on the Distribution of the Industrial Population (1940), the Uthwatt Report on Compensation and Betterment (1942), the Scott Report on Land Utilisation in Rural Areas (1942) and the Abercrombie Report on planning Greater London (1944).

The Barlow Report was a Royal Commission which examined factors affecting the distribution of the industrial population and possible future trends. In its remit it was to consider the social, economic and strategic disadvantages of having the industrial population concentrated into a relatively small number of large urban centres in particular parts of the country; and to suggest possible remedial measures that would be appropriate to the national interest. Whilst noting regional variations in both economic and population growth, the Report concentrated upon problems affecting the major conurbations and especially London. Differences were noted between the old, resource-based industries and new ones which were more market oriented. Also, the growing problems of urban sprawl and loss of agricultural land were commented upon, although the methods for regulating this and other problems were not agreed. Perhaps the chief value of this wide-ranging Report was that it stimulated further investigation into a variety of problems, most of which were related to the development of more effective planning and control. In particular, it recommended closer investigation of public control of the use of land, and made suggestions concerning the dispersal of the industrial population. Both these recommendations were subsequently taken up – in the guise of the New Town Committee, set up in 1945, and the Uthwatt Committee which reported in 1942.

The latter was the Expert Committee on Compensation and Betterment. These terms referred to payment of compensation upon public control of land being assumed and to the recovery of betterment (or profit) consequent upon public control of land. The Committee considered the possible nationalisation of land but recommended that state control should only apply to development rights of undeveloped land. It felt that land for development should be purchased by the state at its existing use value and then leased back at full market value. In this way the state would recoup betterment. It also recommended that development gains made from the granting of planning consents should be taxed. These views and those on betterment and compensation have not been viewed favourably by post-war Conservative governments, but the Committee's recommendations helped shape the 1947 Town and Country Planning Act through its championing of the principles of central control over planning, state control over development rights and the need for planning permission prior to development.

The Scott Committee was asked to consider the conditions that should govern building development in the countryside. This also meant reviewing the factors affecting the location of industry in rural areas, consideration of rural employment, the well-being of rural communities and the preservation of rural amenities. One major theme of the Report was the threat posed to good agricultural land by the continued expansion of urban sprawl. This concern is not so surprising given that one of the Committee's members was Dudley Stamp, the geographer who organised the First Land Utilisation Survey in the 1930s at a time when the rate of loss of agricultural land to urban uses was at its peak. The Report duly stressed the importance of the maintenance of good farmland, suggesting that measures should be taken to ensure injections of capital to agriculture and maintain stability within the farming industry. Indeed, it also suggested that the state should acquire agricultural land compulsorily to protect farmland 'in the national interest'. The Report assumed that a central planning authority would ensure that various desirable objectives would be met. This planning was specified in some detail, for example national land planning; national zoning of land; and the principle of 'onus of proof' in which the onus was upon developers to show that it was in the national interest for 'good land to be alienated from its present use.'

The emphasis upon agriculture and the control of development, especially on good agricultural land, was well to the fore in the 1947 Town and Country Planning Act as well as the Agriculture Act of the same year. Planning focused heavily upon control of physical land use patterns and the agricultural lobby was prominent. It is interesting to note that in the minority report of the Scott Committee, Dennison argued that agriculture need not receive such a favoured status. He felt that improvements in agricultural activity would sufficiently offset losses of farmland to urban sprawl. These views have often been used subsequently by those seeking to convert agricultural land to other uses, not least of all in 'sensitive' areas on the fringe of major conurbations.

It is important to recognise the limitations of the planning processes that arose from the war-time reports and the great wave of socialist legislation enacted by the post-war Labour government:

We must . . . recognise that the planning frameworks developed for rural Britain after the War were derived from urban values with their pre-requisites of living standards and amenities for an urban population (Cherry, 1978: 364).

Rural planning was seen in terms of preservation of the countryside for urban enjoyment. In these terms, reviving agriculture and stabilising rural communities were secondary objectives, but, given the strong farming lobby, agricultural land use was left outside planning control. Unfortunately, preservation did not mean the introduction of a regulatory system able to deal adequately with a range of issues related to the new agricultural revolution.

9.2.2 Development plans and structure plans

The 1947 Town and Country Planning Act brought into being a system of development planning under which a number of developments could only take place after planning permission had been obtained. Thus ownership of land didn't necessarily carry with it an automatic right to develop. County councils and borough councils assumed the role of planning authorities with duties to survey their area and examine a range of factors which could affect the implementation of proposals for development. Each of the planning authorities was required to produce a Development Plan which would be based on the results of their survey. This Plan would indicate proposals for zoning of land use and indicate the stages by which changes to existing uses were to be effected. In addition to the formulation of this Plan, the fundamental concept of 'planning permission' was enshrined in the legislation. The local planning authority was vested with control over building development. For any proposed private development scheme the planning authority could refuse planning permission if it felt the scheme was inappropriate after taking into consideration its own Development Plan and any other relevant factors. Rejection of a proposed scheme did not entitle a landowner to compensation unless land became incapable of 'any reasonable beneficial use'. In these cases, local authorities were empowered to pay compensation, and a fund of £300 million was established for this compensation – in effect, repaying landowners for the expropriation by the state of their land's development value.

The various defects of Development Plans were addressed by the 1968 Town and Country Planning Act. After 20 years of operation it was generally agreed that the planning procedures established in 1947 were too cumbersome and slow and that the system was dominated by negative control on undesirable development rather than upon emphasising positive planning. In the 1968 legislation some of the suggestions of a Planning Advisory Group report were implemented so that the Development Plan was replaced by a Structure Plan and subordinate Local Plans. As before, each planning authority was to prepare a survey of its area upon which the Structure Plan would be based. The Structure Plan was to consist of written statements of policy and proposals for the area, formulated after widespread public consultation and discussion in order to generate more individual participation in the planning process (Bruton and Nicholson, 1985; Elson, 1981).

The 1968 Act was amended slightly by the 1971 Town and Country Planning Act, although this largely reinforced the system of Structure and Local Plans, setting out detailed procedures for the formulation and adoption of these plans. The implementation of Structure Plans, in effect, became the province of newly created planning authorities. These were instituted following the reform of local government after reviews in the Redcliffe-Maud Report (for England and Wales) and the Wheatley Report (for Scotland). The new structures for local government in

England and Wales came into being in 1974 whilst those for Scotland followed a year later. For England and Wales, under the 1972 Local Government Act, a two-tier system of County and District Councils was introduced. The latter were larger than the old County Boroughs and Rural Districts, and were given direct responsibility for planning matters. The major conurbations were administered by Metropolitan Counties which also assumed the responsibilities of planning authorities. Counties acted as county planning authorities and became responsible for district planning, which involved the preparation of Local Plans, and for the control of development under the 1971 Town and Country Planning Act. Provision was made for consultation between districts and counties, especially where applications for planning permission concerned development conflicting with policies or proposals in the Structure Plan.

Unfortunately, the framework of Structure Plans and Local Plans was frequently pre-empted through central government decision-making (Herington, 1982). This decision-making too suffered from the discontinuity within central government concerning ministerial responsibility for rural planning. Post-1945, in addition to the MAFF, the following ministries all played influential roles in rural affairs: The Ministry of Town and Country Planning (formed in 1947); The Ministry of Housing and Local Government (1951); The Ministry of Economic Affairs (1964, withdrawn in 1969); The Ministry of Land and Natural Resources (1965, withdrawn in 1967); and the Department of Local Government and Regional Planning (1969, withdrawn the following year).

These various bodies were superseded in 1970 by the Department of the Environment (DoE) which has a Minister of Cabinet rank. However, its responsibilities are very similar to those laid down in 1943 for the newly established Ministry of Town and Country Planning. Its prime aim is to establish continuity with regard to national policy for the use and development of land. The DoE is responsible for the approval of Structure Plans, the determination of planning appeals, planning co-ordination at the regional level, local government expenditure, and a variety of matters pertaining to countryside affairs, historic buildings and ancient monuments, sport and recreation, housing policy and finance and policy and finance for the water industry.

Within the DoE there are three areas which each have their own (non-Cabinet) Minister: planning and local government; housing and construction; and transport, which has been partially independent from the DoE since 1976. However, this list does not include agriculture and forestry, revealing one major weakness – its lack of direct control over these two major land uses. Other criticisms can be levelled at the policy discontinuity caused by continual reorganisation of the government environmental brief and the policy fragmentation and doubtful use of resources caused by the proliferation of agencies advising the DoE (Rogers, 1985).

The chief legacy of 40 years of Development Plans and Structure Plans is the system of development control that continues to operate. In rural areas this system has functioned principally to enforce the blueprint for settlement growth laid out in the Plans, the nature of which are considered further in Chapter 12. Hence new housing has been permitted in certain villages and not in others, thereby producing some sharp discontinuities between rural settlements and profoundly affecting the distribution of some of the processes referred to in the first part of this book, e.g. gentrification and geriatrification. It could be argued that this differentiation of communities has been the major type of government-directed rural development, although much of the response to the framework of the Plans has been in the hands of the private sector, e.g. builders, building societies, individual employers. Local government has responded to the Plans through the location of local authority housing, but initiatives related to the creation of employment and stimulation of the rural economy have been dealt with by a range of agencies independent of the formulation of Development and Structure Plans. This has made 'integrated' planning almost impossible (MacLeary, 1981; Masser, 1980).

9.2.3 Rural development planning

One of the earliest examples of a concern for rural welfare being translated into direct government action was the Liberal government's creation of a Development Fund through the 1909 Development and Road Improvement Funds Act. This established a Development Commission capable of making grants or loans to non-profit-making bodies and also of formulating their own schemes. Amongst the areas intended to receive assistance from the Fund were rural transport, forestry, land drainage and research likely to benefit agriculture and rural industries. Subsequently the Development Commission came to focus upon two major sectors of rural life – rural industries and community development. The initial budget for the Fund was £500,000 per annum. By the early 1980s payments per annum were £11.5 million. However, by this time the Commission had reorganised, principally in 1974 when it assumed the function of being the main agency for promoting social and economic development in rural England.

The Development Fund may be used for any project likely to benefit the rural economy but which does not have access to other statutory provision. Until 1968, grants to rural industry operated through the Rural Industries Bureau which has since become the Council for Small Industries in Rural Areas (CoSIRA). Community development has been encouraged partly through the work of Rural Community Councils which are independent, country-based and receive most of their funding through the Development Commission. This funding was rationalised in the 1983 Miscellaneous Financial Provisions Act, making the Development Commission a grant-in-aid body responsible for its own budget. The other body assisting community work is the National Council for Voluntary Organisations (NCVOs).

Rural industry is promoted by the Commission largely through efforts to create and maintain employment by the Commission's agents, CoSIRA and the English Estates Corporation. Formed in 1960 under the Local Employment Act and modified under the 1972 Industrial Development Act, the Corporation helps fund factory provision especially in special areas designated by the Development Commission. These areas, known as Rural Development Areas (RDAs) (Figure 8.6), are selected on the basis of criteria including above average unemployment, lack of local employment opportunities and poor access to services, and a declining and aging population.

CoSIRA merged with the Development Commission in 1988 to form the Rural Development Commission. Previously it took approximately one-third of the finance from the Commission's Development Fund, primarily to improve employment opportunities in rural areas (towns and villages of less than 15,000 population) through the establishment of small firms with less than 20 skilled workers. Although initially it built its own factories, now it is restricted to building conversion, and giving credit and advice (e.g. Bowler, 1988). For example, it can pay up to 35 per cent of the cost of converting farm buildings for craft and light industrial use in the EEC's Less Favoured Areas. CoSIRA also receives funding from both the European Social and Regional Funds. Half of CoSIRA's staff are in county offices where there is a Small Industries Adviser, usually working in collaboration with the Department of Industry's Small Firms Advisory Service.

The assistance in the area of community development is more diverse. It covers the provision of housing; support of rural services; encouragement and sponsorship of economic and social surveys and research; publicity, lobbying and information collection and dissemination; and support for the Rural Development of the NCVOs and the independent Rural Community Councils.

In 1982 the Development Commission designated new priority areas, the RDAs, replacing previously designated areas, the Special Investment Areas (SIAs). To qualify as an RDA an area, at least as large as an average sized rural district (5 to 30 parishes) but lacking towns with more than 10,000 population, had to have a rate of unemployment above the British average plus some of the following criteria:

(a) inadequate or unsatisfactory range of unemployment opportunities;
(b) population decline or sparsity of population creating adverse effects;
(c) net out-migration of people of working age;
(d) an age structure 'top heavy' with elderly;
(e) poor access to services and facilities (DC, 1984).

Identification of the RDAs was based on the experience of the earlier designation of the SIAs and those areas within them which were recognised as having severe problems, the *Pockets of Needs*. A database using census data and information from local authorities, the Rural Community Councils and local CoSIRA offices was prepared prior to final designation in 1984 (Wrathall, 1988b).

The area of the RDAs covered 35 per cent of the land area, 5 per cent of the total population and 23 per cent of the rural population of England. This was only 5 per cent less than that of the area of the SIAs, but the RDAs were more widely dispersed and included some areas in the South-East, West Yorkshire and South Yorkshire (Figure 8.6).

Within the RDAs the Development Commission's intent is to implement Rural Development Programmes (RDPs) to tackle social and economic problems, with two principal elements:

(a) setting objectives and priorities to deal with problems over a five to ten year period;
(b) producing a programme based on specific projects with detailed resource requirements from the Commission and other bodies.

However, given relatively limited funding (£25.7 million in 1985–6), they will be acting primarily as 'a power of pumps, a catalyst for self-help and enterprise in the countryside' (DC, 1980).

After their initial phases of operation, Green (1986) has identified three types of benefit accruing from the RDPs:

(a) improved co-operation and co-ordination between local interests and national agencies, especially in the areas with the greatest problems (e.g. Wrathall, 1986b);
(b) the introduction of new social and community developments;
(c) the focusing of resources upon rural problems, with local people making a contribution to the allocation of these resources.

Undoubtedly, though, resource limitations may hinder the RDPs whilst concerns over the designation of RDAs highlights the existence of problem areas which they exclude.

9.2.4   Development boards

The 1967 Agriculture Act made provision for a new type of rural agency, the Rural Development Board, which would attempt to balance the needs of agriculture against those of other elements in the rural economy. In particular the Boards were to address the needs of upland areas where special problems affected the economic and social development. The Boards were to introduce grants and loan assistance to implement their development programmes, paying special attention to the promotion of farm amalgamations and adjustments to farm boundaries designed to create more viable agricultural units. Under the auspices of the Act only one such Board, that for the North Pennines, was ever established. This Board was killed off following the election of a new government in 1969 before it had chance to implement its programme. However, two Development Boards, established under different legislation, have played important roles in upland districts of Scotland and Wales.

The Development Board for Rural Wales (DBRW) was established by the Secretary of State for Wales under the 1976 Development of Rural Wales Act. Its role was to carry out measures to promote the social and economic development of Mid-Wales, an area covering 40 per cent of Wales but containing only 7 per cent of its population, most of which live in small hamlets and isolated farms. Its budget in the early 1980s was £8.5 million, spent primarily on industrial construction, housing subsidies and housing, commerce and on social and economic grants which included amenity and leisure. Although over £4 million in the repayment of loan charges can be added to this budget, the finances available to the DBRW are much smaller than those received by Mid-Wales in the form of agricultural support.

The Highlands and Islands Development Board (HIDB) was established in 1965 with the basic task of 'assisting the people of the highlands and hills to improve the economic and social conditions and of enabling the Highlands and Islands to play a very effective part in the economic and social development of the Nation.' With an initial budget of £11 million, it was given the task of improving economic and social conditions in an area of 3.6 million ha (8.9 million acres) of which 88 per cent was under rough grazing for sheep and deer. In this attempt to bring the area more into the mainstream of the national economy, the Board was invested with wide powers enabling it to influence almost all aspects of upland life and the use of resources. It can carry out its own projects, acquire and own land and buildings, promote development through the provision of grants and loans as well as giving advice and training. Despite the wide-ranging remit, the Board's limited budget and cumbersome compulsory purchase procedures have ruled out major land transactions. Instead, it has operated as a stimulator of new private enterprises, especially in the areas of fishing, craft industry, tourism and large-scale capital intensive industry (Spaven, 1979). It has also concentrated on three 'growth areas' – Fort William, Inverness and Caithness (Figure 9.1).

The policies of the HIDB have been criticised on a variety of grounds (e.g. Carter, 1975; MacGregor, 1978), but it cannot be denied that since its formation economic growth and a degree of prosperity have occurred in the region albeit often in conjunction with oil-related development outside the initial considerations of the Board. Its chief problem has been the difficulty in dealing with the contrasting situations in the region, and the close co-operation required with local government. What has proved feasible in the growth areas has been impractical elsewhere. For example, plans for the reorganisation of the Harris Tweed industry were rejected by the local community whose strong cohesion and distrust of change would not permit a wide-ranging disruption of the industry's existing structure.

Despite these reservations the Board's policies have met with some success, mainly attributable to the selective nature of its work. For exam-ple, its emphasis upon tourism brought a growth in the numbers of tourists of 5 per cent annum from 1966 to 1978, the establishment of

% Population change
1971-81

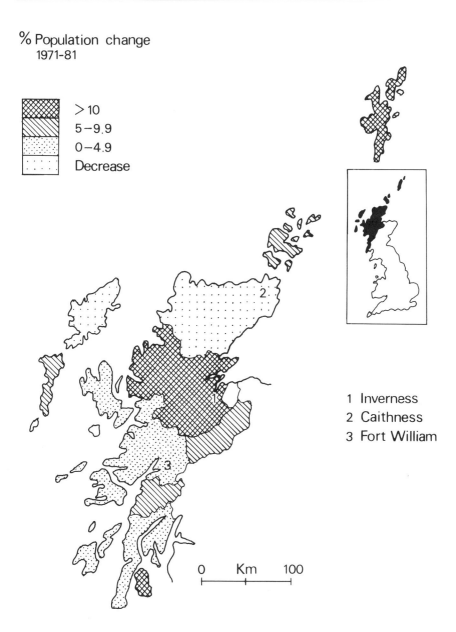

> 10
5-9.9
0-4.9
Decrease

1 Inverness
2 Caithness
3 Fort William

0   Km   100

Figure 9.1 'Growth Poles' and population change in Scotland's Highlands and Islands, 1971–81

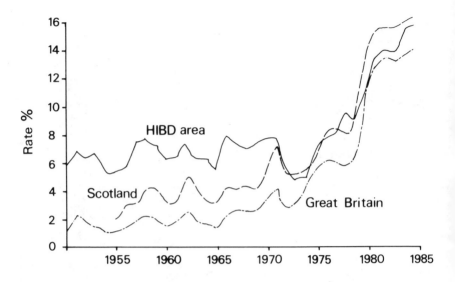

Figure 9.2 Unemployment in Scotland's Highlands and Islands, 1950–85 (based on McCleery, 1988)

a wide range of tourist accommodation in the region and employment in tourist-related industries accounting for over 10 per cent of the labour force. Between 1971 and 1982 the Board gave £59.6 million in grants, 47.6 per cent of which went towards tourism and 22.2 per cent to manufacturing and processing. The corresponding proportions for loans and shares from the Board during the same period were 14.9 per cent and 27.3 per cent, with an emphasis upon fisheries which received 32.4 per cent of the total of £82.2 million under this form of assistance. In all, 16,045 jobs were created by Board-assisted developments in these ten years and 4,777 retained (HIDB, 1983: 125–8).

One measure of the success of the HIDB policies is unemployment, as a central pillar of policy has been to curb out-migration of the young and most enterprising members of the community by providing sufficient jobs in the local area. At the time the HIDB was brought into being un-employment rates in Scotland as a whole were around 7.5 per cent, but in some parts of the Highlands and Islands they were over 20 per cent, Stornoway for example. The unemployment rate in the Highlands and Islands was between 3 and 4 per cent above the Scottish average. However, during the 1970s there was a marked convergence in the two rates (see Figure 9.2), and Scotland's rate has been higher since 1980, though with both rates exceeding 10 per cent.

A government review of the first two decades of the HIDB concluded that many of its activities and procedures were soundly based (Industry Department for Scotland, 1987), and that it had made a substantial contribution to the economic and social well-being of the Highlands and

Table 9.1 Employment structure of the Highlands and Islands, 1951–81

	1951	1961	1971 a	1971 b	1981	1981 Scotland
Primary	18.0	15.3	9.5	12.2	8.9	(4.6)
Manufacturing	13.3	10.0	13.9	11.9	13.6	(24.8)
Construction	14.1	16.0	14.0	11.7	10.3	(7.4)
Public utilities	1.0	1.1	1.5	1.5	1.3	(1.5)
Services	53.5	57.4	61.1	62.8	65.8	(61.6)
Total	100.0	100.0	100.0	100.0	100.0	(100.0)

Source: Department of Employment, DAFS
Note: Figures for 1951–71a are based on National Insurance card count and include the unemployed. They are therefore not directly comparable to figures for 1971b–1981 which are based on the Census of Employment and DAFS estimates for self-employed in fishing

Islands (Walker and McCleery, 1987). It is possible to measure this well-being in several ways, but McCleery (1988) argues that three criteria in particular can be used to support the generally favourable views of the review which also made no proposals to reduce the scope of the HIDB's functions.

The three 'indicators' are population, employment and quality of life. The Highlands and Islands have long exhibited characteristics typical of remote rural areas with respect to these three: long-term population decline, unemployment above national levels and a poor quality of life in terms of service provision, housing and income levels. There have been positive changes in the indicators during the lifetime of the HIDB, although some of these can be attributed to the effects of the oil industry rather than to HIDB initiatives. Thus, the growth of population has been 'patchy', reflecting a general redistribution of people in the region: losses from remoter parts counterbalanced by in-migration in other areas both from elsewhere in Scotland and the rest of Great Britain. The employment structure of the region has moved towards the overall Scottish pattern (Table 9.1), though with a clear under-representation of manufacturing and a bias towards the primary sector. Self-employment remains important, especially in the remoter crofting districts. Services, too, with two-thirds of all employment, account for a higher proportion of employment than in the rest of the country.

The demise of the oil industry is indicated by a 75 per cent decrease in direct employment between 1981 and 1987 (Plate 18). This reflects both the end of terminal construction on the Shetlands and the sharp fall in oil prices in 1986. This has reversed the healthy trend of the 1970s when unemployment rates in the region fell below those of Scotland as a whole for the first time since before World War Two. Since 1981, unemployment in the Highlands and Islands has been rising faster than for Scotland as a whole, though from 1980 the unemployment rate has

Plate 18 Highlands and Islands Development Board assistance for new enterprises, plus oil-related development has led to population increase in Stromness, Orkney, but there are fears that eventual loss of the oil industry will bring about a return to significant depopulation

remained lower than that for Scotland (Plate 18). In some parts, notably Skye and Wester Ross, the rate has exceeded 20 per cent for some time.

Quality of life indicators such as housing, incomes, vehicle ownership and cost of living suggest that the 'Highland Problem' still remains – higher costs but lower per capita incomes; 'a dislocation in space of demand relative to supply' (McCleery, 1988: 174). This dislocation would probably have been much greater but for the influence of the oil industry. With the gradual run-down of this industry in the 1990s and next century will come the real test for the HIDB. Essentially, the basic problems of the region remain intact, but a higher economic baseline has been established compared with 1965 and this gives some grounds for optimism concerning continued investment and development of infrastructure under the HIDB's guidance.

9.2.5   Green Belts

Green Belts are areas of open and low density land use surrounding existing urban settlements where further continuous urban sprawl is deemed to be in need of strict control. Whilst notions of the 'green girdle' or cordon sanitaire around major cities can be traced back at least to the

16th century, the first formally proposed Green Belt was in conjunction with Ebenezer Howard's *garden city scheme* in England in the 1890s. This incorporated an encircling Green Belt as protection against unrestrained expansion of the garden city and was also to act as a source area for agriculture and recreation, providing rural contact to counter-balance the urban setting of the residents in the garden city. Subsequently, several Green Belt schemes were advocated for London, but, until after World War Two, land acquisition costs for municipal authorities were prohibitive (Thomas, 1970).

In Britain one reaction of central government to the unprecedented expansion of urban areas in the 1920s and 1930s was to introduce the Restriction of Ribbon Development Act in 1935. Sheial (1979: 501) comments,

In its attempts to balance the interests of the individual with those of the community, and in its appreciation of the need to reconcile house and road construction with the preservation of rural amenity, the Bill highlighted some of the wider issues of land-use planning.

The Act was the forerunner of post-war attempts to control urban sprawl, especially in south-east England. The idea was applied to London in the Greater London Plan of 1944, which was adopted by the Ministry of Town and Country Planning in 1947 and then embodied in a ministerial circular in 1955. The circular specified three purposes for Green Belts:

(a) to check the spread of further urban development;
(b) to prevent neighbouring towns from coalescing;
(c) to preserve the special character of towns.

A circular in 1984 added a fourth purpose: to assist in urban regeneration. Significantly, no mention was made of possible amenity values of Green Belts. Local authorities are responsible for proposing areas of land which they consider should be given Green Belt status. These areas are then incorporated in the appropriate Structure Plans and approved by central government. Within the designated areas housebuilding has been strictly controlled and largely limited to infilling within existing settlements.

The aims of designated Green Belts that have generally received the greatest attention from planners have been the negative ones of attempting to prevent the growth of large urban areas, stopping the merger of neighbouring towns and preserving agricultural land. Possible positive aspects, such as the provision of scenic areas and recreational space close to urban areas for the benefit of city dwellers, have been largely neglected. Also, the Green Belt designation has carried no extra planning powers, so that it exists as an indication that, in principle, planning authorities will follow certain guidelines concerning development applications in the designated areas (Elson, 1986).

In the post-war period in England and Wales, the area designated as

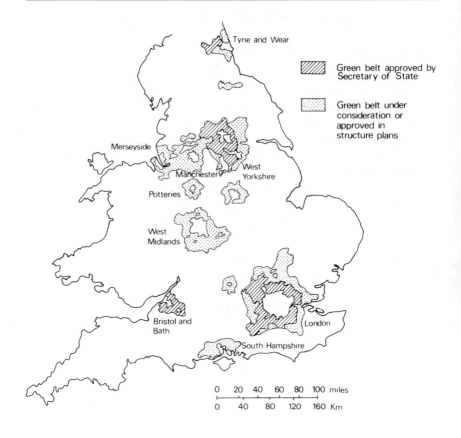

Figure 9.3 Green Belts in England and Wales

Green Belt has been extended to cover 6,070 sq miles (15,540 sq km), with proposals outstanding for further expansions to push the area of Green Belts beyond 11 per cent of England's land surface (Figure 9.3). It seems therefore that such policies have proved popular with planners as a means of controlling urban sprawl. This popularity undoubtedly stems in part from the high degree of success of Green Belts in halting losses of agricultural land (Munton, 1981). Against this must be set the leap-frogging effects that the policy has upon development. In effect, the Green Belt pushes the urban sprawl into other areas, both back into the cities in the form of higher density housing there, or beyond the Green Belt into other rural areas. In addition, its preventative effects upon development within the Green Belt have significantly raised both prices of land and housing inside the Green Belt. Unfortunately, as most Green Belt designation is effectively only in place as a 'guideline' to planners, its presence has only helped to foster speculative landholding, sometimes proving highly profitable as in the case of former Green Belt land sold for development on the fringes of the West Midlands conurbation.

Cloke (1983: 310–15) argues that Green Belt policies have maintained inequalities within the rural–urban fringe by favouring middle-class commuters and high income groups who can more easily afford the rising property prices. Meanwhile, the limited growth encouraged by the policies has helped to limit services for the less affluent, placing them at a greater disadvantage. Unfortunately, even the wave of Structure Plans introduced in the 1970s failed to acknowledge this inequity despite provision for relaxation of restrictions to meet 'local needs' (Gault, 1981). These local needs have varied from allowance of no form of development, even excluding settlement infilling, to provision of housing for low income groups and maintenance of a balanced age and socio-economic structure (e.g. Healey, 1980).

It has sometimes been felt that the Green Belt policies conflict with the wish of expanding firms to locate within a desirable area and thereby also contribute to national economic growth (Hall, 1985). For example, Fothergill (1986) suggests an estimated loss of between 14,000 and 23,000 manufacturing jobs in the London Green Belt attributable to a 'restraint impact' during the period 1974–81. These jobs were probably diverted beyond the Green Belt to the outer parts of the South-East Region and adjacent counties, notably in East Anglia. In addition, the restraint may have simply postponed or curtailed expansion plans, especially for small and medium-sized firms whose growth is not so easily diverted to alternative locations. The severe land use restraints in the Green Belt zone have placed a particular onus on established industrial estates to accommodate industrial growth, and hence the importance of 'infilling, refurbishment and rehabilitation' on these estates (Towse, 1988: 329). However, the pressure upon such estates has favoured particular types of firms, especially those able to pay higher rentals and not requiring large amounts of space. Towse (1988) argues that more flexible planning is required to limit the restriction effect upon industry of the Green Belt policies, e.g. by reclaiming derelict land permitting selective release of land parcels for development in metropolitan areas.

The search for greenfield sites on the outskirts of London for out-of-town retailing, residential and industrial development has naturally led to proposals to develop on land within the Green Belt. This search and pressure for development has intensified following the opening of the M25 motorway, which runs through the metropolitan Green Belt. By the end of 1988, proposals had been made for major projects in a range of locations, reflecting especially the growing shortage of housing in the South-East (e.g. Institute of Manpower Studies, 1986). However, in early 1989 the proposals for the first of these sites, in the 'golden triangle' near the junction of the M1 and M25, was rejected by the Secretary of State for the Environment. This proposal involved a two-storey 70,000 sq m shopping centre, a cinema complex, exhibition space, a health and fitness centre, a 250-bed hotel and parking for at least 5,500 cars. The developers, Town and City Properties (Development) Ltd., were charged the appeal costs of St Albans city council who opposed the scheme. Such charges (around £250,000) may act as a deterrent for other developers with development schemes within the Green Belt, but there remains both the need for new housing and a recognition that both the Green Belt concept and its method of implementation need to be re-examined in the light of changed circumstances.

9.3 Planning for the development of the rural periphery

9.3.1 The rural periphery and internal colonialism

The examples of Mid-Wales and Scotland's Highlands and Islands given in Section 9.2.4 are an indication of the special attention that planners have tended to focus upon rural peripheries in the Developed World. In this focus, notions of how regional planning should promote balanced economic development throughout countries have been combined with concerns having a particular rural dimension. In particular, it has been recognised that there are notable similarities between the relationships of peripheral rural areas to more prosperous 'heartlands' within the same country and those between the Developed Countries and the Developing World. Williams (1977) refers to this situation as *internal colonialism,* which he recognises as having the seven features listed in Table 9.2.

This view holds that despite the spread of new service-based industries to rural locations, the new developments only continue patterns of inequality already laid down. In effect, rural peripheral areas have tended to remain subservient to the demands of the urban core. Marx (1961) described an early stage of this subservience in terms of emigration of surplus agricultural labour being directed to industrial areas where wages and demand for labour were rising. This converted a part of the agricultural population into the urban or manufacturing proletariat and created a distinct depopulating (and under-capitalised) peripheral hinterland and an urban, industrial core with an 'enclave' of capital accumulation as the engine of uneven development in national terms (Holland, 1976). This hinterland–enclave development has persisted in the 'post-industrial' era (Bell, 1973; Hechter, 1975).

Economic relationships between urban and rural areas are often conceptualised in terms of one of three situations:

(a) the diffusion of growth (and development) from urban to rural, thereby associating 'rural' with backwardness;
(b) internal colonialism whereby core area, urban institutions control the economic affairs of the countryside;
(c) uneven development will occur through competitive processes within capitalist societies. Such development will tend to favour core areas but may produce certain kinds of growth in rural areas as illustrated by the ruralisation of industry.

Of these ideas the first two tend to focus upon distinctions between a

Table 9.2 The characteristics of internal colonialism

(a) A commercial, trading and credit monopoly by 'the centre';
(b) Commerce is dominated by recruits from the core;
(c) An economic dependency on external markets arises due to complementary development with the core, and often rests on a single commodity;
(d) The movement of peripheral labour is brought about by forces outside the periphery and is due mainly to variations in the price of a single commodity;
(e) Economic dependence is enforced through legal, political and, sometimes, military measures;
(f) The 'colony' lacks services;
(g) There may be national discrimination on the basis of language, religion or aspects of culture.

Source: Williams (1977)

core and a periphery or between metropolitan and non-metropolitan. So, for example, much of the 'rural' areas in the Paris Basin could be contrasted with rural areas in the rest of France. Under the diffusion argument, the notion of 'trickle down' is the critical mechanism for transferring growth from core to periphery. The periphery 'lags' because of limitations imposed by small numbers of population, remoteness, weaker innovative and competitive behaviour and the preservation of values less receptive to change. There is evidence for aspects of these characteristics in many peripheral rural areas (e.g. Caird and Moisley, 1964; Dunkle *et al.*, 1983; Berry 1973). Yet these ideas ignore situations where such characteristics do not appear in peripheral areas, e.g. the rural orientation of radical social movements in Portugal and Spain or the existence of rural enterprise equally as innovative as that within core areas (e.g. Gould and Keeble, 1984; Pellenberg and Kok, 1985). The diffusionist case rests heavily on core–periphery distinctions of the 1950s and 1960s, and largely ignores the importance of social and economic structures accommodated in the other two concepts albeit in different ways. However, much government policy intending to reduce regional imbalances has made an implicit incorporation of the diffusionist approach in seeking to encourage the spread of growth from core to periphery.

There are various ways in which policies have been developed to spread growth to peripheral rural areas. For example, in New South Wales, Australia, attempts to reduce congestion and concentration upon Sydney and the industrial centres of Newcastle and Wollongong invoked the establishment in 1958 of a Decentralisation Fund to provide assistance on a discretionary basis to industries outside these three centres. Subsequently, the scope of the Fund has increased as Sydney's domination of the state's economy and population numbers has grown. The Fund has helped firms relocate in rural areas, taking advantage of the need for industries processing rural raw materials to utilise cheaper rural labour and land. The greater locational freedom of capital from the large skilled labour pools of the cities through the rise in automation and standardisation has also been a factor (Searle, 1981; Massey, 1978: 118). Special 'maximum assistance areas' were introduced in the 1960s, including the Upper North Coast, the main dairying area outside the Sydney Milk Zone. Growth pole strategies have also been followed in joint state and federal programmes in the 1970s.

Two further examples, from France and Eire, illustrate the combination of *regional planning* with particular measures applied to rural peripheries. They can be compared with Britain's emphasis upon employment creation in peripheral industrial centres plus some specific measures for the rural periphery, e.g. the Development Boards for Mid-Wales and the Highlands and Islands of Scotland.

In France attempts to decentralise metropolitan functions from Paris to eight regional metropoli were accompanied by promotion of 'regional balance' and attempts to prevent disequilibria between core and periphery. The mechanism used was a system of relay centres to link the regional metropoli with their hinterlands. These centres (or second-order towns) were a central theme of the Fifth National Plan (1966–70). But, subsequently, the Sixth National Plan (1971–5) stressed medium-sized towns, *les villes moyennes* (Scargill, 1983), to act as rapid growth centres within a regional context, especially if close relationships with a rural hinterland could be developed (Comby, 1973). This designation also focused upon local participation and the large number of 'medium-sized' towns within large areas of France: 165 towns between 20,000 and 100,000 in 1968. This scheme was strengthened during the Seventh Plan (1976–80).

From 1975 the policy towards medium-sized towns has been extended by promotion of aid for smaller towns and villages as part of a comprehensive rural planning policy (Plate 23). This recognises the need to sustain the farm population and limit rural depopulation. Various designations have been introduced:

(a) *Contrats de pays*: small towns of 5–15,000 and their rural environs;
(b) *Contrats d'amenagement rural*;
(c) *Contrats regionaux d'amenagement rural*.

In 1967 Rural Renovation Zones were identified covering one-quarter of France and one-third of all farms (Figure 9.4). Brittany, western Normandy and the Massif Central were covered by this designation which therefore applied to areas with low average incomes from agriculture, high average age of farmers, poor communications and lack of non-farm employment. Special funds were made available to improve basic infrastructure, e.g. improved roads and water supplies to Brittany. Some funds also went towards rationalisation of stock rearing and developing facilities to attract tourists.

From 1975 more wide-ranging assistance was developed through the establishment of rural planning corporations in Aquitaine, Limousin,

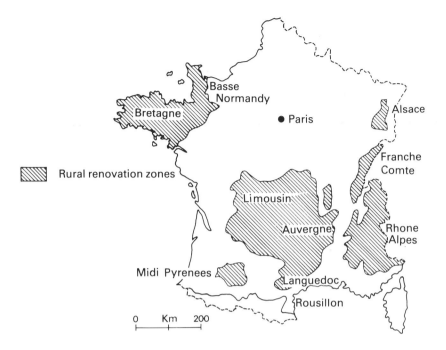

Figure 9.4 Rural Renovation Zones in France

Auvergne, Languedoc, North-Eastern France, Corsica and Alpine districts in Provence-Côte d'Azur. The basic aim of the corporations is modernisation and diversification of the local economy, though their approaches have varied. For example, in north-eastern France there have been attempts to bring substantial areas of heathland into productive pasture whilst irrigation has been extended in Gascony and crop diversification encouraged in the Durance Valley. Detailed planning to meet the needs of specific areas has brought some improvements to farm structure and presented farmers with alternatives to their traditional enterprises. However, given the extent of price guarantees operating as part of the CAP, the rate of change has been slower than desired.

Other planning initiatives have been taken both for specific areas and on a national basis. Of the former one of the most wide-ranging in France was the special development programme for the Massif Central begun in 1975. This was 'to improve communications, encourage small industries and rural tourism, modernise forestry, crafts and farming, and enhance service facilities in small towns' (Clout, 1987b: 185). A similar regional scheme was that of the Mission du Grand Sud-Ouest, commencing in 1979, to improve agriculture, forestry and stock-rearing in Aquitaine, Languedoc-Roussillon and the Midi-Pyrenees. Of the national schemes, there have been special grants awarded by the Fonds

Interministériel de Développement et d'Aménagement Rural (FIDAR) for innovations assisting rural transport, services, education and employment in areas of depopulation. Other sources of government funds have been applied to the preparation of rural management plans for small areas, covering job creation, landscape conservation and provision of services.

In Eire special attention has been paid to the Gaeltacht, those parts of the country where Irish is still the vernacular (Figure 8.5). Its area covers just 7 per cent of the country's land area and has only 80,000 people or 2.3 per cent of the national total. However, it comprises several scattered districts and so affects several parts of the western seaboard. The political and cultural significance of the maintenance and restoration of the Irish language within Eire has promoted successive governments to devote more state aid to the Gaeltacht than to other rural areas. From 1958 this aid has been focused upon a development agency, Gaeltarra Eireann, established to give a new impetus to industrialisation in the Gaeltacht. The agency's statutory powers for industrial development were extended in 1965 and in 1979 it was reconstituted as Udaras an Gaeltachta, with a more democratic structure and more extensive powers. Yet, despite this broader remit, it does not have direct control of statutory activities such as local planning, education, agricultural development, social services, tourism or fishing (O'Cinneide, 1987).

The role of Udaras has been to foster the development of the Irish language whilst working in conjunction with other agencies to stimulate economic development, e.g. the Industrial Development Authority (IDA) which provides grants for industry. Udaras has concentrated upon locally induced economic growth in which endogenous factors play a principal role. In particular it has attempted to help overcome lack of access to information and capital by promoting education and the engendering of positive attitudes towards development. So, in addition to a standard range of financial inducements, it has sponsored a series of courses, seminars and workshops to promote 'social animation'. In support of commercial projects it has collaborated with local community co-operatives. These have helped to provide important infrastructure to assist local development.

The small scale of the projects promoted by Udaras is indicated in Table 9.3, showing an average of under 7 full-time jobs created per project. However, it has played a significant role in revitalising the Gaeltacht and has contributed to the reversal of the long trend of out-migration (Horner, 1986; Keane et al., 1983; O'Cinneide and Keane, 1983).

9.3.2 *Integrated development in the European Economic Community (EEC)*

Since 1973 British regional and social policy has existed within a loose framework provided by EEC policy as reflected in its Regional and Social

Table 9.3 Classification of Udaras-assisted industrial projects by origin of promoters (as at end of 1984)

Origin of promoter	No. of projects	Employment (full-time)
Gaeltacht	576	3,061
Elsewhere in Ireland	35	445
Overseas	44	899
Total	655	4,405

Source: O'Cinneide (1987)

Funds. Although neither fund accounts for a large proportion of the EEC budget (the Regional Fund accounted for 4.9 per cent in 1985 and the Social Fund 4.1 per cent), they both represent substantial sums of money, from which numerous payments have been made to rural areas in Britain, for example the Keilder Reservoir scheme in Northumberland attracted a grant of £65 million from the Regional Fund.

The Social Fund was established under Article 123–8 of the Treaty of Rome in 1958 to provide assistance to certain groups of workers, especially within areas experiencing structural and technical changes. Since its inception it has been reformed in 1971 to help counter unemployment within the EEC and to address particular problems. Further reforms in 1978 have led to a greater focus upon poorer areas of the Community. Those groups which have received the most attention from the Fund in terms of assistance with training or retraining have been migrant workers, textile workers, the handicapped, workers leaving agriculture and those new to the job market.

The Regional Development Fund was created in 1975 to address the problems of regional imbalances produced by industrial change, structural underdevelopment and as a result of agricultural policy. Initially, the Fund concentrated mainly on awarding grants to projects in poorer regions of the EEC, stressing industrial regeneration or improvements of infrastructure. However, following reforms in 1979, the Fund was divided into quota and quota-free sections, and this has resulted in the assistance of some specifically rural-based projects receiving grant aid. The quota section of the Fund is project-specific, awarding up to 50 per cent of a project's total costs, provided it has been given priority by the constituent government, and applies to a development programme region. The non-quota section is aimed at more general development programmes but is small in comparison with the quota sector, accounting for only 5 per cent of the Fund as a whole. In Britain the first major receipt of finance from the non-quota section was for £20 million spread over a five-year period for the Integrated Development Programme for the Western Isles.

Within the EEC one recent attempt to restructure rural areas has involved an initiative from the Common Agricultural Policy for an integrated land use policy to be supported by both national and Community funding. This scheme is the Integrated Development Programme (IDP) aimed at developing least-favoured areas with use of Community funding through the EEC's Agricultural Guidance and Guarantee Fund (FEOGA), the European Regional Development Fund and the European Social Fund. These sources will be supplemented by national governments in sectors indicated by the Programme. The initial three schemes were in Luxembourg Province, Belgium, Lozère département, southern France, and the Outer Hebrides (Western Isles), Scotland.

The aim of the IDP for the Western Isles was to promote socio-economic development in a region where agriculture on its own is unlikely to provide the population with a reasonable standard of living. As in the other two projects, the area selected represents a major problem region. The Western Isles has an unemployment rate nearly double that for the United Kingdom as a whole, has a great dependence upon public sector jobs and has many farmers who are unable to survive from farming alone. Hence, the IDP aims to boost both agriculture and other industries which can provide more part-time work for farmers. Of the holdings in the Western Isles 88.6 per cent are under 20 ha (50 acres), emphasising the dominance of crofting. It is towards the improvement of this large crofting sector that part of the IDP is aimed.

The Western Isles have a population of just over 30,000, 8,000 of whom live in the main town, Stornoway. The land use of the islands is mainly pastoral with livestock providing over 90 per cent of the income. There are 5,974 crofts, but only 1,563 are recorded by the Department of Agriculture and Fisheries for Scotland as being statistically significant. On the rest less than 40 standard man-days of work is done each year. The link between this and the pattern of land use is clear. Out of 212,000 ha (523,850 acres) in crofting, 188,000 ha (464,550 acres) (88.7 per cent) are under common grazing. The individual crofters have a grazing share on the latter plus some 'inbye' land attached to their croft on which they cultivate grass for mowing, root crops and cereals. Although the main activities are the breeding and rearing of beef cattle (4,000) and sheep (150,000), it is very difficult for even the larger farmers to produce adequate amounts of fodder for use in winter. This means that calves and lambs are sold in the autumn for slaughter or for fattening elsewhere. Hence, stocking rates are limited by the availability of winter feed, and the Western Isles are only 40 per cent self-sufficient in beef and 50 per cent in sheep-meat.

One of the central proposals of the IDP is to alleviate this problem of shortage of fodder by improving both the quality of pasture and of winter-feed to promote increased self-sufficiency. A second element is the introduction of high quality livestock to raise the value of farm output. These aims are to be met by an investment of £20 million over five years from the UK government, of which 40 per cent will be refunded by the EEC. A series of grants to farmers will encourage desirable practices, with the highest grants, 85 per cent of costs, going to:

(a) improvement of common grazings involving a wide variety of land improvement works;
(b) improvement of the machair, the fertile coastal sand plain, consisting of a highly calcareous shell sand which is very light in humus and suffers from drought and wind erosion (Adams, 1976: 156; Ritchie, 1967);
(c) improvement of the inbye (land within the head dyke or dry stone/turf wall separating the arable and meadow land from rough grazing) through drainage, fencing, reseeding, application of lime and fertiliser, and the upgrading of access tracks;
(d) the provision of shelter-belts and necessary fencing.

If higher quality livestock are introduced they will qualify for grants: £30 per approved calf, £50 per approved heifer with an extra £100 per approved gimmer (young ewe). Grants will also be available to promote livestock marketing, especially through co-operative ventures.

Whilst it is the aim of the IDP to improve the viability of crofting, more money from the EEC will actually be spent on trying to provide alternative employment and so reduce the extremely high levels of unemployment. So £9 million out of £20 million to be invested by EEC sources will go to fisheries and fish farming (see Table 9.4). Grants ranging from 50 per cent to 75 per cent will be available for processing facilities, new fish farms and marketing.

A crucial factor inhibiting development is the nature of the current infrastructure in the islands. It is here that the IDP'S weakest point occurs. The Programme has identified a need for £35,632,000 to be spent upon infrastructure and socio-economic measures, but this funding will not be met by FEOGA. Reliance for this investment is placed upon the EEC Regional and Social Funds and the United Kingdom government. The most crucial weakness relates to the cost of transport from the Western Isles to the Scottish mainland which adds considerably to the price of goods exported from the islands. Additional investment is required to improve ferry services, air links and roads on the islands themselves. Many of the islands' 703 miles (1,125 km) of roads are single-lane and in need of up-grading as are access tracks to farms. It is recommended that the largest single item of expenditure on infrastructure be on improving the ferry link between both Tarbert and Lochmaddy to Uig, two hours away.

There are several economic activities in the islands that fall outside the scope of finance from FEOGA. The most well known of these are the Harris Tweed and knitwear industries. The production of Harris Tweed increased substantially in the late 1970s through exploitation of growing markets for quality cloth. There are around 650 weavers producing the tweed in their homes, but this workforce is aging, with 47 per cent over 50 years of age. There is a need for more training to assist young people to enter the industry and for re-equipping as looms wear out.

Two other aspects of the IDP are worth noting. These are the elements

Table 9.4 Costs for the Outer Hebrides (Western Isles) Development
Programme

A. UK public expenditure (40% to be refunded by EEC)

	£ million	%
Fisheries and fish farming	9.0	45.0
Land development	3.4	17.0
Livestock development	3.4	17.0
Agriculture and fisheries infrastructure	1.9	9.5
Education, training and advisory services	1.5	7.5
Comprehensive development scheme	0.5	2.5
Project team	0.3	1.5
	20.0	100.0

B. Socio-economic measures identified as necessary but not to be funded under
Common Agricultural Policy

	£million	%
Communications	23.890	67.20
Tourism	2,740	7.71
Water and sewerage	2.670	7.51
Energy	1.735	4.88
Mineral exploitation	1.000	2.81
Seaweed industry	0.800	2.25
Harris Tweed	0.800	2.25
Land and buildings	0.587	1.65
Community co-operatives	0.400	1.13
Miscellaneous activities	0.400	1.13
Knitwear	0.270	0.76
Craft industry	0.200	0.56
Protection of the environment	0.060	0.16
	35.552	100.00

Source: EEC Council (1981)

of the scheme relating to conservation and diffusion of information.
Firstly, developments under the IDP must be compatible with the protection of the environment. However, draining and re-seeding the machair
and using locks for fish farming would undoubtedly place at risk valued
fauna such as the red-neck phalarope. This potential conflict may be
resolved by the Nature Conservancy Council compensating farmers for
not proceeding with improvement schemes, though this could severely
weaken the effects of the IDP. Secondly, it is intended to provide better
education on farming and fish farming to enable the local community to
benefit more readily from the various grants available under the
Programme. It is proposed to establish a network of community
workshops in the Isles to teach basic skills. To this end FEOGA is

contributing £600,000 with £920,000 to come from other sources.

It is this provision of large amounts of additional funding that may prove a severe limitation on the operations of the IDPs. The EEC's Less Favoured Areas (LFAs) scheme has had very limited success because farmers have found it difficult to obtain loans or credit to provide necessary supplements to EEC funding. However, for the Western Isles a crucial factor may be the presence of the HIDB which has announced its willingness to assist with loans.

In practice, after nearly five years the expenditure on agriculture and fish farming within the IDP was £28 million, roughly twice the original estimates, and included around £7 million of private investment. Approximately one-quarter of this amount remained within the local economy, boosting local incomes (Houston, 1987). The direct impact of this investment is not easily assessed, especially in the agricultural sector. Census data suggest the IDP has halted the decline in cattle numbers whilst producing a 10 per cent rise in the sheep flock and seven-fold increase in sheep 'exported' from the Isles. This suggests that the sheep improvement scheme, taken up by 20 per cent of crofters, has been influential. Nearly 90 per cent of crofters have participated in one of the IDP schemes, but under 6 per cent have opted for the Comprehensive Agricultural Development Scheme giving special financial incentives to those adopting a variety of improvements. A more gradual and small-scale adoption of particular measures has been more typical. The same applies to fish farming where 50 new units have been established, but with over half the output coming from just four enterprises. Even so, 131 full-time and 55 part-time jobs are being supported in fish farming as a direct result of the scheme.

To further the notions of integrated rural development, the EEC has developed an integrated plan for its Mediterranean regions. This is a six-year plan, commencing in 1985, designed to raise income levels and improve job opportunities in Southern France and large parts of Italy and Greece (see Figure 6.10). Over £6 million has been allocated to this project, three-fifths coming from member countries and two-fifths from the three beneficiaries. Two types of development will be financed:

(a) improvement of agricultural production to bring it in line with market requirements.
(b) measures to create alternative rural employment, better infrastructure and improved rural services.

Of the budget, 40 per cent will go to agricultural projects, one-third for tourism projects and off-farm employment in industry and craft activities. The remainder is for afforestation, fishing and training schemes (Winchester and Ilbery, 1988: 24). The largest grants are intended for mountain and upland areas and for those lowland areas with the poorest infrastructure, e.g. Crete, eastern Sardinia, and Pyrenees Orientales. To date, significant inputs from this project have gone towards irrigation measures in the Mezzogiorno, flood protection in the Herault Valley near

Pezanas, and afforestation and improved infrastructure in upland areas both in southern France and the Appenines.

This concept of a rural management plan combining objectives that extend beyond purely economic concerns to embrace conservation issues, social considerations and the needs of a range of competing land uses is one which has been referred to by many rural planners. It has often been termed *integrated rural development*, meaning policies reflecting broad long-term concerns for development and the environment in which local objectives are identified and considered, important interrelationships determined and funding provided for specific projects (e.g. Parker, 1984).

Although approximated in the EEC's programme referred to above, it has been difficult to put into practice in national contexts, such as the British and the French, because of the plethora of competing interests and the divided forms of central and local government (Frost, 1986). Lack of policy co-ordination amongst numerous ministries, organisations, lobby groups and factions has severely limited many economic initiatives for rural areas. It is not surprising therefore that the nirvana of integrated rural development awaits concerted rationalisation of planning agencies and interest groups plus clearer recognition of how theories of regional development can be applied to specific rural areas.

10

Planning for conservation

10.1 Environmental protection and the conservation ethic

Conservation is a latent political issue, as yet only rising to the fore occasionally at national level and generating specific political representation only in fringe 'Green' parties, most notably in West Germany. This is not to say that conservation issues have not held centre stage elsewhere in public attention or brought about government action (e.g. Lowe and Goyder, 1982). The passing of Britain's Clean Air Act of 1956 is one such example. Others include the rejection of the Foulness site as the location of London's third international airport, controversies over destructive development plans in Tasmania and Queensland, New Zealand's 'Think Big' policies of the early 1980s and concerted movements against increased use of nuclear energy, e.g. in the United States, Scandinavia and Switzerland. However, 'environmentalist' policies as central pillars of national government policy have been rare, generally being perceived by government as being contrary to their notions of economic growth and increased GNP per capita.

The publication in 1988 of the fifth report on *British Social Attitudes* suggests that such views might be changed by a general shift in public opinions which could prompt new political initiatives both locally and nationally. For example, the report concludes,

Any government which seeks to alleviate housing pressures in the southern half of England by releasing swathes of land for development would, on the evidence of this survey, be taking a considerable political risk. The overall conclusion from our findings is that the countryside is a latent political issue which has not yet quite arrived. It does not, however appear to need much of a push to become a cause and, as other countries have discovered, it is an awkward cause for established political parties to contend with.

The report revealed that 82 per cent of those surveyed felt new houses should be built as part of existing settlements instead of 'in the country', with 78 per cent thinking it more important to maintain the existing state of Green Belts rather than building new homes there. However this represented only part of a general view of how the countryside should be both used and preserved:

> Conservation today seems to imply opening up the countryside to people rather than protecting it from them. And this vision of the proposed future for the countryside places heavy emphasis on the visual, rather than on the economic environment.

It is a reinforcement of the urban dweller's view of the countryside, conveniently ignoring the needs of people living or wishing to live in that countryside. This view has been enshrined in much of the development control legislation of planning systems. According to the Centre for Policy Studies (1988), although such systems have fostered some unimaginative, ugly developments themselves, plus immobility of labour and inflated house prices, they have focused upon control of development by the private sector. Unfortunately, in some cases this control has hindered what many would regard as desirable diversifications by farmers who wish to create more farm woodland as part of 'agricultural extensification'.

O'Riordan (1985) claims the mid-1980s marks the beginning of the 'fourth environmental revolution' (see Table 10.1). This 'revolution' brings together aspects of three earlier phases, combining romantic ideals, technical expertise and scientific understanding with political lobbying and creation of institutions. It has both ecocentric and technocentric elements based upon different environmental ideologies, but these elements are diverging, bringing a gulf between the 'green movement' and governments which have generally tended towards technocentrism (Nicholson, 1987; O'Riordan, 1976; Wibberley, 1976).

The 'revolutions' described by O'Riordan can be seen as part of a gradually increasing concern for preservation and protection of the Earth's resources, but this concern is one which extends beyond the start of his 'first environmental revolution'. Examples of individual attempts to protect both flora and fauna can be found throughout history in both the Old World and the New World: each of which represents one type of 'first', are the proclamations by the government of Bermuda to protect the cahow and the green turtle in the early 17th century; the seasonal prohibition on deer-hunting in Newport, Rhode Island in 1639; and the attempt during the reign of Elizabeth I to establish a cordon sanitaire or 'green girdle' around the City of London (Green, 1981). Two millennia earlier, nobles in Mesopotamia were establishing special areas away from Ur and Babylon where they would camp to escape the heat of the towns in summer. The idea of such parks and special gardens was spread throughout the Mediterranean and Europe by the Greeks and the Romans. Despite the long history of conservationist ideas, their germination and impact was generally limited so that much of the resource use over time in the Developed World must be seen as being essentially exploitative and destructive.

Even in parts of the Developed World only settled within the last two

Table 10.1 Environmental ideologies

Revolution	Date	Characteristics
I	early C18	The growth of romanticism – captured subsequently by poets and writers, e.g. Wordsworth, Emerson, Muir, Chekov, Tolstoy. Human activity judged to be ethically linked to natural processes and purposes.
II	late C19-1930s	The era of the environmental manager. Manipulation of new-found understanding of natural processes in the interests of humankind. Alteration and intervention to improve human well-being.
III	mid-1960s	Environmental concern. Creation of institutions, regulation, growth of politically articulate and legally active pressure groups. New environmental legislation, regulations, 'environmentalism'.
IV	mid-1980s	New environmentalism. Coalescence of romanticism, technical expertise, political lobbying and institution building. Has both ecocentric and technocentric strands.

Source: O'Riordan, 1985

centuries there has often been a tremendous amount of destruction of flora and fauna, the extent of damage only just being realised and acted upon (O'Riordan, 1979). For example, parts of Western Australia have lost over 90 per cent of their flowering species or six times as many as in Britain. In Australia, as elsewhere, the commonest solution to halt this destruction has been to designate a number of different types of *protected areas*. Yet this has not always produced the desired effects. For example, it has been found that in the remaining pockets of Australia's temperate forests the small surviving populations of each species are easily wiped out simply by virtue of their isolation from a larger extent of their natural habitat. The precise needs of particular fauna and flora are often not known adequately, and so even careful management can have adverse effects. Also, the favouring of specific 'national park' designations has usually ignored ideas of including patches of forest between which there can be corridors wide enough to permit migration.

Attempts at conserving particular valued environments have brought conflicts with economic concerns of resource development. In contrast, the association between recreation and conservation has often been seen as symbiotic (e.g. Cloke and Park, 1985). But it is the habitats with the highest conservation value that are most likely to attract visitors, and there is a very fine line between preservation of unspoiled landscapes and significant alteration of the object of attraction. As Revelle (1966) noted for the National Parks in the United States prior to the major growth in outdoor recreation of the 1970s:

> although these treasures of the continent are, in part, sites for active recreation, skiing, mountain climbing, fishing and camping, they are in essence great natural wonders, things of joy and beauty, places for an individual to lose himself in contemplation. While their values cannot be enhanced by human action, they can easily be destroyed by it.

10.2 The National Parks

In North America National Parks have served two principal functions: protection in an unchanged state of areas representative of major natural environments, and promotion of public appreciation of the national heritage (Zaslowsky, et al., 1986) (Plate 19). For example, in Canada the Parks Canada Policy of 1979 states that the National Parks are designed 'to protect for all time representative natural areas of Canadian significance in a system of National Parks, and to encourage public understanding, appreciation and enjoyment of this natural heritage so as to leave it unimpaired for future generations.' Consequently, National Parks are protected by federal legislation from all forms of extractive resource use such as agriculture, forestry, mining and hunting for sport (Nelson, 1987).

Canada has developed a set of criteria to designate potential National Park sites conforming to IUCN standards. Parks Canada has formulated the *natural region concept*, areas containing a unique set of biological and ecological characteristics, with the intention of having at least one park in each region (Foresta, 1985; Nelson, 1973). To date Canada has defined 48 natural regions, 22 with National Parks.

This concept represents more than a century of the evolution of National Parks' designation since Yellowstone National Park was established in 1872. This example was followed in Canada in November 1885 when an area of 26 sq km on the northern slope of Sulphur Mountain, Alberta, was set aside for public use. Hot springs had been discovered three years earlier and were already attracting tourists in the way in which the volcanic areas of Yellowstone had also done. The Rocky Mountain Act, establishing what is now Banff National Park, was passed in June 1887. The North American lead was soon taken up in the Southern Hemisphere, though Europe lagged behind.

Although land for New Zealand's first National Park, Tongariro, was set aside in 1887, only 15 years after Yellowstone was created, it was not until 1909, 15 years after Tongariro's formal dedication, that the first European National Parks were established in Sweden (Abisto, Peljekaise, Sarek and Sonfallets). In 1914 the Swiss National Park was created, then in 1918 Spain created two National Parks and Italy followed with dedications in 1922 and 1923. However, it was not until

Plate 19 National Parks – 1. Rocky Mountain National Park: Tourists can enjoy wilderness where mineral workings, commercial agriculture, forestry and settlement are either prohibited or strictly controlled

the 1950s that large numbers of parks were established (Figure 10.1). Indeed, France had no such parks until 1963, when Vanoise and Port Cros National Parks were created, and West Germany's first National Park, Bayerischer Wald, was not created until 1969. Also, post-1945 several 'National Parks' were designated which, subsequently, have not been regarded by the International Union for the Conservation of Nature (IUCN) as conforming to their international standard (IUCN, 1985). None of Britain's 'National Parks' conform to the standard set out in Table 10.2.

Britain's National Parks come under the IUCN's category of 'protected landscape or seascape' in which the imprint of human occupation is extremely important and where recreation and tourism plays a significant role (Plate 20). The IUCN definitions also exclude Austria, Belgium, Denmark, Luxembourg and Portugal from the list of countries with National Parks. A comparison of national statistics on National Parks shows that the European countries have a much smaller proportion of their total area under National Parks than North America and Australasia. However, they also have a large number of other forms of protected areas. The Scandinavian countries, with comparable low population densities to those in many parts of the New World, have the highest ratios of Parks to surface area. It is not surprising, perhaps, that the first National Parks in Europe were in a remote part of northern Sweden, and that 40 per cent of Europe's National Parks are in polar

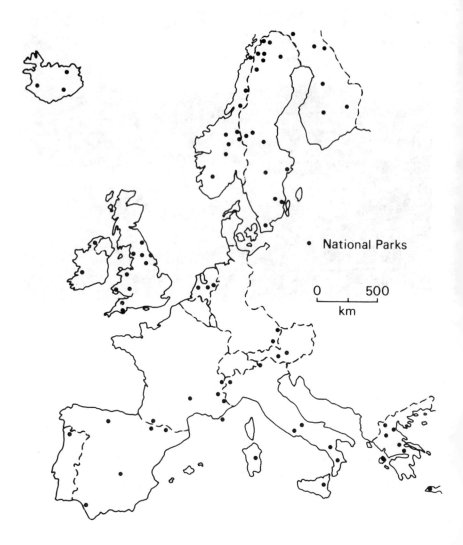

Figure 10.1 National Parks in Western Europe

Table 10.2 IUCN standards for National Parks

1. Large areas where one or several eocsystems are not materially altered by human activity.
2. Contents of outstanding plant, animal or geomorphic features of value for special scientific, educative and recreative interest.
3. Managed by highest competent authority.
4. No exploitation or occupation.
5. Visitors allowed for educative, cultural and recreational purposes.

Source: IUCN (1969)

Plate 20 National Parks – 2. Peak District National Park (Edale): Although within the National Park, the landscape is far from 'wilderness' and the hill farms and small hamlets resemble many found beyond the designated Park boundaries

and sub-polar regions where pressure from competing land uses is considerably reduced. One-quarter also abut onto an international frontier, e.g. Borgefell (Norway), Lemmenjoki (Finland) and Mercantour (France) (Pearce and Richez, 1987).

Land use competition has continued to be a feature of the European National Parks, with growing pressures from tourism and recreation competing with agriculture, forestry, mineral extraction and power generation. Approximately one-third of the European National Parks have agricultural activity within their boundaries, though some, such as Italy's Circco National Park, have now been withdrawn from the IUCN list because of the extent of farming activity. The extension of hydro-electric power (HEP) schemes within National Parks has been an emotive issue. For example, in New Zealand, the 'Save Manapouri' campaign, against the 1960 proposal to raise the level of Lake Manapouri in Fiordland National Park, became a major national environmental issue. Although a power station was built, the lake level was not raised (Henwood, 1982). Even so, Fiordland has around two-thirds of its area subject to exploration, prospecting and mining licences which pose a threat to protection of flora and fauna should significant mineral deposits be discovered (Wallace, 1981).

In France the legislation in 1960 permitting the establishment of

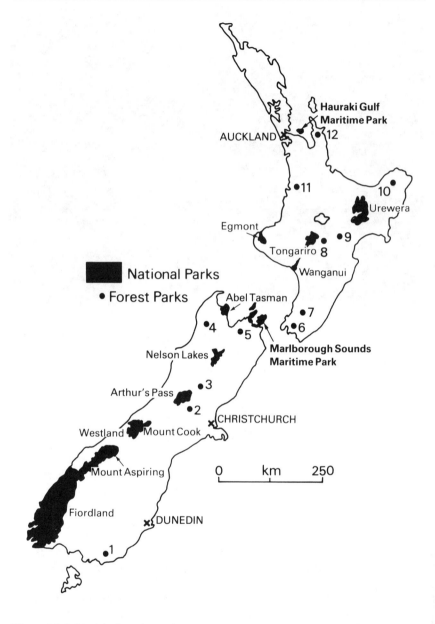

Figure 10.2 National Parks and major Forest Parks in New Zealand (1 –
Catlins; 2 – Craigieburn; 3 – Lake Sumner; 4 – North-West Nelson; 5 – Mount
Richmond; 6 – Rimutoka; 7 – Tararua; 8 – Kaimanawa; 9 – Koweka; 10 –
Raukumare; 11 – Pirongla; 12 – Coromandel)

National Parks states contradictory objectives for conserving the environment and meeting recreational needs (Clout, 1975). To reconcile these
conflicting aims an ingenious system of zones within the Parks has been
designated permitting different degrees of access. A central philosophy of
the designation of National Parks in France has been the need to maintain a population of sufficient size to support the traditional landscape.
To this end more control has been exerted over land use than in the
parks in England and Wales. For example, in the Parc National des
Cevennes, mechanised forestry is not permitted as one of the Parc's
objectives is to manage the area in perpetuity as a semi-natural mixed
broadleafed/coniferous forest. Similarly, although agriculture faces
difficulties imposed by the climate and physiography, special assistance is
given to farmers in order to maintain smallholdings and therefore retain
a significant level of population on the land (Blunden and Turner, 1985:
134–51).

New Zealand has a long-established system of National Parks, which
now covers one-thirteenth of its land area (Figure 10.2; Table 10.3). The
model for the earliest of these, Tongariro, in 1887, was Yosemite, and
the second, Egmont, in 1900, followed similar lines. The land for these
and subsequent parks was publicly owned, although without any
national co-ordination until the National Parks Act of 1952. This
formally recognised the National Parks as areas to be preserved for the
benefit and enjoyment of the public, as far as possible in their natural
state. Priority was given to preservation, with a National Parks Authority
to keep in a state of nature what were wilderness areas.

The French and New Zealand examples illustrate the contrast that can
be drawn between designation of Parks in areas of wilderness and
designation in areas already occupied by settlement and economic
activity. For countries where National Parks have included commercial
agriculture, forestry and other 'non-wilderness' land uses, conflicting aims
between conservation, recreation and resource use have been heightened.
In such cases legislation attempting to resolve the conflicts has often been
complex, with recognition of different types of designated conservation
and recreational areas. Britain provides some of the best examples of this
complexity and a range of designated areas extending beyond the
National Parks.

10.3   Legislation for conservation and conservation agencies in Britain

10.3.1   Agencies and legislation for conservation pre-1950

The concept of conservation in Britain and much of Europe developed
along different lines to those of North America (see Adams, 1986).
Conservation was championed much more as a form of planned resource
use and followed a long-established tradition related to maintaining

Table 10.3 National Parks in New Zealand and England and Wales

A. New Zealand

Name	Designated	Area (ha)	Name	Designated	Area (ha)
Tongariro	1887	78,651	Urewera	1954	212,672
Egmont	1900	33,543	Nelson Lakes	1956	96,121
Fiordland	1905	1,252,297	Westland	1960	117,547
Arthur's Pass	1929	99,270	Mount Aspiring	1964	289,505
Abel Tasman	1942	22,530			
Mount Cook	1953	69,923			
					2,272,059

Other reserve land: 935 Scenic Reserves (255,864 ha), 63 Historic Reserves (1,414 ha); 45 Scientific and Special Reserves (187,010 ha), 886 Public Parks (25,156 ha), Forest and Water Reserves (7.6 million ha).
Total reserve land covers 28 per cent of the country. A new National Park, Wanganui, was designated in 1989.
Source: Pearce and Richez (1987)

B. England and Wales

Name	Designated	Population	Visitors per day p.a.	Land use (%) Enclosed farmland	Open moors	forest	Area (ha)
Peak District	1951	38,000	20 mill.	49	39	7	140,400
Lake District	1951	40,300	20 mill.	33	50	11	224,300
Snowdonia	1951	25,000	9 mill.	21	60	14	217,100
Dartmoor	1951	30,000	8 mill.	35	52	9	94,500
Pembrokeshire Coast	1952	21,500	1.5 mill.	84	12	4	58,300
North York Moors	1952	24,000	11 mill.	4	35	23	143,200
Yorkshire Dales	1954	17,250	7.5 mill.	41	56	3	176,100
Exmoor	1954	10,000	2.5 mill.	57	28.5	10	68,600
Northumberland	1956	2,000	1 mill.	9.5	71	19.5	103,100
Brecon Beacons	1957	32,000	7 mill.	46	42	12	134,400

Source: Council for National Parks

game, fisheries, forests and other natural resources. An early example of this type of concern is the New Forest, in Hampshire, established during the reign of William the Conqueror as a hunting chase nearly 1,000 years ago. The maintenance of sanctuaries for game has had a major impact upon preserving small coppices and thickets in the English Midlands and open moorland in upland areas. Paradoxically, it was the reaction to hunting and shooting performed by the upper classes that provided the stimulus for the formation of protection agencies in Britain such as the Royal Society for the Protection of Birds (RSPB) (in 1889) and the National Trust for Places of Historic Interest or Natural Beauty (1895).

The latter has roots in the Commons, Open Spaces and Footpaths Preservation Society (COSFPS), formed in the mid-1860s to contest the enclosure of the remaining common lands for agriculture. John Stuart Mill was a leading member of this society and he, together with leading figures in the romantic movement, such as John Ruskin, Holman Hunt and William Morris, was influential in the emergence of a conservation ethic. They can be compared with men in the USA like George Perkins Marsh and Edwin Muir who were motivated by ethical and aesthetic reasoning in promoting conservation. Robert Hunter, secretary of COSFPS, was instrumental in the formation of the National Trust as a body that could hold and manage land to preserve areas for the benefit of the nation. In 1907, under the first National Trust Act, the Trust was given power to hold land in perpetuity (Fedden, 1974). The central aim of the Trust was the preservation of the natural aspect of the countryside and its animal and plant life. However, although its early acquisitions included *nature reserves* such as Blakeney Point (Norfolk) and Wicken Fen (Cambs), it became more concerned with landscape, ancient monuments and buildings than wildlife.

The preservation of wildlife was the expressed concern of the Society for the Promotion of Nature Reserves, formed in 1912 to help the National Trust to create nature reserves. In 1915 this Society published a list of potential nature reserves and was subsequently active with COSFPS and the RSPB, in lobbying the government to form a state-controlled organisation to protect the countryside. They advocated the establishment of National Parks along similar lines to those in North America and Australasia. This desire was also expressed by the Councils for the Preservation of Rural England, Scotland and Wales, founded in 1926, 1927 and 1928 respectively.

Government response to this lobbying came in 1929 in the form of a committee of inquiry into the need for National Parks. The report of this committee laid the basis of future legislation as it recommended the establishment of such parks to be administered by a statutory authority. It also foresaw one of the future conflicts within the National Parks – that of the need to provide access to members of the public versus the aim of protecting wildlife and landscape. There was even a suggested solution to this conflict in terms of designating national reserves and regional reserves, the former to be mainly for nature protection and the latter mainly for outdoor recreation.

Unfortunately these recommendations were not implemented in the 1930s, a decade in which 247,000 ha (610,350 acres) of agricultural land was lost to urban development in England and Wales and when thoughts of establishing specifically protected areas of the countryside were secondary to the need to strengthen agriculture and then prepare for the ensuing World War. Various Royal Commissions were forced to investigate these problems, but it was not until 1942 that the Scott Committee on Land Utilisation in Rural Areas recommended the establishment of national and regional parks and nature reserves. These proposals were further developed in the Dower Report in 1945 in which the need for two different types of park was not accepted. Dower saw National Parks as areas in which the countryside could be protected as well as being made available to the public for informal recreation. This parks system, including nature reserves, would be managed by an autonomous authority operating in a similar manner to the United States National Parks Service. Detailed recommendations were then formulated in the Hobhouse and Ramsey Committees, published as white papers, and effected in the 1949 National Parks and Access to the Countryside Act.

One of the outcomes of the 1949 National Parks and Access to the Countryside Act was the formation of the Nature Conservancy, reconstituted in the 1960s as the Nature Conservancy Council (NCC). Its prime functions are the establishment and maintenance of National Nature Reserves (NNRs) and Sites of Special Scientific Interest (SSSIs). This means that it is responsible for 44,000 ha (108,725 acres) of land in 123 NNRs in England and Wales (Figure 10.3) and 800,000 ha (1,976,800 acres) in 3,171 SSSIs. In addition, it provides advice to government on policies of nature conservation, and to all interested parties on a variety of conservation issues. It also commissions and supports research, though its main function was placed in the hands of the Institute of Terrestrial Ecology (ITE) in 1973.

The NCC is a most influential lobby group. Under the 1973 Nature Conservancy Council Act the Council, which had become part of the Natural Environment Research Council, was made an independent body funded by the Department of the Environment (DoE). It was given three main functions:

(a) to advise Ministers on nature conservation policy and to provide advice and information generally on nature conservation;
(b) to establish and manage nature reserves;
(c) to encourage research into conservation.

However, in practice, its research function was taken by the newly formed ITE so that the NCC's main roles have become managerial and advisory.

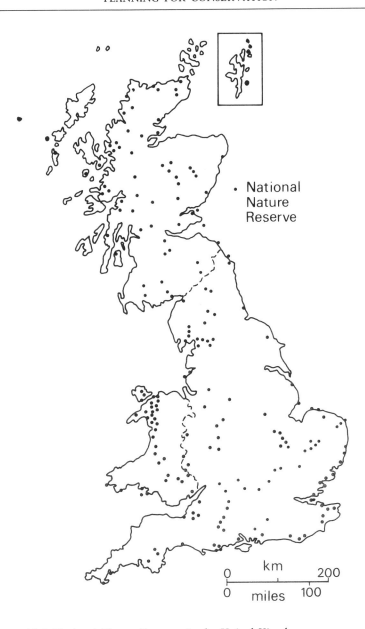

Figure 10.3 National Nature Reserves in the United Kingdom

10.3.2 National Parks

The 1949 National Parks and Access to the Countryside Act listed five specific functions of a National Parks Commission:

(a) to have an overall duty for preservation and enhancement of natural beauty throughout the country, although especially in relation to National Parks and Areas of Outstanding Natural Beauty (AONBs);
(b) to encourage the provision of facilities for the enjoyment of National Parks;
(c) to designate National Parks;
(d) to designate AONBs;
(e) to give advice and assistance to local authorities.

Under the Act any land in a National Park could be acquired by agreement or compulsory purchase order to give the public access to it for outdoor recreation. Indeed, special attention was given to public access, so following the recommendations of the Dower Report. For example, provision was to be made for the establishment of special long-distance routes for walkers, cyclists and horse-riders. Routes could be over either existing rights-of-way or new ones for which land could be acquired compulsorily if necessary. County councils were to determine the extent of open country and to decide whether it was necessary to secure public access to it. If this was deemed to be so then access agreements were to be reached with landowners or access orders issued, with landowners receiving compensation. Certain areas were exempted from access orders, though not from access agreements. Amongst these areas were agricultural land other than rough grazing, parks, gardens, golf courses and nature reserves – a range of land uses which symbolises some of the conflict which has remained in the National Parks between public access, conservation and economic activity.

The Act established a National Parks Commission with powers to prepare a programme for the establishment of National Parks. But the Commission was only to play an advisory role in guiding local planning authorities who would establish and manage the Parks. These authorities had been given wide powers under the 1947 Town and Country Planning Act, and felt it unnecessary to have an executive body for the National Parks that would usurp their role in certain parts of the country. No Parks were deemed necessary in Scotland as here it was felt the pressure of population upon the countryside was much less than in England and Wales, and so the need for extra protection was not necessary at that time.

The 1949 Act permitted the establishment of *National Parks Authorities* which were 'to preserve and enhance the natural beauty of the Parks, and to promote their enjoyment by the public, executing both with due regard to the needs of agriculture and forestry.'

National Parks Authorities were given a responsibility which has since promoted conflicts of interest. On the one hand the authorities were to preserve and enhance the natural beauty of the Parks whilst on the other promoting the enjoyment of the Parks by the public. Frequently, public access to the Parks has been at variance with conservation ideas. Although this has also been true in other National Parks throughout the world, those in Britain have faced a special problem because they are not wilderness areas. Villages, hamlets and farmland play a major role in the scenic attraction of the landscape, and there has generally been little opportunity of influencing the activities of many of the individual landowners in the way certain conservation groups would have wished. This is another problem peculiar to National Parks in Britain – the majority of land (74 per cent) is privately owned. Regulations upon what these landowners can do are relatively strict but have not prevented certain important changes influencing the character of the landscape (Blacksell, 1979; MacEwan and MacEwan, 1981; 1987).

Of land over 240 metres in England and Wales, the area defined as 'upland' by the Countryside Commission, nearly 70 per cent is within National Parks or AONBs. One of the main landscape types within these uplands are moorlands. Although these rarely represent species of climax vegetation, their present characteristics have evolved over many centuries and their retention is felt to be desirable by many conservation groups. Yet since the creation of National Parks in England and Wales, and especially during the past decade, there have been notable declines in the amount of open moorland and the numbers of deciduous trees in the Parks.

One of the main factors in the loss of moorland, which is averaging 5,000 ha (12,350 acres) per annum, is the extent of reclamation and new enclosures made by farmers. These are centuries old processes, but have been encouraged in recent years by grants from both the EEC and the MAFF for conversion of moorland to improved pasture. One estimate of £400 million given by the MAFF as grants to farmers in National Parks contrasts with an overall budget of just £7 million for the National Parks Authorities. Given these priorities in spending it is little wonder that natural and semi-natural vegetation is replaced by cultivated pasture. There is also another threat to the retention of moorland. This is posed by coniferous plantations. Of the Forestry Commission's 1.16 million ha (2.87 million acres) of land in Great Britain 20 per cent are within the National Parks. Their policy has been geared towards the planting of coniferous trees at the expense of both moorland and native deciduous woodland.

Perhaps the greatest controversy over the loss of moorland has concerned the enclosure of the moorland edge in Exmoor National Park (Curtis, 1983). Following publication of the Porchester Report in 1977, Park authorities produced maps of moorland with 'high amenity value'. This designation has been used in conjunction with management agreements based on a voluntary notification system in which farmers notify the authorities of proposed alterations they intend to make to the moorland. Under such agreements the farmers receive compensation for profits lost by retaining moorland as rough grazing.

The agreements reached in the case of moorland in Exmoor are similar to those laid down by statute in the Wildlife and Countryside Act of 1981. Under this Act if farmers within the National Parks are seeking grant aid from the MAFF for agricultural changes, they must notify the appropriate Parks Authority. If the Authority feels the proposed changes are not in the best interests of conservation it is to attempt to reach an agreement with the farmer. This agreement is to include compensation for the loss of income derived from the proposed change and, as in the case of Exmoor, also includes compensation to help finance change. Hence, the implementation of such management agreements will be costly and the authorities are likely to require finance well above current levels if they are to meet this demand. Yet no proposals have been made for substantial increases in the budget of the National Parks Authorities or the Nature Conservancy Council.

One solution to the conflict between conservationists and the various agents of change would be for more land to be controlled directly by the National Parks Authorities. Since the first Parks were created in 1949, only 1.6 per cent of the land in the Parks has been acquired by the Parks Authorities. However, funds for more outright purchase have always been inadequate. The alternative lies in the formulation of management schemes, of which several imaginative ones have been implemented in other countries (e.g. Coulmin, 1986). Within the wider context of the EEC, 11 studies have been made exploring a package of policies and public support measures involving local communities and covering all aspects of rural life. This scheme includes extra finance of £70,000 per year to supplement existing grants available for the encouragement of 'desirable' management practices. One example is the Peak District Integrated Rural Development Programme. Though this has covered only a small area, it has improved employment prospects and addressed environmental issues in its targeted parishes (Parker, 1984).

There are various other types of management schemes within the National Parks. The most long established are the Upland Management Experiments (UMEX) in the Lake District and Snowdonia. The first of these was introduced in 1969 and the most recent in 1978. Their aim has been to encourage farmers to improve both the landscape and recreational facilities. Such projects include the promotion of self-catering accommodation through the conversion of redundant farm buildings. In Snowdonia there has also been a five-year scheme, started in 1975, to halt erosion and to repair damage done by growing numbers of tourists. This had an annual expenditure of over £200,000, focused especially upon the various paths to the summit of Mount Snowdon (Taylor, 1978). Similarly, 27 erosion 'blackspots' have been identified on paths in the vicinity of Kinder Scout, Bleaklow and Featherbed Moss in the Peak District National Park. The laying of an impermeable textile base covered with gritstone has been used in special repairs. Meanwhile, similar problems in the Cairngorms and along the Pennine Way have been treated with daggings, hard-wearing fibres from the hind parts of sheep, mixed with gravel aggregate. Grease and nitrogenous matter in the daggings will encourage plant growth and so increase the longevity of the footpaths' foundation.

10.3.3 Areas of Outstanding Natural Beauty (AONBs)

The 1949 Act also catered for the establishment of AONBs, following recommendations by Dower and Hobhouse that certain areas of high landscape, wildlife and recreational value merited conservation, if not quite the same degree of management to be accorded the National Parks. As a result, during the 1950s ten National Parks, twelve AONBs and three long-distance footpaths were established (Figure 10.4). However, as a consequence of the limited powers the National Parks Commission could exert in the form of development control, there was a good deal of activity in the Parks that conflicted with conservation ideals (MacEwan and Sinclair, 1983). In addition to agriculture and forestry enterprises, the Parks also included land devoted to military use (see Figure 10.5), two nuclear power stations and a radar early-warning station. These Parks were far removed from the North American model of the wilderness park, and with no specially created management authority for all but the Lake District and Peak District National Parks, they did not meet the international definition of 'National Parks' laid down subsequently by the IUCN (1969) (Brotherton, 1985).

AONBs were first mooted in the Dower Report which identified 'other amenity areas' not suitable for designation as National Parks due to their limited size and lack of wilder countryside, but requiring some degree of protection in order to provide recreation, especially in south-west England. The National Parks Commission endorsed this view in their definition of 'conservation areas' which were then designated AONBs in the 1949 Act. These were 'any area in England and Wales not being a National Park but of such outstanding natural beauty that some provision of National Parks should apply'. The designation meant that local authorities within AONBs had similar powers to those in National Parks to preserve and enhance local beauty, and with an obligation to consult the National Parks Commission when preparing development plans. Limited grant aid powers were given for landscape improvement, provision of public access to open country and warden services. Subsequently, 35 AONBs were designated, covering nearly 6,640 sq miles (17,000 sq km), primarily in the lowlands.

Reviews of the AONBs in the 1980s have noted that objectives of 'enhancing beauty' were not always being achieved because of inadequate resources and lack of clear planning and management strategies (Countryside Commission, 1980; 1983). New policies are now seeking to improve co-ordination between local authorities, landowners and land managers. However, statutory management plans for AONBs have not

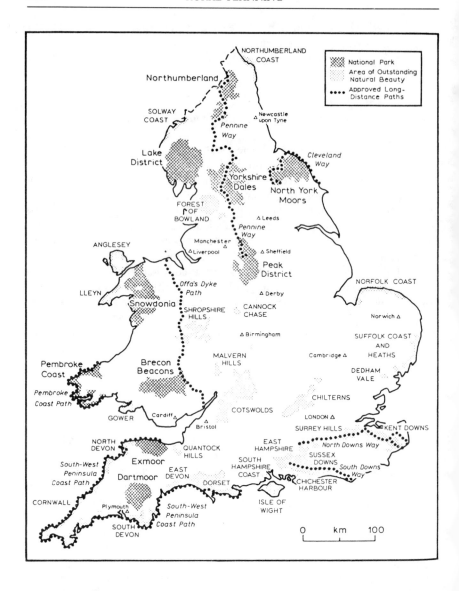

Figure 10.4 National Parks, Areas of Outstanding Natural Beauty (AONBs) and Long-distance Footpaths in England and Wales (*Source:* Clout, 1972: 88)

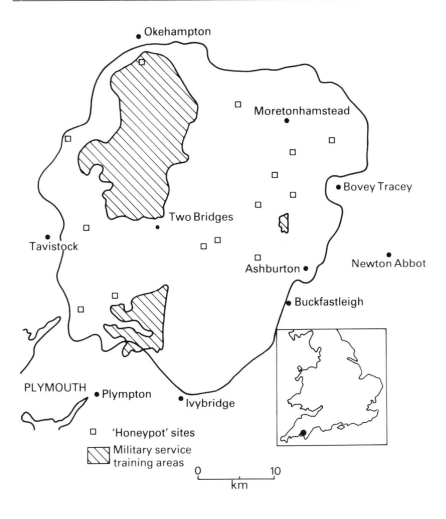

Figure 10.5 Dartmoor National Park – Land designated for military training and 'honey-pot' sites for car-borne tourists

been introduced, and major weaknesses still exist concerning the implementation of the AONB concept (e.g. Anderson, 1981; Cooper *et al.*, 1989). Critical ones include:

(a) The lack of explicit criteria for designating AONBs, so that boundaries of the AONBs have not marked distinct variation in landscape quality (Preece, 1980);
(b) Development control has not always proved an appropriate mechanism for regulating building development (Preece, 1981; Sellgren, 1986);

(c) Policies for nature conservation and landscape conservation have not proved complementary in the absence of detailed management plans (Anderson, 1981; Sellgren, 1988).

Despite the many conflicting pressures within the British National Parks and AONBs, control policies have been followed quite strictly. For example, Blacksell and Gilg (1975) reported that rates of refusal of planning applications were higher in the Dartmoor Park than elsewhere in rural Devon whilst TRRU (1981) made a similar finding with respect to the restrictive attitude towards residential development in the Parks in general. This control has been apparent despite the fact that, in the Parks, planning applications per 1,000 population have been two and a half times the national average (Brotherton, 1982: 455).

10.3.4 The Countryside Commission and new views on rural conservation

A Government White Paper in 1966, Leisure in the Countryside, recognised the vast increase in recreation in the countryside and proposed various steps to deal with the expected further rise in such pressures. In particular it was suggested that the functions of the National Parks Commission should be widened so that the whole countryside was covered rather than just that within National Parks. This was duly carried out in two Acts: the 1968 Countryside Act which replaced the National Parks Commission with the Countryside Commission, and a parallel statute in Scotland, the 1967 Countryside Commission (Scotland) Act, establishing the Scottish Countryside Commission. The roles of the Countryside Commission were to conserve the landscape beauty of the countryside, to develop and improve facilities for recreation and access in the countryside, and to advise government on matters of countryside interest.

In effect, the Countryside Commission is an advisory civil service body to the Department of the Environment (DoE) and is not an executive body. It has no landowning interests, though as well as giving advice it can sponsor research and practical projects and give grant aid. It also designates and advises on policies for both National Parks and AONBs. In its original remit it seemed that, once again, the conflicts between recreation and conservation appeared to have been overlooked. However, in the subsequent development of the provision of Country Parks and picnic sites, clear differences can be seen between their functions of these new designated areas and those of the National Parks (Curry, 1985). The 1967 and 1968 Acts gave powers to both local authorities and private individuals to provide these picnic sites and Country Parks in rural areas for the enjoyment of the countryside by the public, and there are now numerous Country Parks providing opportunity for outdoor recreation in relatively close proximity to major urban areas (Figure 10.6).

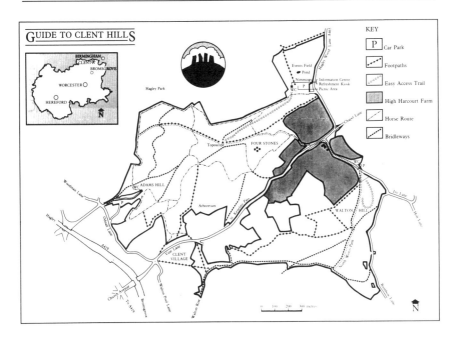

Figure 10.6 Clent Hills Country Park, Hereford and Worcester (*Source: Countryside Service, Hereford and Worcester County*)

The Countryside Commission has also played a major role in developing policy on rural issues, its broad remit enabling it to play a pivotal function in dealing with the growing pressures upon the countryside. In particular, amongst the Countryside Commission's work on the formulation of policy towards and recommendations on major issues affecting the countryside have been a series of major reports, e.g. *New Agricultural Landscapes* (1974), *The Lake District Upland Management Experiment* (1976), *Areas of Outstanding Natural Beauty* (1980), *The future of the uplands* (1981a) and *Countryside management in the urban fringe* (1981b).

More recently these have been followed by one focusing upon recreation, *Enjoying the countryside* (1987a), a general review of rural policy proposals, *New opportunities for the countryside* (Countryside Policy Review Panel, 1987), and consideration of multiple forestry goals, *Forestry in the countryside* (1987c). Within this work is the recurrent theme of a need for competing land uses to be controlled more effectively by policies and systems of management that curb the limited sectional interests of particular 'lobbies', e.g. landowners, MAFF, the Forestry Commission, conservation groups. Yet the implementation of 'integrated' management still seems remote in the current administrative spider's web that provides the framework for countryside management in Britain. Perhaps the best illustration of this remoteness lies in the conflicts promoted by legislation enacted in the 1980s. Although intended to promote agreed forms of rural land management, many feel it has been insufficient to curb exploitative development of rural resources (see Lowe *et al.*, 1986).

10.3.5 The 1981 Wildlife and Countryside Act

Recent conservation legislation in Britain has been controversial. Much debate preceded the passing of the Wildlife and Countryside Act in 1981, and subsequently there have been a variety of complaints about its likely ramifications (e.g. Adams, 1984b; Cox and Lowe, 1983; Lowe *et al.*, 1986: 133–55). The Act addressed the question of improved protection and conservation of the landscape, with three principal areas of concern. These were the protection of birds, the conservation of certain habitats, and public rights of way in the countryside. It was the second of these issues that provided the most substantive part of the Act and which has occasioned the fiercest debate. It refers to new provisions which apply primarily to SSSIs and to the National Parks.

The Act emphasised the 'voluntary principle' favoured by farming organisations (Leonard, 1982).

This approach allows that a farmer prevented from receiving grant aid from MAFF because of the NCC's objections on nature conservation grounds will be offered a management agreement by the NCC based on set government Financial Guidelines (Adams, 1984a: 275).

In other words, the farmer will receive the income he forgoes in not developing the land. The land concerned is a designated SSSI, of which there are 4,048 covering 1.377 million ha (3.4 million acres) or 6 per cent of the land surface of the United Kingdom.

SSSIs are identified by the NCC. They are sites of biological and/or geological/physiographical interest selected for purposes of conservation. Certain types of habitat such as peatlands, heaths and woodlands together with geomorphic and geological features have been selected by the NCC who have deemed them to be suitable for maintenance in their present form. Not all these habitats lack human interference and may depend upon certain management practices for their continued existence. It is this question of management that is considered by the Act.

Under the Act, once the NCC designates an SSSI its owners or occupiers must give notice of any operations on the site that the NCC considers might be damaging, for example the draining of a water meadow or felling of ancient woodland. If such action is deemed potentially harmful to conservation and if the MAFF is willing to provide grant aid for the action, the DoE must be consulted before the grant can be awarded. If the DoE supports the NCC's view, the NCC must then seek a management agreement with the landowner or occupier within a period of three months. According to guidelines issued in 1983, in such an agreement the farmer concerned must be compensated for the loss of

Table 10.4 An example of costs incurred in compensation to farmers following the 1981 Wildlife and Countryside Act

For drainage of 25 ha of pasture in order to plant winter wheat:

Present net income from this land = £2450 p.a.
Income under winter wheat   = £9600 p.a.

Potential extra profit   = £7150 p.a.

Cost of drainage = £16,250
(reduced by grants to £10,150) – to be paid for over 20 years @ 10% =
annual cost of £1200

Extra income from winter wheat less annual cost of drainage =
£7150 – £1200 = £5950

The £5950 could be paid by the Nature Conservancy Council to the farmer as compensation for preserving the pasture, but 45% of the value of the winter wheat comes from artificially high EEC price levels:

i.e. Grants not received   = £ 500
     Consumer subsidies
        farmer forgoes   = £3150
     'True' cost of
        compensation   = £2300
                               ────────
                                 £5950

anticipated future profits. If no agreement is forthcoming it is possible that a Nature Conservancy Order may be imposed requiring the applicant to delay any action for a further period of up to nine months during which an agreement may be reached.

The Act aims to retain the co-operation of farmers in the conservation of wildlife habitats, but various shortcomings have been highlighted. The main concern voiced by several conservation organisations has been the likely cost of the management agreements. An illustration of this cost is shown in Table 10.4. This indicates that compensation for the loss of anticipated future profits includes an element of recompense for subsidies not received from the Guarantee Fund of the CAP. Thus, there is a difference between the 'true' cost of compensation and that which also includes money relating to grants not received and consumer subsidies that the farmer will miss. In some cases this 'true' cost may be less than half that of the total agreed in the management arrangement. This aspect of the prospective compensation has provoked much criticism of the Act and recognition of the legislation's weaknesses by the government itself (e.g. Caulfield, 1984). Other problems already encountered have been a

Table 10.5 The estimated cost of possible management agreements on the whole SSSI system

Habitat	Estimated area (m ha)	Estimated annual cost per hectare £	Estimated total annual cost of agreements £
Woodland	0.3	50	15 m
Lowland grassland	0.21	100	21 m
Peatlands	0.14	30	4.2 m
Uplands	0.13	20	2.6 m
Annual total			42.8 m

Source: Adams (1984a: 279)

lack of NCC staff to deal with the administration of the agreements and insufficient time to formulate them. But the principal stumbling block to its successful implementation is likely to be insufficient money for implementation. The way in which the financial arrangements are inflated by the effect of grants and subsidies will at least double the cost involved. Therefore the NCC's minimum estimate of a cost of £20 million over the first ten years of the scheme's operation may well be short of the amount required.

The Act has proved to have crucial loopholes, most notably in terms of notification to landowners of the existence of an SSSI, the limited time allotted to preparing management agreements and lack of finance with which to pay compensation. If voluntary agreements were concluded for every SSSI, the budget of the NCC would be exceeded by a considerable amount, e.g. agreements to prevent conversion of grazing marsh to arable in the Somerset Levels and the Swale Estuary, Kent, may cost £200 per ha (£81 per acre) (Adams, 1984b; Nix, 1983). Adams's (1984a) estimates for the costs of agreements for the whole system of SSSIs are shown in Table 10.5. Also, some new solutions have been found to ensure greater protection for certain endangered sites, e.g. in the Halvergate Marshes, Norfolk, a General Development Order was used to require planning permission to be granted for drainage work (Adams, 1984a: 279).

Similar problems exist in Scotland where there have been complaints over the lack of funding to protect the National Scenic Areas (NSAs). The Countryside Commission (Scotland) received a grant of £5.5 million in 1987 compared with £22.7 million for its counterpart in England and Wales. Yet, Scotland's NSAs cover six times the area of the National Parks south of the border.

10.3.6   Agriculture and rural development versus conservation

The contribution of both agricultural and forestry policies throughout the Developed World, and especially within the EEC, towards dramatic changes in rural landscapes has become increasingly a topic brought before the attention of the general public (Robinson and Blackman, 1989). Concern has been expressed by a growing number of organisations over the destruction of the variation within the landscape caused by the erosion of certain types of habitat at the hands of modern farming and commercial forestry. Farms have lost more of their natural and seminatural features to make way for more land under tillage or pasture, supported by fertilisers and pesticides permitting the growth of a monospecies of grass. Examples given below of the conflict between modern farming methods and landscape conservation illustrate the difficulty of producing legislation that can balance the needs for food production on the one hand and protection of the environment on the other (see O'Riordan, 1987).

In Britain one of the chief symbols of the conflict between farming and conservation of the countryside has been the hedgerow. Although many hedgerows were only brought into being as a result of the enclosure movement in the 18th and 19th centuries, they are viewed by many as representing an essential component of rural landscape in lowland Britain, perhaps being matched in the uplands by dry-stone walls. The removal of hedges and walls post-1945 has been fostered both by the changing nature of agricultural activity and by government policies driving these changes and giving special grants for field enlargements (e.g. Robinson, 1983b: 115). These enlargements have permitted the more effective operation of large-scale equipment used especially at harvest time and have produced scale economies, e.g. effective working time increases by two-thirds if field size is increased from 2 ha (5 acres) to 10 ha (25 acres) and by nearly 75 per cent in a 40 ha (100 acres) field. It is not surprising therefore that of the 494,000 miles (790,400 km) of hedgerow in England and Wales in 1947, 109,000 miles (174,000 km) or 22 per cent had been lost by 1985. The rate of loss increased in the early years of the 1980s, as favourable prices for cereals and some break crops encouraged arable farmers to strive harder for production increases. This has meant that East Anglia and the eastern counties in particular have lost hedgerows. Other measures of destruction are the loss of four-fifths of chalk downland, over half of the fens and mires, and half the ancient lowland woods since 1945, mainly to the extensions of the arable acreage. Destruction of flora and fauna has also been great through ploughing of grasslands, poisoning by pesticides, drainage of marshes, pollution of waterways and overgrazing of moorland.

There was a change in policy by MAFF in 1985 that encouraged some reversal in the destruction of hedgerows. Farmers were able to recoup 40 per cent of the cost of planting or repairing dry-stone walling. The MAFF claim this has produced an addition of 2,770 miles (4,432 km)

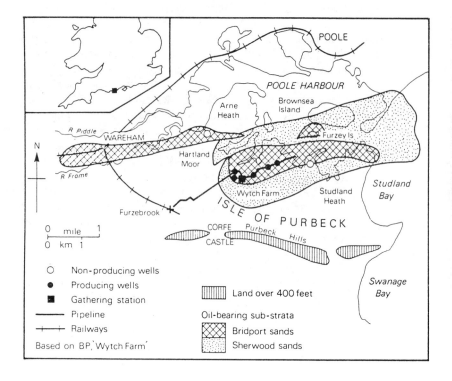

Figure 10.7 The Wytch Farm Oilfield, Dorset (*Source:* Robinson, 1986b)

in four years, but this must be set against the annual rate of removal of
3,700 miles (5,920 km) between 1980 and 1985. Similarly, the creation
of Environmentally Sensitive Areas (see Chapter 6) and the work of Farm
and Wildlife Advisory Groups (Cox *et al.,* 1985) are positive forces for
conservation but are working against market forces and farm support
policies encouraging increased production. In addition, farming's
contribution to environmental destruction is not the only one that has
produced a dramatic decline in flora and fauna. Other major contri-
butors include continued urban sprawl, rural industrialisation and the
growing use of the countryside as the playground of the urban population (Blacksell, 1982). An example from south-western England, where
there is concern over the planned expansion of the Wytch Farm oilfield,
illustrates the cumulative impact of these destructive forces.

The oilfield underlies heathland, saltmarshes and distinctive flora and
fauna. Appraisal wells for the planned expansion are located within an
SSSI. Also in the immediate vicinity of the oilfield is an AONB, the Dorset
Heritage Coastline, National Trust Land, and significant wader and
wildfowl populations around Poole Harbour, part of which is a reserve
held by the Royal Society for the Protection of Birds (Figure 10.7).

The National Trust argue that the decision to permit further oil development in Dorset implies any location in the south of England could now face oil and gas exploration with little regard to the environmental consequences. To other conservationists the Wytch Farm development represents just one more noxious threat to the largest area of heathland in southern England.

Thomas Hardy vividly portrayed the distinctive character of the Dorset heaths in his Wessex novels, but during the last two hundred years the area of the heath has diminished from 40,000 ha to just 6,000 ha as numerous competing uses have contributed to dramatic changes. Conifer plantations of the Forestry Commission, an army tank range, gravel extraction and the requisitioning of heathland by the Atomic Energy Authority to build prototype nuclear reactors all have contributed to a diminution of the variety of species found in the area. Special concern has been voiced over the possible extinction of species such as the Dartford Warbler and the aquamarine marsh gentian, but a more general conservationist argument questions the general perception of the heathland as an 'unattractive' landscape suitable only for land uses that are frequently both noxious and unsightly (Robinson, 1986b: 357).

The conflict at Wytch Farm represents that between what can be seen as a desirable economic proposition on the one hand and as an environmentally destructive project on the other.

One type of site which has aroused great controversy within the overall context of modern farming versus conservation has been the 'wetlands' (e.g. NCC, 1989; Turner et al., 1983; Turner, 1987). For example, in Spain an inventory of wetlands in 1945 recorded 60 wetlands in reasonable condition, but by 1980 all but one had been seriously affected by draining, introduction of exotic species, pollution and recreation. There are 19 'Ramsar' Wetland Sites covering 66,663 ha (164,725 acres) and several other important wetlands in the United Kingdom (see Figure 10.8). These sites were designated by governments in accordance with the provisions of the Convention on Wetlands of International Importance, especially as waterfowl habitats, signed at Ramsar, Iran, in 1971. Such sites and many others designated as SSSIs, National Nature Reserves and other types of conservation area depend on the maintenance of existing farming practices to preserve their current ecological equilibrium. Thus changes in farming methods can immediately be seen as a threat to the very existence of wetland sites. That such threats are real is evident from the extent of both river canalisation schemes and the increase in pump-drainage. However, the complexity of the conflict between farming and conservation must not be ignored, as illustrated with respect to the recent debate over the future of Broadland, East Anglia (Figure 10.9)

As has been the case for many wetlands, the low-flying, flood-prone meadows of Broadland have been under pasture for several centuries. Only during the past 20 years has there been a threat to change the traditional usage of the land. This has been brought about by the development of deep draining equipment and the availability of subsidies from

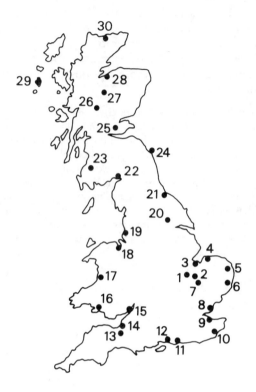

Figure 10.8 Major Wetlands in Great Britain (based on Maltby, 1986) (1 –
Woodwalton Fen; 2 – Ouse Washes; 3 – The Wash; 4 – North Norfolk Coast;
5 – The Norfolk Broads; 6 – Minsmere; 7 – Wicken Fen; 8 – Foulness and
Maplin Sands; 9 – North Kent Marshes; 10 – Romney Marsh; 11 – Amberley
Wild Brooks; 12 – Chichester Harbour; 13 – Somerset Levels; 14 – Bridgwater
By; 15 – Severn Estuary; 16 – Crymlyn Bog; 17 – Borth Bog; 18 – Dee
Estuary; 19 – Ribble Estuary; 20 – Derwent Ings; 21 – Teesmouth Flats and
Marshes; 22 – Solway Firth; 23 – Silver Flowe; 24 – Lindisfarne; 25 – Loch
Leven; 26 – Rannock Moor; 27 – Insh Marshes; 28 – Moray Firth; 29 –
Hebridean machair; 30 – Strathy River Bogs

the EEC for grain production (Morris and Hess, 1986). Together these
have made it far more profitable for farmers to drain the marshy pasture
and plant grain crops. Hence, local drainage boards in Broadland have
submitted proposals to improve drainage in various marshes alongside
the principal rivers. When notified of these plans, the Broads Authority,
established in the 1970s as an arbitrator between a range of diverse
interest groups, objected on the grounds that such action would be
detrimental to the landscape of the area. One of the most recent sites
involved in the dispute has been Halvergate Marshes where voluntary
agreements to safeguard 487 ha (1,203 acres) of marshland were reached
in early 1982. Subsequent requests for grant aid towards drainage costs

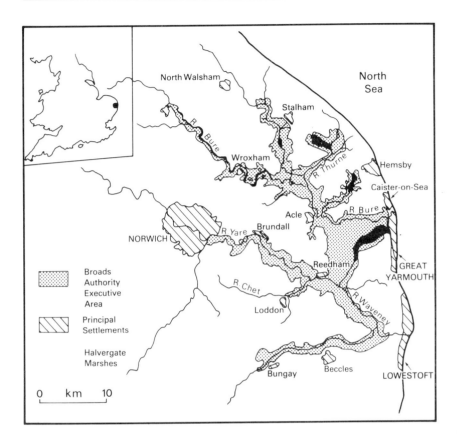

Figure 10.9 The Norfolk Broads

with respect to other marshes in the vicinity were refused, but with a compromise that will retain some marshland whilst also permitting some drainage. On the Acle/Tunstall Level 700 ha (1,730 acres) will be drained whilst 500 ha (1,235 acres) will be preserved in its existing state. Meanwhile, 1,200 ha (2,965 acres) on the Seven Mile/Berney Level has been the subject of further negotiations (Lowe *et al.*, 1986: 265–300).

Further policy developments to bring about greater emphasis upon conservation in the Broads are the introduction of the Broads Grazing Marshes Conservation Scheme in 1985 and the establishment of controls which will effectively give the Broads the same status as a National Park. The conservation scheme has three main aims:

(a) to keep permanent grassland with livestock grazing on the Broads marshes;
(b) to encourage the management of grazing marshes in ways which support both farming and conservation;

(c) to re-establish permanent grassland in some areas formerly grazed but now arable.

These will apply in the Halvergate Marshes and will extend along parts of the rivers Bure, Waveney and Yare.

It is the high cost of providing compensation to the farmers that has been the main stumbling block to many of the negotiations between the Broads Authority and farmers. Yet this is by no means the straight-forward conflict of interests between farmers and conservationists as it is sometimes presented. This is illustrated by the Broads Authority's (1982) own schematic representation of the environmental impacts on the rivers and broads under their jurisdiction (see Figure 10.10). Farming is only one element in the changing pressures upon the landscape, with recreational use, discharge of urban wastes and unintentional modification of flora and fauna also playing major roles. This implies that government policy towards the management of areas such as the Broads must be much more intricate in its formulation for some sort of balance to be attained between the various factors involved (see Moseley, 1987; Turner and Brooke, 1988). In the case of the Broads this policy in 1987 established the Broads as a national park in all but name by giving the Broads Authority statutory control over conflicting interests, with the chance to co-ordinate sectoral activities in an attempt to conserve the area for the benefit of all its users (Blunden and Curry, 1988: 200–1).

The controversy surrounding the Halvergate Marshes exposed the viability of the Broads Authority to finance a full suite of management agreements with farmers who had land on the marshes. However, it did lead to further provision for compensation for farmers through the introduction of financial incentives to encourage 'appropriate farming practices' in newly designated Environmentally Sensitive Areas (ESAs), introduced by the EEC where 'the maintenance or adoption of particular agricultural methods is likely to facilitate conservation, enhancement or protection of the nature conservation, amenity or archaeological and historic interest of an area'. Twelve such areas were designated in England and Wales (Figure 6.5), with a total of 738,000 ha (1,823,600 acres) in the United Kingdom. The majority of these areas had landscapes and ecology dependent upon the continuance of extensive livestock production. To maintain the desired form of agriculture, two-tier compensatory payments were introduced. By early 1988 in the English ESAs, over 1,400 farmers had applied to join the scheme, putting forward 32,000 ha (79,075 acres) or 78 per cent of land considered suitable for inclusion by the MAFF.

Blunden and Curry (1988: 77) see the ESA programme as a significant departure for agricultural policy as it represents payment to farmers to 'produce countryside' rather than simply to stop undesirable practices. Yet incentives encouraging overstocking and grassland improvement still continue, and farmers are not compelled to enter into management agreements. The same weakness can be applied to recent attempts to

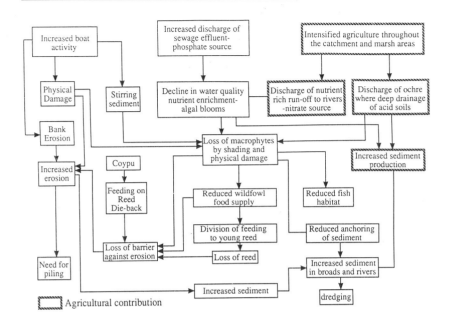

Figure 10.10 Environmental impacts on the rivers and Broads (*Source:* Broads Authority, 1982)

reduce pollution of watercourses, in the form of a dozen pilot water protection zones. In these '*nitrate sensitive areas*', covering 15,375 ha (38,000 acres), farmers will be offered compensation for restricting their use of nitrate fertilisers. The National Rivers Authority (NRA) chose the twelve areas and a further nine 'advisory areas' where farmers will be encouraged, without compensation, to restrict nitrate use. However, these relatively mild voluntary controls are below EEC requirements. A House of Lords Committee has produced a report, *Nitrate in Water*, that suggests that as much as half the land under cereals in East Anglia would have to be taken out of production if draft EEC directives on nitrate use were followed. High nitrate levels in drinking water have become an emotive issue, and the issue of the farmers' requirements versus public health considerations seems likely to be one of many 'agriculture versus conservation' battles in the 1990s.

10.4 Landscape evaluation

In the second half of the 20th century it has become common for the rural environment to be assessed with a view to designating certain areas for special protection and others for special types of use by the general public. In some cases the designation is a fairly straightforward affair, for example designating an area for conservation as it contains the last surviving occurrence of a particular plant species or bird colony. However, there are many other instances where the designation has to meet both conservationist and recreational criteria, and where what is being evaluated is not the existence of one particular species or type of habitat, but the aesthetic characteristics of the whole landscape. The implicit question being asked by planners designating 'scenic areas', or 'areas of outstanding beauty', or even National Parks, is 'are scenic qualities of this area sufficiently outstanding for them to merit special protection from exploitation and development?' There may be a subsidiary question to this in the form 'does this area merit the provision of legislation to restrict development and/or to encourage recreational activity, perhaps via the necessary provision of special recreational facilities?' Hence both conservationist and recreational considerations demand an evaluation of the aesthetics of environment and have given rise to a wide range of literature under the general heading of 'landscape evaluation'.

Techniques for landscape evaluation abound. Often they are closely comparable, differentiated by factors of merely peripheral importance, occasionally methods are flatly contradictory . . . The state of the art, then, is one of confusion (Dunn, 1974: 935).

Despite a large volume of literature dealing with the evaluation of scenic resources, the over-riding subjectivity of such evaluation has invoked a plethora of approaches. Some have added a quantitative element to their approach, but subjectivity and an underlying theoretical vacuum have limited the usefulness of much work in this field. Yet despite these glaring limitations, landscape evaluations in one form or another have formed a vital ingredient in the production of designated areas by planning agencies, and also in improving inventories, environmental impact measurements and determining carrying capacities. It is this widespread use of evaluation methods that make them a worthy subject of attention despite the often highly unsatisfactory nature of the methodology them employ.

It is not surprising that criticism has been voiced concerning the way in which landscapes have been evaluated. How can the beauty of scenery be defined? How can the majesty of snow-clad peaks be compared in quantitative fashion with the desolate beauty of an open moorland or the sunset over cliff-backed sands? What elements in the complexity of a landscape are being evaluated? and how may the complexity be simplified in order to simplify the task of the investigator? Despite the difficulties inherent in answering the first two questions raised in this paragraph, the third and fourth questions have been answered through studies that, according to Mitchell (1989: 126), have adopted one of the three general approaches referred to below.

One of the most dominant characteristics of landscape evaluation studies is the wide range of different purposes for which such evaluations are undertaken. This was demonstrated very clearly in Penning-Rowsell's (1975) survey of the different purposes and units of data collection for landscape evaluation developed by a sample of county planning departments within England: 47 per cent had used an evaluation to decide between the relative merits of preservation and development; 28 per cent were for assessing suitability for landscape improvement schemes; and 25 per cent were for recreational purposes. Just over half of the sample had used the kilometre square of the National Grid as the basic unit of data collection. The different aims of the evaluations has assisted the development of varying methodologies.

10.4.1 Evaluation by consensus

This is the oldest and most frequently used method adopted by planning agencies confronted with the task of delimiting an area for conservation purposes (Turner, 1975). A variety of methods may be employed by 'experts' who designate areas of high aesthetic value using field reconnaissance and/or maps, aerial photographs and other available material. There is rarely any systematic database so that the justification for the designations produced is not easily defeated. Such designation, based on an agreed demarcation by various planners, has been the basis of many management decisions throughout the world. It is not surprising then that there are major disagreements over the details of the boundaries of areas delimited in such fashion. National Parks and AONBs in Britain have been designated by this method, for which Blacksell and Gilg (1975: 135) propose the substitution of more systematic methods which could possibly reduce conflicts of opinion. Similarly, designation of National Parks and Scenic Areas in the United States has relied largely upon expert judgements (Henderson, 1976: 282).

A management team pooling its expertise to arrive at a decision has some merit, especially in minimising the time spent and exercise involved. However, in many cases such expertise has made use of a more rigorous methodology in which the aim has been to incorporate an objective or more systematic approach to evaluation. Such approaches have the virtue of being replicable and subject to a more systematic cross-examination by interested parties, more especially by members of the public on whose behalf the evaluations are made.

10.4.2 Evaluation by description

Landscapes may be described in many ways. In one type of description a landscape may be regarded as 'very beautiful' or 'unremarkable', and in another it might be awarded a numerical score so that its relative value can be assessed against other landscapes. Descriptive evaluations have often combined numerical and adjectival assessments in the

Table 10.6 Linton's composite assessment categories

Category	Score
Urbanised and industrialised	− 5
Continuous forest	− 2
Lowland	0
Treeless farmland	+ 1
Low uplands	+ 2
Moorland	+ 3
Plateau uplands	+ 3
Varied forest and moorland	+ 4
Hill country	+ 5
Richly varied farmland	+ 5
Bold hills	+ 6
Wild landscape	+ 6
Mountains	+ 8

Source: Linton (1968)

production of an overall scale of values. An example of such a scale, produced by a geographer rather than a planner, is that developed by Linton (1968) in an assessment of scenery as a natural resource. His work, and that by others based on the formulation of scales, demonstrates some of the complexity and great subjectivity.

Linton attempted to appraise landscapes in Scotland on the basis of two variables, landforms and land use. He divided landforms into six types on the basis of relative relief, in conjunction with steepness of slopes, abruptness of acceleration, frequency and depth of dissecting valleys, and isolation of hill masses from their neighbours. Arbitrary numerical values were assigned to the six categories, with additional points scored if water was present as this was regarded as increasing scenic quality (Table 10.6). Land use categories were dealt with in a similar fashion to landforms. His seven categories of land use were assigned numerical scores, again in purely arbitrary fashion. This scoring system was then transferred to a map which therefore represented a composite assessment of scenery based on an evaluation of landforms and land use. Linton put forward the strength of this approach as being one which others could repeat, albeit perhaps with modifications to the scales employed (see for example, Gilg, 1974; 1975; 1976b; Duffield and Owen, 1970).

As an eminent geomorphologist, Linton based his numerical scale upon his long experience of research in the field and did not regard his final map as anything more than an indicator of the 'extent and location of our scenic resources' (Linton, 1968: 238). In contrast, Fines (1968) developed his numerical scale for scenery as part of a review of development control in East Sussex. Indeed, Linton's technique was offered as an alternative to Fines' which he regarded as overly complicated, time-

Table 10.7 Fines' scale of landscape values

Category	Score	Lowland landscape types
Unsightly	0–1	Countryside spoilt by excessive
Undistinguished	1–2	clutter; flat unrelieved plains;
Pleasant	2–4	flat or gently undulating
Distinguished	4–8	'humanized' countryside; woods
Superb	8–16	and forest (interior); coastal
Spectacular	16–32	marshes, creeks, dunes; flat or
		gently undulating heaths and
		commons; landscaped parks; low
		hills; coastal cliffs

Highland landscape types	Townscape types
High hills and moors; lower mountains; Great mountains, canyons, waterfalls	Slums and derelict areas; modern industrial and commercial areas; modern suburbia; towns of architectural and historic interest; classic towns

Source: Fines (1968)

consuming and based on unverified measurement. Fines' study certainly had greater complexity, but it was notable for its attempt to include in the formulation of the scale the general public's preference for certain types of landscape (Dearden, 1987). This preference was ascertained by asking 45 'representative' persons to rank 20 colour photographs of landscapes. One of the photographs was assigned a score of one so that the rest were given scores relative to this 'control view'. However, the attempt to include public preference was very limited as most of the 'representative' group had design experience, and the final range of scores (Table 10.7) was based primarily upon the decisions of those interviewed who had the greatest design experience.

Fines' procedure for allocating scores to his study area in East Sussex was much more complex than Linton's method. Brancher (1969) queried the validity of the scale based on the views of the 'representative' sample, but it was largely the time involved and need for experienced, skilled personnel carrying out an extensive amount of field-work that prompted Linton to produce an alternative approach. One of the chief differences between the work by Fines and Linton illustrates the different emphases placed upon maps and field-work in many landscape evaluations (Dearden, 1980a).

The question of the arbitrary scale of values adopted in the evaluation of landscape is a crucial one. Gilg's (1978a) survey of landscape evaluations employed by planning authorities throughout England revealed several arbitrary scaling systems and demonstrated the need for the establishment of some clearly recognised criteria for the production of standardised scales. For 17 planning bodies and individual studies he indicated that 13 different elements had been analysed within the landscape evaluations, although of these elements, only woodland and water had featured in more than half the studies. Other common features examined included relative relief, type of relief and the presence of settlement. Only three of the 17 referred specifically to 'views' of the landscape, often a vital consideration when considering the importance of a site's tourist potential. The disaggregation of a landscape into individual component parts which are then given a score is questionable, though, given the critical interrelationships that exist between the components and the 'totality' of any given landscape (Dearden, 1980b).

In North America one of the chief pioneers of landscape evaluations with a descriptive approach was Litton (1968: 1972; 1974), a landscape architect whose aim was to assess the likely visual impact of proposed development schemes. Here the link between landscape assessment and aesthetics went further than Linton's. He referred to the 'compositional type' of the components of a landscape, namely the panoramic landscape, feature landscape, enclosed landscape, focal landscape, forest or canopied landscape and detail landscapes. The form, variability and spacing and/or timing of these compositional types within the landscape may be recognised by an observer. The evaluation of these types was dependent upon their unity, vividness and variety rather than an attempt to introduce personal preference. Such characteristics may also be applied to music or a painting. This is a more 'artistic' approach to evaluation, with Litton attempting to indicate which landscapes would be most sensitive to development. Examples included those with contrasting elements and vivid vegetation. When tested by Craik (1972) in the same way that Gilg had tested Linton's method, a substantial consensus was obtained concerning the recognition of the various compositional types (see also Zube, 1974).

A good example of the way in which landscape evaluations may be linked to planning decisions is provided by the descriptive evaluation produced by Leopold (1969) in an assessment of the aesthetic value of the Hell's Canyon area of the Snake River, Idaho. This evaluation was to assist the enquiry concerning a proposed hydropower dam in the area. The initial part of Leopold's methodology was similar to those already described: deciding upon the main factors contributing to landscape aesthetics. He selected three: physical features, biological attributes and the human use of the landscape. These three components were translated into 46 different criteria which were applied to 12 river valleys in Central Idaho and, for comparative purposes, to four river valleys in National Parks. The criteria varied from some measurable characteristics, such as river width, depth, turbidity, to non-numerical ones, such as land use, amount of algae, and flora, but were all described in terms of categories. Observations for each criteria were expressed in terms of five classes in order to obtain a common measurement scale.

Using this data Leopold focused upon how unique each of his river valleys was with respect to the 46 chosen criteria. So, if there was only one river in category five with respect to the criterion of river depth at low flow, this would give that particular site a *uniqueness ratio* of one. Two rivers in the same category would give a ratio of 1:2, three rivers 1:3 and so on. In this fashion he was able to calculate a uniqueness ratio for each criterion at each site and, by adding the ratios for all 46, obtained a total uniqueness ratio for each site. The higher the ratio the more unique the site. Within this ratio it was also possible to determine which of the three basic components (physical, biological, human interest) was contributing most to the status of the site.

As with Fines' work in England, Leopold's evaluations provoked criticism (Hamill, 1975) and attempts at modification or development of an alternative approach (see also Leopold and Marchand, 1968). Two examples of simplifications of the basic Leopold technique are Knudson's (1976) evaluation of rivers in Indiana and Chubb and Bauman's (1977) work in Michigan. They both employed fewer criteria and variables. Leopold himself identified various problems with his method, the chief of which related to his concept of uniqueness. For example, uniqueness does not indicate the degree of attractiveness of a site and is only one of the characteristics of a site worthy of attention. The measure of uniqueness is based on three equally weighted factors. Yet these are not directly comparable and are not the only valid indicators of landscape aesthetics (Dearden, 1985).

10.4.3 Evaluation by preference

An example of an evaluation including people's preferences for a particular type of landscape was illustrated above in Fines' (1968) work in East Sussex. A good background to this idea of preferred landscapes is provided by Lowenthal and Prince's (1964; 1965; 1976; Lowenthal, 1978) work on people's perception of 'the English landscape'. The basis of their studies is indicated in the following statement:

Landscapes are formed by landscape tastes. People in any country see their terrain through preferred and accustomed spectacles, and tend to make it over as they see it. The English landscape, as much as any other, mirrors a long succession of such idealised images and visual prejudices (Lowenthal and Prince, 1965: 186).

It is this 'visual preference' that may be tapped as part of a landscape evaluation, and which Prince and Lowenthal attempted to indicate in an English context. For example, they noted the variety and openness associated with many recurrent features in the landscape, such as parklands and fields, and demonstrated preferences for the 'homely', picturesque, deciduous trees, antiquarianism and individualism. Lowenthal (1968) pursued this idea in the United States, judging that Americans preferred landscapes with elements of wildness, formlessness, extremes and size. Though he did detect some marked changes towards preferred landscape in a later study (Lowenthal, 1975).

The work of Lowenthal and Prince has sparked many studies focusing upon people's preference for certain types of landscape. For example, the study of landscape painting has been one area of investigation (Prince, 1984; Rees, 1973; 1976; Zaring 1977). However, such studies are not attempts at a formalised survey of public opinion. They use the character of the landscape or a landscape painting as the surrogate for the aesthetic preferences of the population. In fact, they are obtaining only a guide to the preferences of a very small part of that population – artists, landowners and people such as politicians, clergy and teachers whose views may have appeared in print. Contemporary evaluations are able to go beyond this directly to the population itself with questionnaires and statistical sampling techniques.

There have been numerous such attempts to incorporate public opinion in an evaluation, with an intriguing variety of approaches, aims, degree of quantification and end result. Perhaps the most popular method has been to ask respondents to indicate their preference for various landscapes depicted in a series of photographs, e.g. Daniel and Boster's (1976) *Scenic Beauty Estimation Method*. Whether photographs are a valid simulation of the real world is open to question, but the cost and logistics of transporting people to a variety of landscapes has often ruled this out as a practical alternative. Photographs also have limitations, for example, there can be significant variations in response according to the weather indicated in the photograph and the time of day.

Photographs have been used in a variety of ways to form evaluations of landscapes (Calvin et al., 1972; Dunn, 1976; Hendrix and Fabos, 1975). One group of studies is typified by Peterson and Neumann (1969) in an examination of the visual appeal of beaches along Lake Michigan within the Chicago Metropolitan area. A sample of users of the beaches were shown 14 black-and-white photographs depicting a variety of beaches. The respondents were asked to allocate each scene a score whilst also describing what was liked and disliked about it. The results suggested a grouping of respondents into those who preferred city beaches and those who preferred beaches in a more natural setting, whilst the descriptions revealed certain characteristics that were felt to be important, for example sand texture, degree of crowding and water safety.

A British example of the use of the preference method is the work by Penning-Rowsell and Searle (1977). They conducted a household survey in North London, collecting respondents' rating of landscapes using semantic differential scales. The reasons for the rankings were also investigated to explain sources of variation.

Certain evaluations have made use of relatively complex methodologies with a large element of quantification overlying a basically subjective core (Dearden, 1980b; Dearden and Rosenblood, 1980; Penning-Rowsell, 1981). A good example of this approach is the work of

Coventry City Council *et al.* (1971; Blacksell and Gilg, 1975). The quantitative aspect of the evaluation was incorporated in two parts of a three-part process. Initially, 24 landscape components were measured for one kilometre squares using maps and field surveys. These components were selected to represent a wide range of landscape features and included such things as the number of hedgerow trees, area of woodland, number of farms and area under farmland. Then a subjective assessment of the visual quality was made for each kilometre square. This assessment was linked to the landscape components by means of a step-wise multiple regression analysis using the components as the independent variables and visual quality as the dependent variable. The multiple regression eliminated nine components from the evaluation as these were shown to have no independent effect upon visual quality. It also indicated which components were most important in explaining visual quality. The components were then weighted accordingly and totals obtained for each kilometre square. By varying the number of components selected and the weightings applied, different evaluations were produced based on different assumptions. Such a procedure has the advantage of using a clearly defined methodology with easily replicable statistical analysis. However, there is still a reliance upon a subjective assessment of visual quality and a degree of subjectivity concerning the original choice of components. The latter may be reduced by including components which will increase the overall explanation of the regression model. According to Curtis and Walker (1982: 390), this essentially statistical approach 'takes no account of the significance of the arrangement of the elements, of the belief that the whole may be more than the sum of its parts, nor of the influence of aspects of the landscape such as historical association, which are not observable.'

Much of the designation of National Parks and other protected areas in the countryside was based upon consensus by experts, using maps, aerial photographs and field reconnaissance to evaluate the area being considered as worthy of protection. Yet very rarely did such designation utilise any definite measurements or attempt to elicit views from beyond a small group of 'experts'. Much of the work of landscape evaluation in the 1970s and 1980s has sought to advance the underlying basis for such designations beyond this elementary approach. It can be argued that there have been some notable successes in incorporating landscape evaluations into planning decisions and designations, but the inherent subjectivity underlying evaluation has also meant that arguments over appropriate philosophy, objectivity, techniques and methodology have proliferated (Hamill, 1985). A stronger link between landscape evaluation and environmental protection remains to be developed.

11

Planning rural services

The number and range of services and facilities available to rural dwellers has exhibited a marked and steady decline in the postwar era. In 'deep rural' areas depopulation has undermined the economic threshold for many services while in less remote areas the spreading competitive influence of towns combined with increased personal mobility has had a similar effect on service viability. These difficulties have been compounded by a general trend towards centralization by public and private service providers in an effort to achieve economies of scale (Pacione, 1984: 263).

This statement encapsulates some of the key problems of service provision in rural areas – inadequate access to services, especially for those rural dwellers reliant upon public transport; and the lack of development of policies to prevent further rapid decline of rural services (Smart and Wright, 1983). Issues relating to rural services have become of greater concern during the last two decades as both planners and non-planners have recognised a variety of approaches to these issues. The commonest approaches have focused upon the actual extent of declining service provision, the concept of *rural deprivation* deriving partly from the continued decline in provision, the question of accessibility to services, and planning measures aimed at reducing deprivation and increasing accessibility. These are not necessarily mutually exclusive approaches as will become apparent as they are dealt with systematically below.

11.1 The decline in provision of rural services and the growth of rural deprivation

The loss of rural services has occurred throughout the Developed World, but especially in those areas where prolonged depopulation has occurred. The falling numbers of consumers of a range of rural services has prompted their demise, a wide range of individual services being affected by such a decline, e.g. schools, bus and rail services, health care, retail provision (Plate 21). However, the decline has been exacerbated by two other factors which have extended the area affected well beyond the

Plate 21 The elevators marking grain delivery points have been a symbol of prosperity on the Canadian Prairies. Increasingly, though, there has been rationalisation closing both elevators and railway lines

depopulating periphery (Woollett, 1981a):

(a) The raising of population thresholds by both public and private services in order to achieve economies of scale. This has promoted concentration of services in higher order central places (e.g. Pacione, 1982).

(b) The greater personal mobility produced by the spread of car ownership has enabled the majority of people to travel longer distances to utilise services.

It is clear therefore that reduced levels of service provision are part of a self-reinforcing cycle of decline, perhaps most pronounced in depopulating areas but generally widespread.

In their conceptualisation of the use of public services in rural areas, Joseph and Poyner (1982) distinguished between critical and non-critical services. This implies that fewer factors intervene to prevent the use of 'critical' medical services as opposed to others such as libraries or dental services. However, Taylor and Emerson (1981) suggest that whilst the reduction in provision has affected virtually all services and retail facilities too, one of the most crucial has been the post office which frequently has acted as a multiple service facility, e.g. in Britain most village post offices are sub-post offices of which 80 per cent are run in conjunction with private businesses like grocery shops and newsagents

Plate 22 Maintenance of a post office often symbolises the viability of a rural community, but few surviving ones are as small as this one in the Barossa Valley, South Australia

(Pacione, 1984: 273). The combination of postal service and village store has often been a major focal point for small communities. Hence, any threat to the store or post office poses a significant threat to that community. Rural inhabitants driving into town to buy cheaper groceries from a supermarket can bring about closure of village stores and perhaps post offices too (Shaw, 1976). For example, the number of sub-post offices in 354 remote Scottish rural parishes fell by 42 per cent from 1960 to 1980 whilst the number of shops fell by 46 per cent over the same period.

Whilst the closure of post offices tends to have the greatest effects upon pensioners and young families who obtain a variety of welfare payments via the post office, the closure of another key service, rural schools, affects young families directly but with a profound feedback to the very existence of some communities. The demise of village primary schools removes a vital ingredient in retaining the young and active element in the rural population as it is so often associated with young families moving closer to the location of their children's new school, usually in a higher-order settlement (Forsythe et al., 1983; Jones, 1980). 'To close a primary school may in fact diminish a village in more senses than one and provide a further reason why young married couples will want to leave it' (Central Advisory Council, 1967). Thus the closure of rural primary schools and, perhaps to a lesser extent, post offices and

Village stores, can be seen as a trigger mechanism likely to lead both to depopulation, greater reliance on centralised services, and hardship for those rural inhabitants least able to travel to services elsewhere (Plate 22). This tends to mean it is the elderly and the poorer members of rural society that suffer most from reduced service provision, and therefore are most likely to experience *rural deprivation* (e.g. Knox and Cottam, 1981; Rose *et al.*, 1979).

Two long-continuing characteristics of rural areas give an important insight to the notion of rural deprivation: rural incomes are lower than those in urban areas, and in rural areas there are also greater inequalities in income distribution, the latter partly due to the increased mixture of farm and non-farm populations in the same locality (Gunter and Ellis, 1977). In the United States for over three decades the proportion of the population below the federal poverty level has been higher in non-metropolitan counties and in farm rather than non-farm populations (Bogue, 1985: 608). This rural poverty has been concentrated in the South, especially Alabama, Mississippi and Arkansas. Other concentrations of rural poverty occur in Kentucky, the Dakotas, New Mexico and south-west Texas. These have tended to be areas where poverty has persisted over a long period of time, reflecting factors contributing to uneven development in advanced capitalist economies. Fitchen (1981) summarises the reasons for the continuance of poverty in the same areas and the same families:

(i)   the continued impact of history;
(ii)   the crippling economic situation;
(iii)   inadequacies of the social situation;
(iv)   barriers to upward mobility;
(v)   corrosive stereotypes;
(vi)   constant pressure of too many problems at the same time;
(vii)   difficulty of balancing aspirations and achievements;
(viii) failure syndrome;
(ix)   psychological defects from early childhood;
(x)   the closing horizon.

Although the concern with rural deprivation in the early 1970s largely attributed deprivation to the imbalance in the allocation of government resources between urban and rural areas, it has been argued more recently that deprivation is less a result of unequal allocation of state funds and more a reflection of general socio-economic inequalities (e.g. Bentham and Haynes, 1986; McLaughlin, 1986b). This view implies that specifically spatially-based solutions to the problem of rural deprivation, such as the Development Commission's designation of Rural Development Areas in England and Wales, are likely to be of limited effect.

Various rural problems, for example, restricted housing and employment opportunities, low wages, declining services, poor public transport and lack of accessibility to services and utilities have attracted the attention of government policies in the 1970s and 1980s under the general policy aim of 'reducing rural deprivation' (e.g. Neate, 1981). Unfortunately, the term 'rural deprivation' has not been employed rigorously and has become a 'catch-all category for every negative consequence of the operation of an hierarchical system of social stratification within a market-oriented society' (Lowe et al., 1986: 2). Some observers do not regard it as an appropriate or useful concept (e.g. Rutter and Madge, 1976) whilst others use it as a synonym for inequality or poverty (e.g. Holman, 1978; Townsend, 1979). However, in the 1980s it has been frequently used by those seeking administrative solutions to social problems (e.g. Klein, 1983). In short, it is a concept which has been placed in several social, political and ideological contexts, but with an emphasis placed upon prescriptions rather than analysis or evaluation. Bhasker (1978) refers to deprivation as a 'chaotic concept' and, rather like the term 'rural', it has been part of a long and confusing terminological debate. The confusion seems to arise because of four overlapping problems:

(a) There are numerous theories which have some bearing upon what produces deprivation, underlying causal explanations are by-passed.
(b) Deprivation of an individual or group is expressed in relative terms, but there is agreement neither over how deprivation should be measured nor how the deprived group or areas should then be determined.
(c) Deprivation is multi-dimensional in that, for example, there are a number of services and utilities of which groups can be deprived. Hence the term 'multiple deprivation' has often been employed. However, measurement and assessment of multiple deprivation, especially in terms of formulating suitable remedial policy, has been problematic.
(d) When addressing the causes of multiple deprivation vis-à-vis policy generation, precise causal relationships between relevant variables have been difficult to determine. A major consequence of this has been that urban deprivation, involving larger total numbers of people, has received greater attention from policy-makers especially in the area of infrastructural decay and environmental degradation in inner cities.

Two fundamental paradigms provide the keys to most work on rural deprivation:

(a) Deprivation is seen as a consequence of inequality within rural society. This refers to changes in rural society brought about by rural depopulation, the reduction of the agricultural labour force, and the growth of the adventitious population. These changes have brought an unequal sharing of the resources of economic and political power and social status. In addition, they have influenced the distribution of material resources and brought declines in services, retailing and social amenities, especially affecting lower income and low status

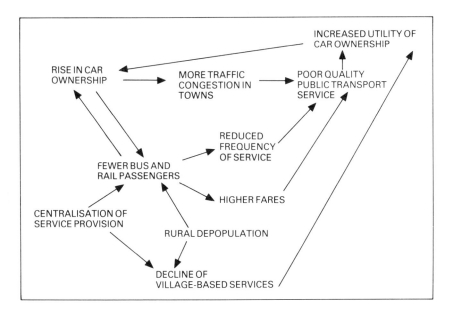

Figure 11.1 The Spiral of Deterioration centred on decreased use of public transport services (based on Moseley, 1979)

elements in the rural population. This approach is concerned with social justice and societal distribution of welfare.

(b) Deprivation is seen as a consequence of decisions made beyond the village. These decisions produce various types of deprivation, illustrated by Shaw (1979) in terms of a three-fold typology of 'household', 'opportunity' and 'mobility' deprivation. This set of problems can be viewed in terms of differential accessibility to job opportunities, services and utilities (e.g. Moseley, 1977; 1979; 1980), the keys being levels of provision set by government and other authorities and the mobility of individuals within rural communities. Hence, this approach is an adjunct to policy-making and suggests that deprivation can be overcome by adequate provision of services to maintain rural communities. In contrast, considerations of 'inequalities' within rural society view such provision as part of a planning process which itself helps to create and maintain inequalities.

A central problem with both the study of and policies towards rural deprivation relates to the extent to which it can be treated as a separate entity from urban deprivation. Structuralists have argued that deprivation in urban and rural areas shares common structural economic roots and common indicators. This is demonstrated too by Moseley (1980) and Knox (1987) who compare deprivation in inner cities and peripheral regions, in which a similar *spiral of deterioration* is characteristic and affects amenities, services, out-migration and investment (Figure 11.1). A major difference between the two types of area, though, are the numbers of people affected by deprivation; and hence the possibility for governments to overlook the rural problem when faced with larger numbers unemployed in the inner city or living in poor housing.

Two of the key factors affecting the continuation of and growth of rural deprivation, as shown in Figure 11.1, are *depopulation* and *centralisation of services*. For example, in Eire, despite attempts to encourage rural economic development, especially in western districts, there has been much centralisation of service provision. For example, under the recommendations of *Investment in Education* (Eire Department of Education, 1965) 1,800 one- and two-teacher schools were phased out between 1965 and 1977, being replaced by new or enlarged schools built in higher order central places (Curry, 1980). This policy closed from one-quarter to half the schools in western Eire (Cawley, 1986). Similar centralisation has affected primary health care and hospital services, providing a significant contrast to the increased dispersion of manufacturing industry. In the west this has worked against the professed aim of improving accessibility to health care. Thus variations between regional health boards have grown, e.g. in the East region (centred on Dublin) one doctor serves an estimated 656 medical cardholders, compared with 1,093 in the West (Galway) and 1,176 in the North-West (Donegal). There are similar variations for per capita access to dentists, district nurses and pharmacies (Eire Department of Health, 1986; Brunt, 1988: 91–3). These contribute to what Cawley (1986) describes as a 'cycle of cumulative deprivation' which she identified for 40 rural districts.

Similarly, in Finland marked reduction in rural services and facilities was apparent in the 1960s as out-migration rose. However, the decline in the Finnish rural population can be traced to the early years of the century in more remote areas where the viability of services was gradually reduced by long-term out-migration, culminating in school closures, loss of bus routes and decline in other services in recent decades. An influx of displaced persons in 1947 slowed the decline in some districts, producing an upsurge in service provision followed by a subsequent decline. Nurminen's (1981) study of school provision in the Kanta-Hame region revealed 68 closures from 1963 to 1977, with a peak rate of closure in 1968–9. Villages remote from urban centres suffered most, largely because of a fall in their numbers of inhabitants of primary school age due to out-migration of young families. The rate of closure was reduced by the influence of government legislation which, in 1977, lowered the state subsidy for a second teacher in small rural schools from 25 pupils to 12. Conversely, rural closures had been encouraged by legislation in 1958 enabling school-children to be bussed for up to 2.5 hours daily to and from school rather than limiting distance between home and school to 3 miles (5 km).

It is this type of reduction in services, often encouraged by legislation, that is at the heart of the deprivation problem. So too is the concept of people having differential access to services. Variations in access constitute a key aspect of deprivation as accessibility provides the link between the consumers of services and the services themselves. For example, Blacksell *et al.* (1988), examining the use of legal services in rural areas, suggest there are special problems of access for those living in remote areas, partly through low levels of awareness as to what constitutes a legal problem and unfamiliarity with the workings of official bureaucracy (Johnsen, 1978: 199), but also because of widely established tendencies towards centralisation of services, through which urban areas benefit at the expense of rural areas (see Lonsdale and Enyedi, 1984).

11.2 Accessibility

At the heart of the issue of rural deprivation and the provision of services is the notion of *accessibility*. This refers to the ease with which people are able to obtain (i.e. gain access to) goods and services that they desire. It is a concept with both a supply side and a demand side, and therefore a full appreciation of the dynamics of accessibility has to consider both aspects (Stanley and Farrington, 1981). However, in general, the supply of goods and services in rural areas has diminished post-1945 as has the demand for a range of items to be supplied in lower-order settlements. As the growth of private ownership of cars has grown so a greater proportion of the rural population has been more easily able to consume goods and services offered at some distance from their own residence. Indeed, the attractions of consuming these goods and services in a centre offering a wide range of 'consumables' has promoted 'distance shopping' and desire to extend personal mobility through families becoming owners of more than one car.

This picture, though, is misleading. First, it ignores the pressure placed upon rural residents to extend car ownership in order to maintain a satisfactory degree of access to goods and services in the face of increased centralisation of outlets. Secondly, it also ignores that portion of the rural population whose personal mobility is limited – most commonly through factors of age and wealth. Indeed, whilst the poor, elderly, young and housewives may actually have a very high demand for particular types of service, they are often the ones who have the least access to those services by virtue of lack of mobility. So it is within certain groups of rural society that the lack of accordance between demand and supply can create critical problems of accessibility, posing critical questions about both the nature of demand and supply. In addition, a significant number of rural inhabitants live well beyond the range of major urban areas so that the factor of distance alone confers a lack of easy access to certain goods and services (e.g. Nutley, 1979; 1980).

In terms of demand, Figure 11.1 illustrates how general increases in personal mobility through rising car ownership have had a direct feed-back to the decline of village-based services. This feedback has been a growing reality post-1945, for example the number of cars per head of population in Britain will have risen from 0.05 in 1951 towards around 0.44 by the end of the century (Moseley, 1979: 17). The highest levels of car ownership have been in rural areas, a fact often regarded as a clear indication of the association between basic accessibility problems and rurality. Concurrent with this rise in car ownership has been the reduction in public transport. For Britain, Oldfield (1979) estimated that between 1952 and the late 1970s the increase in car ownership directly accounted for nearly half of the decline in the patronage of public transport. Mitchell (1976) showed that for the country as a whole the number of journeys per capita by cars first exceeded those by public transport in the late 1960s, and the greater reliance upon cars has increased subsequently. One major effect of this change has been for substantial contractions in rural bus and train services: halving in terms of total patronage in Britain since the early 1950s, with passenger rail services being most severely affected. Similarly, in France there was a 30 per cent reduction in rail track between 1950 and 1970 (Clout, 1988: 114).

11.2.1 Measuring accessibility

Despite the generally rising levels of car ownership in rural areas there are distinct groups in rural areas for whom mobility and accessibility remain limited. Indeed, increased car ownership has contributed to this by helping to undermine both public transport and the local provision of shops and services in rural areas.

In simple terms, there are three basic elements in any problem of accessibility:

(a) the people seeking a particular activity, e.g. shops, buses;
(b) the link between the people and the activity, e.g. private car, train;
(c) the activity.

An alteration to any one of these three produces a change in accessibility and can result in people suffering from lack of access to particular activities. For example, the different range of opportunities associated with various types of transport mode are illustrated in Figure 11.2 (Huigen, 1983).

Unfortunately, accessibility is yet another term used frequently in a wide range of literature but which has been given a number of different definitions (Gould, 1969: 64). Physical accessibility is perhaps most readily recognised as a concept involving an individual's capacity to over-come space to reach a particular object or destination. Thus it links mobility with the opportunities that may or may not present themselves as a result of this mobility (Moseley, 1979: 58). Within the concept is

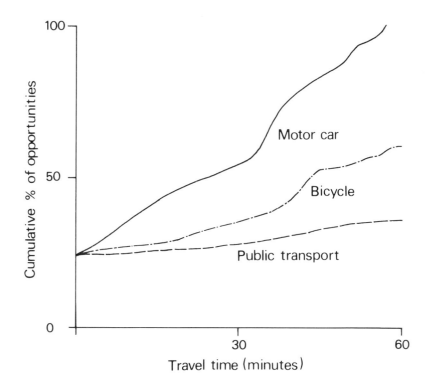

Figure 11.2 'Opportunity profiles' for different modes of transport (based on Huigen, 1983: 91)

also the rider that accessibility can vary significantly from individual to individual – contrast the differential accessibility experienced by a husband and wife to a doctor's surgery several miles from their home if the husband can travel by car to the surgery just five minutes drive from his place of work whereas his wife is reliant on a once-a-day bus service which takes an hour to reach the surgery and then will not return her home for another three hours. However, whilst accessibility to services can vary tremendously between individuals in the same household, it is often measured as a composite term. Alternative measurements have been made on a comparative basis whilst the variation of accessibility amongst individuals can be considered via a time–space approach (Pirie, 1979).

(i) *Composite Measures* – Johnston's (1966b) work on part of the North Riding of Yorkshire gives a good illustration of this type of measure (Figure 11.3). In this work he was able to combine a series of individual measures into one map, e.g. presence of a daily bus service arriving in town before 8 a.m., and a Sunday bus service, giving points to these services in order to produce a composite index

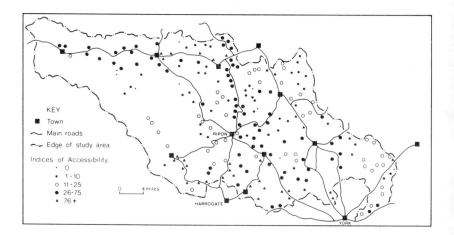

Figure 11.3 Indices of accessibility of villages in North Yorkshire (*Source:* Johnston, 1966b)

of accessibility for settlements in his study area.

(ii) *Comparative Measures* – whilst Johnston's work can be used to make comparisons between services available at particular places, it does not take into account the circumstances of individuals in those places. Yet, it is the variation in these circumstances that could contribute substantially to any development of transport or land use planning in rural areas. For this more demand-related view of accessibility, more focus upon individuals is required, as described below, but some comparative measures can be useful in terms of comparing access by car versus access by different forms of public transport, e.g. for time distance from a given locality to particular services for different groups in society (e.g. Moseley *et al.*, 1977).

(iii) *Time–Space Approaches* – although composite and comparative measures are easier for planners to deal with, because essentially they are presenting an aggregate picture, it is important to recognise that accessibility varies from individual to individual because of a series of constraints that are placed upon individuals' mobility. This is best demonstrated by comparing the *time–space budgets* of two rural inhabitants: a housewife with access to a car only when it is not used by her husband; and a retired pensioner who has to rely on public transport or lifts in a friend's car (Figure 11.4). The representation of their travels through space and time during the day gives a graphic portrayal of the restrictions upon their movements, and can be used to demonstrate the variations in mobility experienced by different groups within the rural population. This considers accessibility in terms of constraints operating at the individual level, but its practical application by planners remains limited (Nutley, 1983; 1984; 1985).

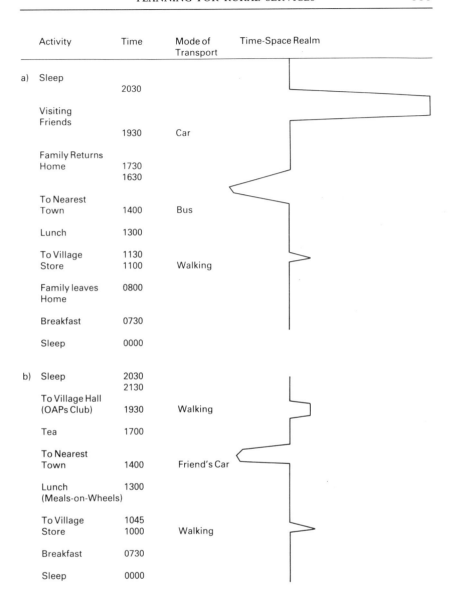

Activity	Time	Mode of Transport	Time-Space Realm
a) Sleep	2030		
Visiting Friends	1930	Car	
Family Returns Home	1730 1630		
To Nearest Town	1400	Bus	
Lunch	1300		
To Village Store	1130 1100	Walking	
Family leaves Home	0800		
Breakfast	0730		
Sleep	0000		
b) Sleep	2030 2130		
To Village Hall (OAPs Club)	1930	Walking	
Tea	1700		
To Nearest Town	1400	Friend's Car	
Lunch (Meals-on-Wheels)	1300		
To Village Store	1045 1000	Walking	
Breakfast	0730		
Sleep	0000		

Figure 11.4 Time-space realms for rural residents. (a) A housewife in a one-car household; (b) A pensioner with no car

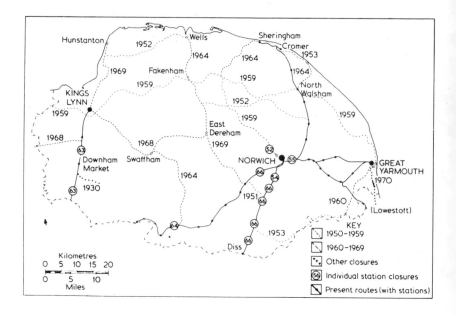

Figure 11.5 Post-war rail closures in Norfolk (*Source:* Moseley, 1979: 24)

11.3   The provision of rural transport

11.3.1   Passenger rail services

Figure 11.5, showing the closure of railway lines post-1945 in Norfolk, England, illustrates the way in which rail closures have left large areas of a quite densely populated countryside without access to railways. This example can be compared with that in Figure 11.6 for the Canadian Prairies where falling demand for both freight and passenger carrying has brought a similar reduction in service and closure of grain delivery points.

   In Britain much of the justification for closure was economic, reflecting steadily diminished passenger demand. However, the Beeching Report of 1963, which prompted a 29 per cent reduction in lines, questioned the economic viability of many rural lines even in earlier periods when they were carrying more passengers. By 1962 this uneconomic nature of many rural lines was clear: 40 per cent of the rail network carried only 3 per cent of the total traffic (Beeching Report, 1963). Whitby *et al.* (1974: 191) state the problem being faced then by many rural branch lines: 'the fact remains that for the majority of rural branch lines and duplicated main lines, traffic generated is not sufficient to justify even single-track operations'. The reduced rail services brought about by the Beeching 'axe' has left large areas of rural Britain without rail services, so that by 1979 only 1 per cent of people in rural areas used the train as the main form

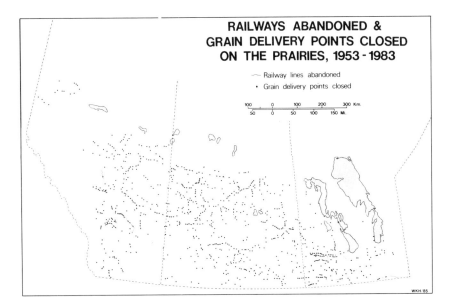

Figure 11.6 Railways abandoned and grain delivery points closed on the Prairies, 1953–1983 (*Source:* Carlyle, 1988: 257)

of commuting transport. Although this proportion may have increased as a result of the continued growth of commuter villages, it is representative of a pattern found in many other parts of the Developed World (e.g. Briggs, 1981).

Given the expense of maintaining track and rolling stock, there has been relatively little attempt to reverse the decline in provision of rural railways except for small private schemes usually associated with attracting tourists and train enthusiasts, e.g. the Severn Valley Railway. The one planning scheme that has been adopted quite widely has been to replace rail services by buses and coaches. Thus coaches replaced many of the rail services closed following the Beeching Report. However, from 1968 to 1978 one-third of these replacement bus services were closed down and a further one-third were substantially modified (Jones, 1985; Keen, 1978). The economics of the replacement services were clearly also questionable. In many cases they failed to be well co-ordinated with the remaining rail services, making suitable connecting services less viable and highlighting a general lack of integrated management of public transport in Britain as well as unwillingness by government to subsidise rural transport provision.

In Britain, effectively, rail closures have been a prime example of state disinvestment in rural areas, though there have been few studies of the effects of such closures (Whitelegg, 1987). However, a study by Hillman and Whalley (1980) did refer to a series of damaging consequences of closure related to reduced accessibility. The extent of the disruption of people's lives by rail closures has perhaps not been fully appreciated:

The impact has been socially regressive, for it is the older people, those without cars and those in blue-collar households, who have cut back most on their activity and thus who are hardest hit. Women have been harder hit than men (Hillman and Whalley, 1980: 111).

Counter views have also been presented in arguments stating that rail travel is largely the preserve of the more affluent who can more easily withstand rail closures by switching to car travel (Dodgson, 1984; Pryke and Dobson, 1975).

11.3.2 Passenger bus services

Given that an increasing number of rural communities have no access to rail services, it is not surprising that a growing amount of attention has been focused upon rural bus services and alternatives to the bus as a form of public transport provision. For example, in Britain between 1968 and 1980 there were six different Acts of Parliament that included substantial legislation which influenced rural bus services, mainly as part of a series of Transport Acts (1968, 1978, 1980) (e.g. Banister, 1983). However, in much of the Developed World, the provision of rural bus services has declined from the 1950s as the demand for such services has fallen. In fact the pattern of falling demand has been complex and has been affected by certain critical changes. These include the drop in numbers of people seeking entertainment in towns following the spread of television; access to cars through informal lift-giving; depopulation; and the spread of mobile services. To add to reduced demand, rising costs of transport have also critically affected bus services, especially via increased petrol prices in the 1970s and rising labour costs. These have tended to limit companies' willingness to cross-subsidise rural routes with money made from urban services (Edwards, 1985; 1987).

The recognition of the need to subsidise rural buses through injections of state funds has been made in many countries, most notably in all the Scandinavian countries (Sjoholt, 1988: 87–8). Indeed subsidisation in various forms constitutes one of the principal forms by which the supply side of public transport has been addressed by policy-makers. For example, in Britain under the 1968 Transport Act, a system of direct grants was introduced with the central government paying half the running costs provided the service covered half its operating costs. Grants were also made to cover losses on unremunerative rail buses. Fuel-tax rebates were made, concessionary fares introduced, and schemes arranged whereby school service contracts could carry fare-paying passengers provided there was spare capacity. A National Bus Company (NBC) was also established (the Scottish Transport Group in Scotland) as a nationalised undertaking, but in the form of a federation of regional bus companies, each with a commercial remit.

It was the operations of the NBC's subsidiaries that largely shaped the provision of rural bus services in Britain prior to *deregulation* following the 1986 Transport Act. However, the NBC co-existed with a range of local authority bus operators and thousands of small private firms (Moseley, 1979: 98–9). Undoubtedly, NBC's use of government subsidies enabled it to maintain services in many rural areas which otherwise would have had them removed or severely diminished. So despite a 38 per cent decline in the number of journeys on services run by the NBC between 1969 and 1980, Moseley (1979: 180) argued that 'as for the operation of conventional buses in rural areas, the NBC subsidiaries appear generally most suited for providing a basic scheduled transport network linking the major villages with each other and with the urban centres'. However, this national system has been transformed following legislation to deregulate the bus industry, creating fears for a worsening of services to rural areas (Cloke and Edwards, 1984).

Deregulation of bus services in South Dakota under the 1982 Bus Regulatory Reform Act contributed to 61 of 159 communities with bus services in 1975 losing them by 1984 (Bailey, 1986). Prior to deregulation, there was cross-subsidisation from profitable contract, charter and inter-city routes to unprofitable sectors of carriers' operations, the latter mainly associated with services to small settlements. Deregulation was intended to promote financial viability in an industry subject to increased costs but falling revenue. However, services to rural areas were already contracting rapidly, especially in the south-east of the state. This contraction was continued after the legislation, with no new services being initiated through the reintroduction of market forces. Indeed, post-1982 'the burden of community abandonments was lifted . . . from the poorest, and most socially disadvantaged settlements to more socio-economically typical towns' (Bailey, 1986: 297; see also Rucker, 1984). Many feel that the same consequences will be felt in Britain as a result of deregulation (Bell and Cloke, 1989; Farrington and Harrison, 1985).

11.3.3 Transport policies for rural areas

In addition to a variety of government subsidies there have been a tremendous range of schemes attempted in order to provide road transport services for rural dwellers. These schemes can be categorised in a variety of ways; five are distinguished here as a guide to some of the principal schemes implemented (Moseley and Packman, 1983).

(a) *Stage Carriage Bus Services* – these are the familiar 'traditional' form of rural transport in which buses operate to a specified timetable. This type of service has to cope with a series of daily peaks and troughs in demand which make it very difficult to provide an economically efficient service. For instance, journeys-to-work and

-to-school focus on early morning and late afternoon periods whilst the other significant element of rural travel, for shopping purposes, is spread throughout the day (Banister, 1983).

There are numerous examples of attempts to maintain a timetabled service, with a rationalisation of routes, but some modifications to existing services have been more imaginative than simple cuts in service. For example, in Belgium the Societé Nationale des Chemins de fer Vieinaux used a series of local user surveys and data on ticket sales to match services more closely with demand. Similar procedures were employed in Britain in the NBC's Market Analysis Project, promoting substitution of one route for two and service rescheduling (White, 1978). A similar scheme in Scotland was the strategy effected by Strathclyde Regional Council, based on bus use, surveys of travel behaviour by both bus and non-bus users, and perceptions of existing bus and rail services. Such approaches are attempting to make services more user-specific and match spare capacity to demand. However, these considerations are often taken further by schemes that extend beyond the bounds of normal stage carriage services.

(b) *Demand-actuated Transport* – 'A demand actuated transport service is one whose route and/or timing responds at least in part to *ad hoc* passenger requests' (Moseley, 1979: 123–4). Whilst the taxi is the most well-known example of such services, transport planners have introduced variants of this to ease rural accessibility problems. *Dial-a-ride schemes* for flexibly routed minibuses have been proposed, but seem a very expensive option for rural areas (e.g. Oxley, 1976). Diversions to existing scheduled bus services may be a more realistic option, though this would forgo the benefits of a precise timetable.

(c) *Multi-purpose Services* – more practical than the concept of the demand actuated schemes have been those using a single vehicle and driver to perform several distinct tasks. Perhaps the most common of these is the idea of the *postbus* in which postal deliveries are combined with the carrying of passengers. Postbuses have been long established in the rural areas of Austria, Norway, Sweden, Switzerland and West Germany, and were first introduced to the United Kingdom in the late 1960s. They have been widely utilised in Scotland, with over 100 routes operational by the 1980s, aided by government grants for new vehicles and a fuel-tax rebate. In parts of Scandinavia the passenger service offered by postbuses has come to resemble that of conventional bus services in terms of speed, routes and passenger capacity. However, a range of trips are offered in conjunction with the collection of mail, and not all of these are likely to appeal to passengers, e.g. contrast a 'direct' run from a village to a nearby town with one which travels through all the neighbouring villages and hamlets. Hence the often circuitous and time-consuming routes can make postbuses unattractive, and timing usually means they are unsuitable for journey-to-work or -to-school.

A similar multi-purpose service to postbuses can be provided by

Table 11.1 The advantages and disadvantages of voluntary services in rural areas

Advantages	Comments
Cost effective	Reliance on the volunteer and/or self-help, plus private funding; lack extensive administrative structures
Demand responsive	Can respond to individual requirements
Flexible	Few bureaucratic constraints; can run services when needed as providers of services do not rely on service for livelihood
Local based	Community based and built upon local knowledge and understanding of local needs
Well integrated	Centred on people or communities rather than types of need
Community catalyst	Schemes can assist in bringing together community activists and can help stimulate action
Public participation	Conforms to and encourages the principles of community involvement

Disadvantages	Comments
Respond unevenly to need	Can accentuate rather than reduce overall inequalities in welfare provision
Instability	Volunteers may be harder to replace than paid employees
Personalised	Can stigmatise the less well off
Unprofessional	Motivation of volunteers often idiosyncratic and specific; not concerned with equity
Employment	Reduce the amount of paid employment available

Source: Banister and Norton (1988)

school buses carrying passengers in addition to school-children. However, there are crucial drawbacks to this type of scheme, related to carrying capacity, timing and lack of year-round service. A compromise between specialist schools' services and conventional stage services has been operated in some areas (e.g. Oxfordshire County Council, 1976).

Other attempts at utilising multi-purpose services have focused on greater use of privately owned cars and the operation of community-run transport services. In the case of the former, this has taken two forms. Firstly, attempting to *subsidise rural car ownership* through lower vehicle excise duty or fuel-tax rebates for rural residents. Generally, this has not proved practicable. Second, extending *car sharing arrangements* via community care organisations, for example using volunteers from groups like the Red Cross, the Women's Institute and the Women's Royal Voluntary Service to provide transport for special purposes, e.g. hospital visits, shopping trips. Ideally, such voluntary provision is complementary to statutory and commercial services, but this is not always so, with those areas in the greatest need not necessarily possessing the greatest potential for voluntary sector intervention (Wolfenden, 1978; Banister and Norton, 1988; Nutley, 1988) (Table 11.1). Organisational and logistical problems have tended to limit voluntary arrangements to all but the highly localised and irregular applications. On the other hand, regular car-sharing arrangements have become more popular, especially for commuters from rural residences to urban places-of-work. In some cases, legislation discouraging single occupance of cars in cities has encouraged such sharing arrangements, e.g. Athens, though payments for such sharing are often illegal.

The logical extension of such sharing schemes has been the use of *minibuses*, either as part of a private scheme, by commercial operators as a cheaper operating alternative to a bus/coach, or by local authorities for specific purposes. However, commercially, savings on the operation of a minibus compared with a coach are small as overall operating costs are dominated by labour costs which do not diminish with the size of vehicle.

(d) *Mobile Services* – mobile shops and libraries are a familiar feature in many rural areas. However, there are a range of other services that can also be brought to rural customers in this mobile form, e.g. banks, meals-on-wheels for the elderly, play school facilities and community/social welfare information. The most recent extension of this idea is to use telecommunications to transmit services direct to people's homes via radio (e.g. schools' broadcasts), television (e.g. teletext data, Britain's Open University) and distributed computing (e.g. 'home' banking). None of these alone, though, seems to be a complete answer to overcoming rural isolation and the need to obtain certain services requiring person-to-person contact.

(e) *Relocate Services and People* – Two other policy alternatives considered by Moseley (1979: 144–62) involve planning decisions to change the locations of either the rural population or rural services. For the former, the *key village policy* (see Chapter 12) in which population tends to become more concentrated in particular villages and small towns would be a 'controlled' version of the tendency towards the centralisation that has helped to create major accessibility problems for rural residents. Under the key village option, though, this centralisation is encouraged within rural areas rather than outside and may encourage both people and services to concentrate in designated centres (Cloke, 1979). In practice, many public utilities in developed countries conform to their own notions of concentration within rural areas as a reflection of economic expediency.

The 'concentration' of people in particular locations can be encouraged by housing policies facilitating building in certain places and by the promotion of factories/offices in designated centres. However, there are also policy options which can make shops and services in certain locations more easily accessible to a wider range of people by catering to people's time-budgets. This assumes that people are able to go in person to a particular location to derive a service, but introduces a certain flexibility in opening hours. This flexibility can take a number of forms:

(i) vary the consumers' work-time to create larger blocks of time available to travel to and from a service, e.g. the introduction of a four-day working week or the instigation of a child-minding service to permit mothers to be more mobile;

(ii) clustering activities in time as well as space, e.g. the principle of a single evening opening for shops;

(iii) staggering the timing of activities that make intensive use of peak-hour transport facilities. This tends to be impractical, especially as it may weaken the need for a viable peak-hour transport system;

(iv) provide better co-ordination between public transport and particular services, e.g. surgery hours of general practitioners;

(v) move towards the mobile service option by having part-time outlets in several locations; almost resembling the traditional periodic markets or stances in Scotland (Carlyle, 1975).

Nutley's (1985) evaluation of planning options for the improvement of rural accessibility in Central Wales considered 40 options based on removal of public transport services, use of conventional bus services and adoption of 'unconventional' modes. That so many possibilities were examined demonstrates the theoretical variety of scenarios available to the planner. However, Nutley shows that even if the concern is social benefit as expressed by the additional accessibility provided by each scheme, costs and benefits can vary tremendously – high access often correlating closely with high cost. Hence, it is the philosophy behind the planning that is the key determinant of policy adoption. The tentative suggestions in the work on Central Wales were for the combination of a social car scheme plus an elementary level of conventional buses, and a community bus 'minimum' service supported by a social car scheme (Nutley, 1983).

11.4 Education

If the closure of schools is recognised by many observers as likely to have 'a significance beyond its ostensible educational one' (Neate, 1981: 17) then it is not surprising that centralisation of educational facilities, which has occurred throughout rural areas of the Developed World, can be recognised as controversial to say the least. Yet centralisation policies have prevailed in many countries for much of the post-1945 period,

Table 11.2 Arguments for and against closure of rural schools

Arguments for closure

1. The range of subjects in a modern curriculum is too wide for one or two teachers to cope with, with the result that pupils in small schools could be disadvantaged by the limited range offered to them compared with their peers in larger schools.
2. All-age classes are difficult to teach, and special needs cannot be adequately catered for.
3. It is difficult to overcome any personality clash between teacher and child when they are together for four to six years.
4. While a school could thrive under the influence of a good teacher, pupils may suffer under a mediocre one.
5. Larger schools are more flexible and can cope with changes in the birth rate by using portable classrooms.
6. Children do not receive the stimulation from varied contact with adults and others of their own age.
7. There is a difficulty in recruiting teachers to isolated schools.
8. Teachers are isolated and lack professional contacts.
9. The buildings are often old and in need of repair, many having outside lavatories and limited space for games.

Arguments against closure

1. Small classes mean more individual attention, a closer relationship between child and teacher, and the generation of a 'family' atmosphere.
2. The advantages of a generous pupil–teacher ratio are linked with a high level of parental involvement and commitment to the local school.
3. Discipine is rarely a problem.
4. Smallness enables a high degree of flexibility in the organization of teaching and the conduct of the school day.
5. Academic achievement is as good as in larger schools.
6. Vertical grouping in primary schools enables younger and older children to learn together.
7. Contrary to the findings of the Plowden Report (1967) teachers are now more eager to work in small rural schools.
8. The demand for a wide range of subjects can be met by peripatetic teachers serving a group of schools.
9. Long journeys (in excess of 30 minutes) to and from centralized schools can affect attainment, and preclude children from after-school activities.

Source: Pacione (1984)

based both on economic and educational grounds (Tricker and Mills, 1987). Closure of one-teacher schools on the Canadian Prairies has been justified as a cost saving exercise in reducing maintenance and running costs of schools plus 'more effective' teaching to be gained from redeployment of teachers to larger schools where relocated pupils will also benefit from a broader curriculum and better facilities (Carlyle, 1988). Indeed, as shown in Table 11.2, a range of educational arguments can be cited against the retention of small local schools. However, many of these can be countered, also on educational grounds, but, in addition, by a consideration of wider community-based issues. Even the diseconomies of small rural schools can be questioned as, although the high cost of building maintenance and payment of teachers per number of pupils taught is high (e.g. Nash *et al.*, 1976), there are savings on pupils' transport costs and on 'social' costs, e.g. fewer demands made on other local authority services such as child guidance and juvenile care (Lewis, 1989).

In Britain the demise of rural primary schools throughout the 20th century has reflected rural depopulation, falling birth rates and periods of economic stringency. In 1926 the Haddow Report recommended that 11 years of age should mark a distinct separation of primary and secondary education. Hence, this should have marked the closure of many rural schools as they lost their over-11s. Lack of funding to build the requisite centralised secondary schools reduced the Report's impact, but effectively the implementation of its recommendations was only delayed until the 1944 Education Act and increased central government finance for educational reorganisation. From the late 1940s village schools meant primary schools, with rural children bussed to a secondary school in a higher-order centre (e.g. Drudy, 1978). Subsequently, though, further reports have encouraged major reductions in the rural primary school network.

Further impetus to closure of rural primary schools in Britain was given by the Gittins and Plowden Reports in 1967, recommending a minimum size of 50 pupils and a range of classes. This 'immediately put 6,500 primary schools at risk throughout England and Wales' (Rogers, 1979: 4) and contributed to an accelerated rate of closures in several counties, e.g. 800 primary schools closed in rural England from 1967 to 1977, the smallest settlements being most likely to suffer school closure (e.g. Stockford, 1978). In this respect and in the general loss of rural schools, a general trend was found throughout the Developed World (e.g. Bradshaw and Blakeley, 1979; Lonsdale and Holmes, 1981). Some of the chief expectations were in Scandinavia where the minimum size of school classes was lowered to permit retention of small schools, thereby minimising bussing of children which was perceived as a major strain on children and communities (D'Aeth, 1981).

In Britain and several other Developed Countries (e.g. West Germany and Sweden), the closure of rural schools has also been closely linked to falling birth rates. In Britain birth rates fell every year from 1964 to 1977, as a result of which the numbers of pupils in primary schools fell by 30 per cent between 1973 and 1985 (Tricker and Mills, 1987: 42). Most closures of schools have actually lagged behind this fall in the school-age population, but, nevertheless, the fall has been a powerful reason for further closures of rural schools in the 1980s (Figure 11.7) (Tricker, 1984). In contrast, in France critical thresholds for schools and post offices were revised downwards, and 85 per cent of rural communities have retained a primary school. This reflects the continued

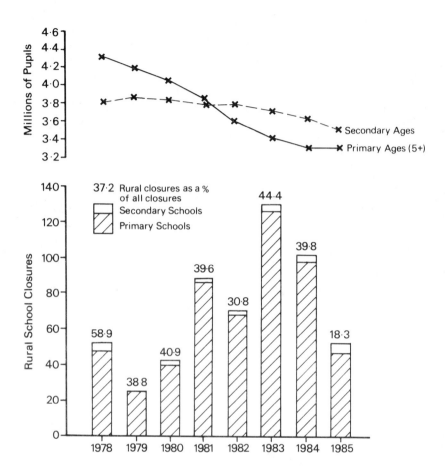

Figure 11.7 The decline in the number of school pupils and rural school closures as a percentage of all closures in England, 1978–85 (*Data source:* Tricker and Mills, 1987)

concern in France for the maintenance of viable rural communities, especially following decentralisation policies in the 1980s (Mackenzie, 1983).

Continued closures have taken place despite the growing awareness of the impact upon the rural community. Increasingly, strident opposition, particularly from parents, has met proposals for school closures (e.g. Rogers, 1979) and has produced considerations of a range of alternative patterns of school reorganisation. Retaining one- and two-teacher rural schools has become an instrument of policy in some areas, but often such maintenance of the existing school system has only been possible through the introduction of additional measures, some of which have also been utilised for other rural services. A number of alternatives can be recognised:

(a) *The 'nuclear' option* – rather than transporting children to an 'area' school with a large catchment areas, some authorities have adopted a 'clustering' arrangement in which a number of schools become annexes of one school under the jurisdiction of a single head teacher or single school administration. In this way most lessons can be carried out in the individual annexes, but with some held in a central school. With a highly flexible pattern of organisation this nuclear option can be much cheaper than the area school approach in terms of initial capital costs, but more expensive for teaching costs. However, the benefits to individual communities might be regarded as outweighing such costs.

(b) *Community schools* – in some cases it has been possible to utilise school buildings for other uses outside normal school hours to serve the needs of the local community. In particular, sporting uses have proved popular as have a range of educational and 'assembly' functions, e.g. for old age pensioners' associations, aerobics classes, Scouts and Guides, the Women's Institute, community councils, DIY classes, arts groups, parent-teachers' associations, ladies' circles, and music clubs. This solution can provide some income to run rural schools but has tended to be followed most by centralised schools serving a wide hinterland rather than by schools serving small communities.

(c) *Mobile resources* – one current solution to the decline of rural population below threshold levels for services has been to maintain the provision of these services but on a periodic basis in the form of a mobile supply. The most common examples of this have been mobile shops and mobile clinics. However, peripatetic teachers, supplementing permanent teaching staff in rural schools, and mobile teaching resource units have been used, especially in areas of widely scattered population in North America and Australasia. In the latter this mobility has been extended by taking virtually all teaching away from fixed-location schools and using telecommunications to transmit 'lessons' to children in their own home (Holmes, 1988: 206–7). This 'school of the air' specifically tailored to the needs of the 'outback', has been copied for tertiary education in some countries, e.g. Britain's Open University. In Australia there has been a strong emphasis upon providing improved transport and communications to the remotest areas, and both daily school access and the School of the Air have benefited from this, e.g. the *Community Access Programme* of Telecom Australia (from 1979) and their *Rural and Remote Areas Programme* for introducing a microwave relay service using digital radio concentrators. In addition, though, financial assistance has been increased progressively to support boarding and tuition costs for rural children attending boarding schools well away from home (Tomlinson and Tannock, 1982).

(d) *Unorthodox funding* – there are a range of solutions that have been adopted to maintain small rural schools through raising additional funding to supplement that derived from state support. These schemes include parent-teacher co-operatives and market-determined salaries under which a lower salary may be offset against attractions of rural living or provision of accommodation.

None of the above solutions has been sufficiently widespread and have not countered 'centralising' legislation in those countries where commitment to maintenance of rural communities has been secondary to other considerations. Yet, there are a range of arguments in favour of retaining small rural schools where a sense of personal and communal responsibility is more easily instilled. 'Maintenance of communities' can be linked to historical and sentimental arguments in favour of retention as well as some cost factors associated with the expense of bussing children to centralised facilities. Ultimately, though, it is a combination of rural 'self-help' and maintenance of numbers that is likely to be the most potent force in saving rural schools from closure.

11.5   Health care

Throughout the rural areas of the Developed World both hospital and general practitioner services have become increasingly centralised. Many small, local hospitals have been closed as hospital services have become concentrated on large general hospitals in major towns (Haynes and Bentham, 1979a). Similar trends have been apparent in general practitioner services through the formation of urban-based health centres and the growth of group practices (Cartwright and Anderson, 1981). Haynes and Bentham (1979b; 1982) argue that these trends have resulted in inhabitants of remoter areas receiving less health care than is typical of the population as a whole.

Despite lower levels of health care, there is evidence for England and Wales to suggest that rural residents enjoy better health than urban residents (e.g. OPCS, 1981), i.e. mortality rates are lower in rural areas than urban areas, and especially in contrast to conurbations and larger towns. However, a study of mortality rates in England and Wales by Bentham (1984) showed these rates to be higher in 'more truly rural areas' than in other rural areas. These more truly rural areas were also experiencing more rapid increases in mortality. As it is these areas which have the poorest levels of access to and lowest levels of use of health services, it appears that rural health services have not been reaching the areas needing them most. The deteriorating mortality rates may be a consequence of continued population decline and its accompanying differential migration. However, the precise causes are not clear.

There is evidence from several other developed countries that there is a disproportionate provision of medical services in urban as opposed to rural areas (e.g. Barnett and Sheerin, 1978; Bentham and Haynes, 1985a; 1985b; Roemar, 1976; Shannon and Dever, 1974). For general practitioner services this has reflected a tendency towards centralisation into group practices despite overseeing authorities such as Britain's Medical Practices Committee which try to encourage an evenly distributed patient–general practitioner ratio (Knox and Pacione, 1980). But this is only one aspect of health services used by the public in a complex set of relationships between services and patients.

A general distinction is usually drawn between primary and secondary health services. The former includes not only general practitioners but also dentists, opticians and pharmacists supported by community health services, especially for expectant and nursing mothers, and health visitor services. The public have direct access to these services whereas they are usually referred from this primary sector to secondary health care in the form of hospital in- and out-patient services. Especially in countries where the state funding and management of health care is limited, centralisation of general practitioner services has often occurred. A more pronounced trend has affected secondary services, with their tendency to demand higher threshold populations.

Fielder (1981) has identified five main difficulties in the provision of primary health care in the rural USA:

(a) the limited opportunities to practise the sophisticated types of medical care in which the doctor has been trained;
(b) doctors' attraction to urban areas where rapid advances in medical research and development are occurring;
(c) the lure of specialisation and other attractions of group practice entice solo rural general practitioners to abandon their rural practice;
(d) workloads in rural practices are heavier, with greater patient:doctor ratios, longer distances for house-calls, and frequently more irregular hours;
(e) training for primary care often fails to prepare general practitioners for the demands of a rural practice, leading to disillusionment and a desire for a different kind of medical work.

There are specific attractions of centralisation that can be added to these 'push' factors for abandonment of rural practice. These include 'the advantages of mutual support, shared costs, a range of technological equipment beyond the reach of a single practice, and close proximity to other members of the primary health team engaged in complementary services to the community' (Pacione, 1984: 264).

Unfortunately, this tendency towards centralisation of primary services in urban areas has coincided with a growing awareness of higher concentrations of need for certain types of medical care in rural areas. For example, in the United States, rural areas have a higher incidence of chronic disease, more days away from work through illness and higher rates of work-related injuries. To the traditionally higher infant mortality rates in rural areas can now be added growing demands on medical care following the growth in numbers of elderly choosing to live in rural areas. The lack of services has its greatest effect upon the poorest members of this elderly group, who do not own a car. They and others in the rural USA who do not have medical insurance to pay for treatment are amongst the most disadvantaged.

Hospitals as well as general medical practices have been subject to centralisation. For example, in Britain, as with education services, hospital centralisation has been a product of planning at the national level. The first national hospital plan in the 1960s recommended the formation of a network of large district general hospitals providing services unable to be offered economically in smaller units. Hence many small hospitals in rural areas were closed (Haynes and Bentham, 1979a), and subsequently access to both hospitals and other health services has declined steadily.

A similar decline in the United States has been met by a series of Federal Government policies, though of relatively limited success, in attempting to reverse the decline in the level of medical care available to rural residents. According to Pacione's (1984: 266-7) summary, these policies and other initiatives have focused on five strategies for improving patient access to health care services:

(i) increasing the number of medical schools and the total number of medical graduates;

(ii) providing incentives for the establishment of general practices in rural areas, e.g. special loans;

(iii) developing and promoting general practice programmes in medical schools;

(iv) establishing a *National Health Service Corps* to help train family (general) practice doctors to be located in rural areas;

(v) use of mid-level health personnel, the so-called '*new health practitioners (NHPs)*' (e.g. Reid, 1975). These are essentially ancillary medical staff trained to perform particular tasks, and thereby reducing the load on more highly qualified personnel, especially general practitioners, and particularly in rural areas (e.g. Blake and Guild, 1978). Despite worries that the NHPs would require close supervision by general practitioners or hospital doctors, the establishment of a set of protocols for the NHPs to follow has enabled this scheme to be followed with some success in some sparsely populated rural areas. Indeed, whilst Dhillon *et al.* (1978) cite as one of the most pressing health care problems in rural America the scarcity of health care workers through the maldistribution of physicians, their suggested remedy is the provision of non-physicians for primary health care (i.e. some preventative, acute and restorative care short of dealing with 'serious illness'). Such provision could be brought about both through use of NHPs and also greater use of satellite clinics staffed by non-physicians, with the use of improved telecommunications permitting consultation between such clinics and physicians.

It is pertinent to note, though, that in Australia 'remote outback locations serviced by the Royal Flying Doctor Service have a higher-quality medical service, for both preventative and emergency treatment, than do some outer suburbs' (Holmes, 1988: 209). This has developed following the introduction of special financial assistance to cover costs of travel and living costs for remote residents for time spent away from home receiving specialist medical treatment.

11.6 Self-help solutions

The preceding discussions referring to a range of different rural services have highlighted the tendency for centralisation of these services, often produced through the operation of scale economies at regional and national levels. Some counters to this trend have also been suggested, in terms of initiatives deriving from rural communities themselves. Such initiatives may be stimulated and supported by government, generally in an attempt to support social equity in peripheral areas but without excessive cost, or they may be generated by communities themselves. The latter can more truly be said to form a *self-help solution* to service provision, but government-assisted schemes with significant local input can also be included under this heading. Indeed, McLaughlin (1987) argues that a variety of self-help policies have proliferated in Britain (e.g. Rogers, 1987; Woollett, 1981b), largely because they help to legitimate the social and economic policies of central and local government. In addition, they help to support numerous rural statutory and non-statutory agencies whose existence depends on political and financial support from government.

In Britain one of the chief stimulators of self-help schemes in rural areas are the Rural Community Councils (RCCs), generally existing as independent voluntary organisations but with support from local authorities. The RCCs usually give advice and information rather than direct financial assistance, limited technical help, and liaise with the statutory services and voluntary organisations to improve local services. The RCCs are overseen by the National Council for Voluntary Organisations (NCVOs), established in 1915 and supported by the Development Commission. The Rural Department of the NCVO deals with promotional, lobbying and information dissemination activities through the RCCs. The RCCs have undertaken a series of community development projects in the 1980s to seek 'low cost solutions, in areas not currently statutorily provided'. However, these schemes have tended to focus upon making further use of existing facilities, e.g. village halls, schools and rural societies, rather than the establishment of new services based on local initiatives. Much of the local-level initiatives in both housing and development of the rural economy in Britain are promoted by RCCs, and co-ordinated at a national level by a charity, Action for Communities in Rural England (ACRE). Through the RCCs, ACRE is linked to 11,000 parish councils plus schools, youth groups, Women's Institutes and local

Table 11.3 The operations of Action with Communities in Rural England (ACRE)

A At national level

1. Raises the profile of rural issues and campaigns for specific action through media and parliamentary contacts.
2. Organises conferences and seminars to highlight initiatives and areas of concern.
3. Publishes books, guides and information to develop good practice and share experience. Recent titles include *Halls for the future* and *Who can afford to live in the countryside?*
4. Provides training and advice to develop staff and managers in RCCs and related organisations.
5. Liaises with other rural agencies in Britain and Europe to share ideas and work together. One important way is through Rural Voice, the alliance of ten leading countryside organisations, representing nearly a million people. ACRE is an active member of Rural Voice and also provides its secretariat.

B At local level

1. RCC fieldworkers help rural communities survey their local needs for housing, jobs and essential services; ACRE provides training and publications in support.
2. RCCs help their local communities fight to save the village school, set up a 'community shop' or provide a new bus service. ACRE's magazine and books share this experience throughout the UK.
3. RCCs themselves fill service gaps, e.g. with mobile information centres and travelling art workshops. ACRES's research identifies good practice from such innovations.
4. RCCs provide an advisory service to village halls on legal obligations, safety, design and good management. ACRE's publications, videos and technical advice enable them to do this.

Source: ACRE, n.d., *Another countryside organisation?* (ACRE, Fairford, Gloucestershire)

voluntary organisations. Through finance from the Rural Development Commission and various major businesses and financial concerns, it attempts to provide information, advice, training and action 'to improve the quality of life of all those living and working in the countryside'. Table 11.3 shows how these general aims have been translated into practicalities. Perhaps the most critical aspect of the work of ACRE and other self-styled 'countryside organisations' is the co-ordinating role, which seeks to develop greater coherence within the plethora of policies being pursued by voluntary, governmental and private-sector agencies. Without such coherence, the rural planners' holy grail of integrated management of rural areas seems unattainable.

Another form of self-help development found in different parts of the Developed World has been the *community co-operative*. This has usually been an example of government 'pump-priming' to encourage local economic development via new business activity in which rural initiatives share in communal enterprise. For example, community co-operatives in both northern Scotland and western Ireland have been based on similar lines, with three main features: a package of financial incentives including a five-year management grant and provision of 'start-up' grants equivalent to the capital raised via local shareholdings; the provision of support services, especially related to marketing and financial advice; and field officers to provide information of community co-operatives (Breathnach, 1984: 109; Brownrigg, 1982; Storey, 1982; Nurminen and Robinson, 1985).

A wide range of activities has been undertaken by such co-operatives, though often it has been retailing that has been the most important aspect, with a village store being maintained or a pooling of local skills to support construction services and social projects (Williams, 1984). The focus upon co-operative formation to maintain rural retailing is indicative of the extent to which, traditionally, government has neglected this aspect of rural life. However, in Scandinavia government assistance for the maintenance of rural services has extended beyond the sphere of the social services to that of private retailing, recognising the centrality of the rural shop to the maintenance of a viable community (Ekhaugen et al., 1980). Thus, from the mid-1970s, in Finland (1975), Norway (1976) and Sweden (1973) there have been measures of support for small retailers located in rural areas. These measures have included investment and management aid as direct subsidy, for example to assist home deliveries in the Swedish case, and consultancy facilities in the Norwegian case (Mackay and Moir, 1980). Whilst these policies seem to have significantly benefited recipients of the aid, it has more often been community self-help schemes that have tackled the problem of maintaining retail outlets in remote areas, e.g. the co-operative outlets of the Inuit in northern Canada (Stager, 1988), community shops in rural East Anglia (Packman and Wallace, 1982), bulk-buy schemes and multi-purpose community centres including retailing as one facility.

Self-help schemes have undoubtedly improved the quality of life in some rural communities. The organisation of more cultural activities, recreation and possibly also improvements to local services have been of immeasurable benefit to some rural dwellers, especially in remote areas. The satisfaction gained from 'grass-roots' developments should not be overlooked, but there are important caveats concerning the degree of need for such local initiatives, the 'abnegation' of responsibilities for service provision by local and central governments, and the conflicting roles being played by a plethora of voluntary and semi-governmental agencies at local level. Self-help remains the ultimate solution of rural inhabitants deprived of services, but generally it signifies underlying defi-ciencies in government and general societal attitudes to rural service provision.

12

Village planning

12.1 Settlement planning in rural areas

Planned change in rural settlements has reflected both the need to correct maladjustments in the current settlement pattern and a desire to implement certain theories and ideologies in practical context to correct perceived economic and social imbalances. It is important to recognise, though, that the use of theory as a basis for settlement planning has been an extremely inexact science in which the correlation between theory and policy has been very loose.

According to Cloke (1983: 55–71) four principal formative theories have underlain settlement planning:

(a) The concept of a *hierarchical settlement pattern*, based initially on the work of Christaller and Losch (Beavon, 1977). A simple example of a settlement hierarchy is shown in Figure 12.1. Within the hierarchy are two concepts which have proved central to settlement planning – *threshold* and *complementary region* or *hinterland*.

The threshold of a particular good or service refers to the number of people required to support it. Certain functions such as hospital services require larger numbers of consumers than other functions, e.g. sales of grocery. The latter may be considered to be lower order goods (convenience goods), and their provision will occur more frequently than that of the higher order goods (shopping goods) and services. Christaller's notion of complementary regions refers to the area from which a service draws its customers. Therefore a service with a high population threshold will tend to draw its custom from a wider area than one with a smaller threshold. The settlement hierarchy reflects the variation in thresholds and complementary regions such that those settlements, or central places, at the top of the hierarchy offer both higher order and lower order goods, thereby serving a wider complementary region or *sphere of influence* or *urban field* (Smailes, 1944), than settlements at the lower end of the hierarchy where only lower order goods are available.

- •   Small village
- o   Rural Service Centre serving small area with daily facilities
- ▲   Self contained village
- △   Self contained village acting as Dormitory Settlement
- ☐   Local Service Centre serving larger area with some weekly facilities
- ■   Small Market Town serving larger area with all weekly facilities
- ▆   Other towns
- ▒   Large towns (County Boroughs)

Figure 12.1 The rural settlement hierarchy for Lindsey, Lincolnshire, as set out in Lindsey County Council's Development Plan, 1955 (*Source:* Cloke, 1983: 91)

Planners have often assumed that such an arrangement of services and settlements in hierarchical form exists and have planned accordingly. For example, in Britain 'the early Development Plans' emphasis on existing central places . . . not only set the pattern for trend planning in rural areas, but also had some considerable bearing on the introduction of key settlement policies which stress the importance of a centre's ability to service its surrounding area' (Cloke, 1979: 42–3). The notions of hierarchy, threshold and complementary region have also been vital to the planning of new settlement on the Dutch Polders (e.g. Van Hulten, 1969). However, the simplicity of the Loschian and Christaller theories breaks down in reality, and the extent and nature of this breakdown needs to be considered if rural planning is to be effective (e.g. Martin, 1976).

(b) *Thresholds* – as described above, central place theory includes the notion of minimum population required to support a particular service, good or facility, i.e. a threshold population. Much rural settlement planning has focused upon this idea of a service threshold by way of attempting to centralise service provision. However, centralisation has placed emphasis upon the social advantages of larger rural settlement at the expense of the needs of smaller settlements and of poorer rural inhabitants unable to travel easily to centralised services. Changes in threshold levels have also not been dealt with very readily (Shaw, 1976), though recognition has grown that the concept of threshold in central place theory is essentially an economic indicator rather than a social consideration to which planners should perhaps be paying more attention (Cloke, 1983: 63–4).

(c) *Economies of scale* – as stated by Ayton (1980), economic arguments have been used by planners in four ways to extend the concept of threshold and link it to general theories of the benefits of scale economies:

(i) individual small villages cannot support education, health and commercial services which need large 'threshold' populations;

(ii) limited and diminishing resources limit public-sector service options;

(iii) both private-sector and some public-sector services will not be provided where they are not profitable, which can therefore severely restrict their operation in some rural areas;

(iv) alternative service provision in mobile form can often incur high running costs whilst offering a low quality of service.

These are points derived from classical scale economies as illustrated in Figure 12.2, which also shows how these economies can be translated into cost curves for particular services. The logical conclusion adopted by many planners in rural areas has been to stress 'the bigger the better' and hence the fostering of centralisation of service provision. Though the demonstrability of this economic argument in practice has been questioned (see Gilder, 1979), there is no doubt that it has played a major part in the prevailing 'wisdom' underlying much rural planning.

(d) *Growth Centre Theory* – this incorporates the notion of two opposing forces governing economic prosperity: *backwash forces* and *spread effects*. Backwash forces, as described by Myrdal (1957) and Hirschmann (1958), refer to the ability of central places to attract factors of production from their surrounding areas. For planners this notion has been used as a basis for concentrating growth at particular centres which can act as poles of growth for a region.

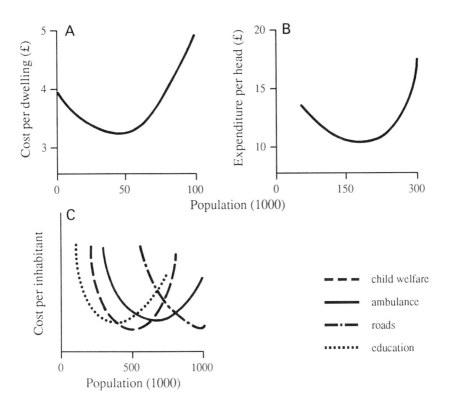

Figure 12.2 Scale effects on service costings (based on Toyne, 1974: 58)

Spread effects are forces which transmit economic prosperity from a central node or region to the periphery. Thus spread effects might be viewed as complementary to backwash forces as growth at designated centres might be transmitted outwards by such effects. This complementarity has been an implicit factor in several rural settlement schemes, utilising growth poles both to concentrate development and spread economic growth to a wider region. Yet it has generally been found that, in rural areas, backwash is more dominant than spread, and so growth pole designation of small service centres can easily drain small neighbouring centres of their economic vitality. For example, in several studies in eastern England and Brittany, Moseley (1973a; 1973b; 1974; 1978) showed that the spread effects of rural growth centres only applied in limited circumstances. In particular, size seems to be a limitation upon spread, with most rural central places falling well below the population level capable of generating spread effects.

It appears that not only have planners often tended to assume that spread effects will operate even from small rural centres but they have also paid insufficient attention to the mechanisms by which backwash forces may promote growth in such centres. Too many ill-conceived plans have attempted to use both growth centre and central place theories as the bases of their operation and, ultimately, have neglected:

(i) to consider how peripheral settlements will be affected by centralisation of resources in growth poles;

(ii) to consider what planning decisions will promote growth by backwash forces in designated poles, and what form this growth will have both in the growth poles and areas affected by backwash.

This common lack of detailed consideration for the mechanisms whereby the four general theories outlined above operate within the context of specific plans can be examined with respect to one of the commonest forms of rural settlement planning, *key settlement policies*. These represented the most widely adopted policies in Britain in the 1950s and 1960s (Cloke, 1980b), being followed subsequently by the gradual implementation of more flexible arrangements. Hence most of the examples used in the following section are taken from plans adopted in British counties after legislation in 1947 and 1968 establishing respectively, Development Plans and Structure Plans.

12.2   Key village policies in Britain

The 1947 Town and Country Planning Act in Britain required each of the 145 planning authorities to carry out a survey in its area and to formulate a development plan indicating how land within its administration would be used over a 20-year period. Much of the detail of the Act was based on the findings of the earlier Barlow, Scott and Uthwatt reports (see Chapter 9) and so placed great store upon preservation of agricultural land and very little upon a settlement's capacity for residential development. Perhaps the first major result of the legislation was that the planning authorities, the county and county borough councils, made detailed catalogues of the educational, health, retail and social services which were provided by individual settlements within their jurisdiction. These catalogues then formed the basis for policy documents, many of which designated '*key settlements*' on the basis of an assumed hierarchy of central places.

Although some of the Development Plans had somewhat diffuse stated objectives, Cloke's (1979) analysis of the key village policies within them lists four prime objectives:

(a) promotion of growth in remote rural areas;

(b) reduction or reversal of rural depopulation through the creation of locations of intervening opportunity;

(c) achievement of the most efficient pattern of rural services;

(d) concentration of resources in centres of greatest need.

These objectives were to be met through plans focused upon an identified settlement hierarchy. The hierarchies in the various counties were recognised in different ways, but with the tiered arrangement suggested in Figure 12.1, for Lindsey, being a common feature. From the hierarchy, selected centres would be designated for a comprehensive concentration of housing, employment, services and facilities in order to maintain a level of rural investment to support both the key settlement and hinterland villages (explicitly recognised in Lindsey) (Cloke, 1983: 91).

With few exceptions little account was taken of possible changes in the settlement pattern and only limited consideration was given to the impact of this approach upon the non-designated settlements (Ash, 1976). Only in a few cases was the key settlement policy extended to become one of explicit planned decline under which a direct rationalisation of the settlement pattern could be pursued, possibly through curtailing of public support for certain settlements and a concentration of population in larger settlements.

As Green (1971: 43) remarked, the notion of 'growth' was viewed as a measure of success and prosperity for all settlements. Hence a planning measure aimed at actually reducing population levels in some settlements strikes at the core of deeply felt rights. Not surprisingly, therefore, settlement rationalisation policies when stated explicitly, as they were in the case of County Durham (Table 12.1), raised great controversy (Barr, 1969; Blowers, 1972). In Durham's case, 114 settlements were considered to have no long-term future, largely being non-viable mining settlements. People in these settlements, with no local mining employment available, were expected to migrate gradually to larger centres, thereby supporting many planners' wishes for greater centralisation. However, the political conflict promoted by this designation in 1951 meant that its implementation was retarded, and by 1970 only eight villages had been completely cleared of population. The attraction of people to particular locations, even if their infrastructure was poor and their appearance unattractive, had not been fully appreciated by the planners. This attraction was displayed again in small mining communities threatened by pit closures in the 1980s and by outlying fishing communities in Newfoundland when the government promoted a resettlement scheme. Generally, though, such explicit designations for decline have been rare, and in Britain have been exceeded in number by plans which have merely classified settlements, classification appearing as an end in itself, e.g. Gloucestershire (in 1955), Hertfordshire (in 1958) and Wiltshire (in 1959).

One of the greatest variations in key village policy has been in the selection of criteria used for designating the key settlements. The lack of any agreed criteria, even between areas facing similar rural problems, has been one of the weaknesses most frequently highlighted by critics of such

Table 12.1 Settlement classification in County Durham's Development Plan

A. Those settlements in which the investment of considerable further amounts of capital is envisaged because of an expected future regrouping of population, or because it is anticipated that the future natural increase in population will be retained (70 settlements).
B. Those settlements in which it is believed that the population will remain at approximately the present level for many years to come. Sufficient capital should be invested in these communities to cater for approximately the present population (143 settlements).
C. Those settlements from which it is believed that there may be an outward movement of population. Sufficient capital should be invested to cater for the needs of a reduced population (30 settlements).
D. Those settlements from which a considerable loss of population may be expected. No further investment of capital on any considerable scale should take place. This generally means that when the existing houses become uninhabitable they should be replaced elsewhere, and that any expenditure on facilities and services in these communities which would involve public money should be limited to conform to what appears to be the possible future life of existing property in the community (114 settlements).

Source: Cloke (1983)

policies. Two examples of criteria adopted by two of the English counties are given in Table 12.2. They show too how criteria have varied according to different intentions of the policies. Cloke (1979) demonstrated this when comparing the selection of key settlements for growth (in Devon) with selection in an area of population pressure where growth was to be channelled carefully (in Warwickshire).

Despite growing criticisms of key settlement policies (Woodruffe, 1976), many authorities utilised this type of policy in their Structure Plans brought into being in the 1970s. Cloke and Shaw's (1983) survey revealed that at least 16 counties had utilised 'a form of key settlement policy', selecting settlements where comprehensive growth of housing, services and employment would be encouraged and which would serve the surrounding area (see Figure 12.3). More emphasis was placed upon housing and active promotion of development than in the early Development Plans, but the overall concept remained clearly recognisable, for example, compare Devon's Development Plan of 1953 with its Structure Plan of 1981. One change, though, was often a greater flexibility towards permission for growth in non-key settlements.

12.3   More flexible policies

As the amount of literature in the field of rural planning has grown in the 1970s and 1980s so the criticism of key village policies has also grown, helping to bring about new and more flexible policies. Cloke

Table 12.2 Criteria used in settlement designation: two examples – Devon and Roxburghshire

A. *Devon*
(a) Existing social facilities, including primary (and in some cases, secondary) schools, shops, village hall and doctor's surgery; and public utilities (gas, water, electricity, sewerage).
(b) Existing sources of employment (excluding agriculture) in, or in the vicinity of, a village.
(c) Their location in relation to principal traffic roads and the possibility that new development may create a need for a by-pass.
(d) Their location in relation to omnibus routes or railways providing adequate services.
(e) Their location in relation to urban centres providing employment, secondary schools (where not provided in the key settlements itself), medical facilities, shops and specialised facilities or services. A town will provide all services and facilities which one would expect to find in a key settlement. Key settlements are not appropriate near main urban centres.
(f) Their location in relation to other villages which will rely on them for some services.
(g) The availability of public utilities capable of extension for new development.
(h) The availability and agricultural value of land capable of development.
(i) The effect of visual amenities.

Source: Cloke (1979)

B. *Roxburghshire*
A. Settlements which are suitable for considerable expansion. In these settlements development will probably substantially alter the existing character of the settlement. This will mean that considerable expenditure will need to be committed, and that cost thresholds will be of less significance than in settlements where more limited growth is proposed.
B. Settlements which are suitable for moderate expansion, but where the basic existing village form will not be affected. These villages should be developed to their existing thresholds, and beyond if the scale of development would justify this.
 1. Settlements where this scale of growth would be desirable, and on the basis of existing and future trends can reasonably be expected to be achieved.
 2. Settlements where this scale of growth would be desirable, but where on the basis of existing trends and future prospects so far as they can be estimated, it cannot at present be considered likely.
C. Settlements where expansion should be mainly in the form of minor infilling, rounding-off and essential redevelopment, and where considerable capital expenditure in relation to the expansion of the village involving the crossing of the thresholds should be avoided. (This does not mean that expenditure considered necessary for the improvement of facilities for the existing population should be avoided.)
 This group may again be divided into two:
 1. Villages which could be allowed to grow by rounding-off, infilling and other limited forms of development within their existing thresholds.
 2. Villages which should be strictly controlled as to their future growth or where even the limited development mentioned above seems unlikely in present circumstances.

Source: Woodruffe (1976)

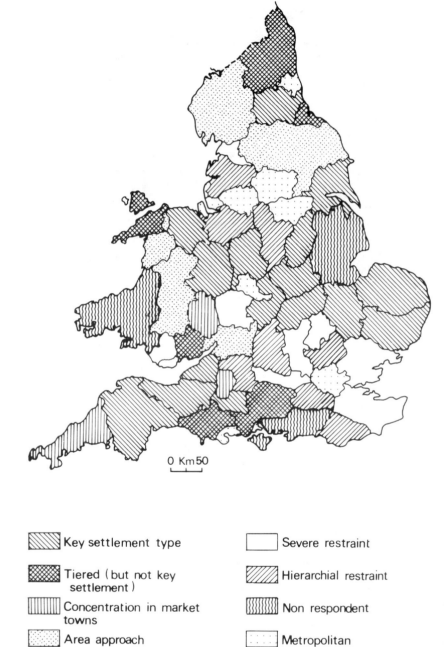

Figure 12.3 Policies adopted in County Structure Plans (based on Cloke and Shaw, 1983)

(1983: 168–75) elaborated criticisms of key village policies under four headings:

(a) The *lack of a proper theoretical basis* – this contrasts the theoretical arguments described above with the realities of practical implementation of plans (Cloke, 1980b). Too often the actual key settlement policies were introduced without any attention paid as to how they were supposed to operate. In other words, the policy itself was subvented by economic expediency and administrative pragmatism. Hence far too little attention was paid to how the theoretical bases of the policy could be translated into practicalities.

(b) The *economic criticism* – that the key settlement policies were not the most effective mechanism for allocating rural resources. Many of the centralisation aspects of the policies assumed that economies of scale would operate, but there must be serious question marks over whether these can have the desired effect when dealing with small levels of population. Indeed, Gilder (1980) argued that only a few services concur with traditional notions of scale economies. Therefore the relationship between size of settlement and the cost of service provision may be limited. His own economic review of policy (Gilder, 1979) suggested that a pattern of dispersal for growth in Suffolk would be more cost effective than a policy of concentration. However, there are many problems involved in trying to assess the effectiveness of planning policies purely in economic terms. Thus Whitby and Willis (1978: 243) urged that a clear distinction needed to be maintained between budgetary costs and social cost when making any policy evaluation. And 'if all costs accruing to both the supplier and beneficiary of a particular service are taken into account, it may be less expensive for a travelling service to visit a scattered population than for the people to travel to a centralised resource' (Pacione, 1984: 61). The weakness of such arguments is that financial assessments by local authorities are based on costs of the authority and generally ignore personal expenditure by rural inhabitants. Given this narrower version of costs, the centralisation option seems more suitable, especially when allied to the economic vitality that can often be demonstrated in key settlements because of their retention of private-sector retailing and services plus provision of a range of public services (Woodruffe, 1976: 26). Indeed, Cloke's (1979) survey of the results of the policies in Devon and Warwickshire in the 1970s showed distinct economic benefits in the key settlements from the greater concentration of activity. The 'flip side', of course, may be growing diseconomies for residents of non-key settlements.

(c) The *social criticism* – that key settlement policies exacerbate social problems experienced by many rural residents. So it has been argued that such policies can lead to increased rural deprivation for residents of the non-key settlement, especially those with less mobility whose access to goods and services may well be diminished by concentration policies (e.g. McLaughlin, 1976). This may present critical social problems counteracting benefits of reduced depopulation from key settlements (Cloke, 1981). Indeed, in some cases it has been relaxations of policy that have helped ease deprivation, e.g. granting permission for housing development in non-key settlements, thereby maintaining or even raising population numbers so that services can be retained. Given its economic base and the type of accounting system of local authorities, key settlement policies seem poorly designed as instruments to deal with social concerns. As these concerns have been pushed to the fore in the 1970s and 1980s, then so different planning policies have been designed and implemented.

(d) *Political criticisms* – that such policies are not compatible with practical political considerations. In Britain critical political problems underlying the implementation of planning policy have been the changing ideologies (from free-market to socialist) and varying relationships between different tiers of planning authorities. Overall, a common view amongst planners themselves seems to be that the scope of rural planning is too narrow to deal with countryside problems. More specifically, key settlement policies have been a part of restrictive planning measures aimed primarily at development control. Thus the policies have by-passed the necessary positive measures required for the various theoretical consequences of key settlement policy to take effect. In short, the planning framework, based on control, restricted or neglected developments that would have promoted spread effects from key settlements, and possibly also retarded backwash effects, e.g. lack of attention to encouraging new housing schemes and new economic stimuli.

Cloke and Shaw's (1983) survey of the key settlement-type resource concentration policies identified four prime problems recognised by the planners themselves:

(i) The assumption that non-key settlements would benefit from investment in the designated settlements was undermined by the decline in rural accessibility.

(ii) The key settlement concept granted respectability to market trends of service rationalisation and centralisation, thereby further discouraging beneficial spread effects in non-key settlements.

(iii) The scale of new public investment was frequently too small to produce benefits from scale economies.

(iv) Planning authorities have had limited control over the provision and allocation of rural resources, and therefore policies have often failed to produce the desired effect.

The many criticisms of key settlement policies have led to a series of alternatives being introduced in the British Structure Plans produced in the 1970s. These alternatives have interpreted the debate of resource allocation and development control in a variety of ways, ranging from substantial approval for resource concentration to rejection in favour of pronounced dispersion. Those counties favouring concentration and centralisation have been the ones continuing to pursue key settlement type policies. However, given the greater emphasis in the 1980s upon market forces and stricter controls over local authority spending, it seems reasonable to expect that some of these counties will have effectively helped to increase divisions between designated and non-designated settlements, as has been the case in France where more emphasis has tended to be placed upon centralisation (Darley, 1978a). Meanwhile, though, a consistent trend amongst the British counties has been to adopt policies more flexible than those reliant totally upon key settlements, usually incorporating an element of greater resource dispersal.

Cloke and Shaw (1983) recognised six categories of rural settlement policies in England and Wales:

(a) place-specific strategies for comprehensive rural development;
(b) area approach policies;
(c) policies dominated by restraint and control of development;
(d) market town policies;
(e) key settlement-type policies;
(f) tiered (but not key settlement) policies.

Yet, when these six were scrutinized more closely, it was apparent that only a small minority of counties had made any concerted attempt to implement policies of resource dispersal. Proposals in the Structure Plan for North Yorkshire would have utilised groups of villages to act as service centres for less prosperous areas in an effort to disperse resources, but objections to the 'uneven distribution of potential investment' (Cloke, 1988b: 35) led to intervention from central government that emasculated the original proposal. The Plan for Gloucestershire was similarly affected (see also Sillince, 1986).

After over 40 years of Development Plans and Structure Plans it is tempting to ask whether the Plans have significantly disrupted the prevailing economic trends favouring concentration and centralisation of resources. In some cases the key settlement policies may have reduced the amount of undesirable sporadic urban development in the countryside, but, especially in peripheral areas, they have not focused upon social needs and may have been detrimental in that centralisation has drained population from non-key settlements more rapidly than would otherwise have happened. It has also been recognised that all the Plans 'nest' within regional, national and even international frameworks, so that decisions far removed from the local level play crucial roles in determining rural settlement plans and their implementation. It is this recognition that has contributed to a much greater academic concern with underlying philosophies of national governments and with power structures and relationships at a variety of levels.

12.4 Comparisons between settlement planning in Britain and other countries

Whilst rural planning in Britain has given rise to a growing literature and there has been some redressing of the previous emphasis upon town planning, this has not been true in the United States. There, rural planning, even in the general context employed in Britain's Development and Structure Plans, has tended to be viewed as unnecessary. Instead it has been the influence of the market and the individual decision-maker rather than the government that has dominated. This domination has only been broken by the growing concern for the environmental well-being of rural areas. Hence rural planning has largely meant environmental planning during the 1980s, though with some significant secondary considerations which have begun to make it more comparable with some of the concerns of British rural planners. The aims recognised in Lassey's (1977) survey of American rural planning, as it extended its remit in the late 1970s, are set out in Table 12.3. Cloke (1983: 114) argues that this growth of rural planning in the USA has led to a more comprehensive approach to problems than the British model, accompanied by a concern to reduce imbalances in services, employment, and housing provision between urban and rural areas.

A similar concern has been voiced in Canada, as illustrated by work on service provision in Ontario. For example, Joseph and Smit (1985) attempted to trace the association between rural residential development and municipal service provision in a township near Guelph, Ontario. Their chosen study, Puslinch Township, had been transformed in three decades by predominantly non-farm residents raising the population over 50 per cent. This growth had increased both the need for public services and the ability to pay for them, though with high costs due to the dispersed nature of settlement. The expansion of settlement had produced increased problems of land fragmentation and competing land uses, with planning constraints addressing this issue and helping to determine the nature and location of new residential development from the early 1960s (Joseph and Smit, 1981; Punter, 1974). Limitations upon new development reflected service-related concerns, though with policies often being unco-ordinated or contradictory.

The statements above about rural planning in the USA must be treated as being highly generalised given the enormous diversity between regions of the country, between states and within states. However, Lassey et al. (1988) do recognise certain general points. The presence of a federal structure instead of Britain's high degree of centralisation is one key feature. At the federal level the US Department of Agriculture (USDA) has had a significant impact on rural communities through its price supports, credit schemes and fostering of research and development. But the role of social planning, as considered in this chapter, has fallen to certain aspects of federal regional policy and to programmes carried out by individual states.

Table 12.3 Fundamental aims of rural planning in the United States

1. Preservation of ecological integrity so as to provide a continuing supply of life-supporting resources.
2. Development of efficient and appropriate land use.
3. Creation of healthy living conditions through the construction of a suitable physical environment.
4. Preservation of aesthetically pleasing environment.
5. Creation of effective social, economic and government institutions.
6. Improvement of human welfare.
7. Development of physical structures and adapted landscapes of pleasing design.
8. Adoption of a comprehensive viewpoint to include physical, biological and human factors in rural regions.

Source: Lassey (1977)

The depression in the farm economy of the American Mid-West has been sufficiently severe to promote renewed rural depopulation, especially from 'towns' of under 2,500 population. The presence of large numbers of such towns in Iowa has posed problems for the state in terms of whether to spread public resources amongst many towns or to concentrate resources on relatively few. It appears that a key settlement policy is emerging in which attempts are made to create selected growth centres promoting economic diversification (Daniels and Lapping, 1987).

An example of the ways in which different types of planning control can exert a major influence over population growth is given in Groenendijk's (1983) comparison of the northern Netherlands and Norfolk, England. The greater uniformity of growth in the former reflects the way in which planning decisions are made at local level and the sharing of population growth between numerous centres. There has been much greater clustering in Norfolk, tending towards a focus upon small market towns, administratively divorced from the rural districts of the county. Public housing has been concentrated in these towns whereas in the northern Netherlands there has been a much more even spread between town and country. To some extent this reflects differences in the perceived population size for 'growth centres'. Such differences are also apparent in Scandinavia. For example, in northern Norway a lower limit of 1,000 inhabitants was proposed in background material for the *North Norway Plan* for settlements to serve as growth centres for employment and services. The selection of this figure reflected the post-war population growth in settlements of this order of magnitude. However, Hansen (1975: 260-1) argued that such growth was the product from rural population reservoirs now empty or greatly reduced. Hence the lower limit may have been too low, and contrasts with a figure of 2,500 to 3,000 inhabitants suggested for similar areas in Sweden (e.g. Bylund and Weissglas, 1970).

The growing academic interest in rural problems and in the institutional

Plate 23 Although emphasis has been placed upon the growth and prosperity of
the major French regional centres, in the 1980s more power has been devoted to
smaller centres such as St Malo (Ille et Vilaine) and from there down to
individual communes

mechanisms through which policy measures are enacted has highlighted
the diversity of approaches within rural planning (Cloke, 1988b; 1989c).
The British system described above and in Chapter 9 has had a pronounced focus upon the physical planning represented by development
control whereas much looser control has prevailed in North America and
also Australia. However, direct international comparisons made in the
1980s have offered prospects of greater exchanges of views between planners of different countries so that lessons learned in one may be applied
in another.

One set of examples offered in the 1980s has been in France, where
development control and environmental conservation have only been a
major part of l'aménagement rural from the late 1970s (Clout, 1988:
116). One facet of this changed emphasis may be the influence of rural
planning elsewhere in Western Europe; another is undoubtedly a central
'driving force' of rural planning – the prevailing political ideology. In
France's case in the 1980s this has been the Socialist administration's
decentralisation of power to the regions and communes (Plate 23). Yet,
overlying this devolution of power, there are the national agricultural
policies that have produced area-specific rural projects and also national
environmental guidelines promoting conservation (Barthelemey and
Barthez, 1978). The development of local initiatives for settlement

planning within this confused framework has led to a bewildering complexity of planning policies, matched by a plethora of agencies often in conflict with one another and the powerful Ministère de Agriculture et Développement Rural.

In terms of settlement planning, since 1983 all communes have been required to produce a plan. This prospect of 36,512 local plans has been reduced in size by some collaboration between communes, usually 15 to 20, in producing either a rural management outline, Plan d'Aménagement Rural, bearing some similarity to the British Development Plans (e.g. Vaudois, 1980), or a more recent version, the Charte Intercommunale. However, the outlines are not legally binding and incorporate no fixed budget (Le Coz, 1984). Co-operation between communes has been encouraged in the form of development controls, *contrat de pay*, signed between groups of communes, small towns and the state as represented by the Délégation pour l'Aménagement du Territoire et à l'Action Régionale (DATAR) (Scargill, 1983). These contracts have incorporated a measure of financial commitment from regional authorities, but have been extremely diverse in their objectives. Perhaps the greatest attention has been given to provision of infrastructure (e.g. piped water and tele-communications) and maintenance of public services. Both of these have also received support via the DATAR and the Ministry of Agriculture through the Fonds Interministériel de Développement et d'Aménagement Rural (FIDAR), especially assisting service provision in remote or disadvantaged areas (Aitchison and Bontron, 1984).

In France and most of the Developed World the co-existence of local, regional and national planning frameworks with the private sector, which has assumed or been given major responsibility for generating jobs and providing housing and even certain rural services, has tended to produce certain 'imbalances'. That of over-centralisation versus dispersion has already been identified as has a tendency, especially in Britain, for planning to focus upon selected aspects of rural life at the expense of others. Although the private sector can be strongly guided by state controls, certain characteristics of the private sector have meant that the 'gap' between planning intervention and state controls has created crucial weaknesses in the composition of rural society and economy. Perhaps the best example of this is *rural housing*.

12.5   The provision of rural housing

As prices inexorably rise, so the population which actually achieves its goal of a house in the country becomes more socially selective. Planning controls on rural housing have therefore become – in effect, if not in intent – instruments of social exclusivity (Newby, 1980b).

When evaluating the extent and severity of rural deprivation, one critical variable to be considered is rural housing – its quality and opportunities of access to the housing stock by different social groups. The notion of the rural idyll has tended to hide the presence of poor rural housing from more widespread attention, but the poor quality of a significant section of rural housing stock has been just one of several housing-related issues that have become a focus both for academic investigation and legislation in many parts of the Developed World (Dunn *et al.*, 1981; Phillips and Williams, 1982).

The quality of rural housing has been a major problem in many countries for decades, though the population turnaround has often helped to turn inferior housing into highly desirable 'rural retreats' (e.g. Duffy, 1986). In the early decades of this century many rural areas in Western Europe had large proportions of their rural housing stock lacking in basic amenities. In parts of the Mediterranean in particular these deficiencies persisted into the post-1945 period. Even in some of the wealthier countries there were sizeable problems, e.g. in the United Kingdom in the 1940s 11.6 per cent of rural houses were found to be 'unfit for occupation' and 33.4 per cent 'in need of repair' (Ministry of Housing, 1944; 1948). Three decades later the proportion deemed unfit had fallen to around 5 per cent, though in proportional terms rural areas were faring worse than their urban counterparts (Gilg, 1985: 60–1; Rogers, 1983).

The improvements in housing conditions post-war have reflected the general increase in living standards. For most rural dwellers the days of no piped water supply, no sewerage, and cramped conditions with no separate cooking accommodation or bathroom have long since disappeared as higher disposable incomes have helped to transform housing quality. Nevertheless, there remain significant pockets of rural poverty as expressed by poor housing. In Britain it is in some of the more remote, less accessible parts of the countryside, e.g. north-east Scotland (MacGregor *et al.*, 1987; Shucksmith, 1984), Mid-Wales and Cornwall, but in some European countries the problem is more widespread (e.g. O'Brien, 1989). Here the deficiencies are partly related to age of dwellings, for example Rogers (1983) reported that in the mid-1970s over 60 per cent of all French rural homes were built pre-1914 and the majority of homes in rural communes lacked inside toilets and baths.

The problems associated with increased demand to build new houses in the countryside to accommodate commuters have provoked an increasingly acrimonious debate (Clark, 1982a). In Britain the argument has focused upon different perceptions of the need to reduce the controls imposed by Green Belt designation, especially around London. Differences have arisen partly through revisions in the estimates for new homes required within the South-East region. For example, in 1988 government estimates were for 610,000 new homes whereas in 1986 the London and South-East Regional Planning Conference estimated 460,000 new homes were needed, at the time criticised as an inflated figure based on unrealistic assumptions. The new estimates have increased fears that further relaxations in Green Belt controls will be encouraged in order to accommodate the large scale of development. The area to the west of

London in the M4 corridor has become a particularly popular target for developers. In early 1988 11 sites south of Reading were targeted for a total of 23,750 houses, bringing into the area a potential extra population of 95,000 people. However, these proposals largely represent a leapfrogging of the existing Green Belt, an inherent weakness in this type of 'green girdle' legislation.

A recent survey by the Rural Development Unit of the Scottish Development Agency (1989) revealed that rural housing conditions in Scotland were deteriorating in that there had been a 40 per cent increase in waiting lists from 1981 to 1986, an 83 per cent increase in homelessness from 1983 to 1987, and higher levels of houses 'below tolerable standards'. These problems were being compounded by problems of access to suitable homes for first-time buyers, the elderly and those requiring sheltered housing. Sales of council houses had reduced the availability of public sector housing (Phillips and Williams, 1981; Williams and Sewel, 1987) whilst private sector property for renting had also declined. Investment in council housing in real terms had fallen by two-thirds from 1976 to 1986 (Shucksmith, 1988).

Within the Structure Plans of the 1970s, now being put into practice, there has continued to be designation of settlements for infill or medium-scale housing schemes in or adjacent to country towns. Increasingly such settlements have seen a growing polarisation between 'desirable' new homes, often bought by newcomers, and traditional rural residents whose children can no longer afford housing in their home village. Planning has fostered social exclusivity whilst permitting market forces to dictate that new houses for the middle class take priority over those for poorer sections of the community (Shucksmith, 1987). In addition, the greater emphasis upon market forces in the 1980s has brought a relaxation of planning constraints, especially in areas under the greatest pressure from urban sprawl, e.g. the area immediately beyond London's Green Belt. It is this relaxation and the continued demand for land by commercial developers that has provoked claims that the South-East is being engulfed in a tide of concrete, bricks and mortar. Blunden and Curry (1988: 92–3) give the example of Consortium Developments Ltd who wish to create 15 to 20 new 'country towns' in the South-East, with a total of 100,000 new houses. Spurred on by real economic growth to be promoted by the Channel Tunnel transport corridor through Kent and the high-tech industries of the M4 and M40 corridors, such schemes may prove difficult to reject. If so, they seem likely to bring the town to the country in the form of increased social segregation and a lack of accommodation for the rural working class.

To overcome the problem of 'no homes for locals' central government encouragement is now being given to councils who set aside land specifically for new low-cost homes for rural residents. The land would effectively become part of village infill, though developers could be prevented from building high-cost housing on land being set aside for locals.

One response to the concern in England about the shortage of affordable homes in villages for people on modest incomes was the establishment in 1975 of the National Agricultural Centre Rural Housing Trust (NACRHT). This is a charity which has the support of the Royal Agricultural Society of England and the Rank Foundation. The NACRHT pursues its objectives by helping *housing associations* to build in villages. It advises and assists village groups (e.g. parish councils) in identifying housing needs and sites for development, working in conjunction with the Rural Development Commission and the House Builders Federation (Rogers, 1985).

In recognition of the housing shortage in rural areas a housing association to provide homes for people who cannot afford soaring village house prices was launched at the 1988 Three Counties Show. It is estimated that rural property prices in the three counties (Gloucestershire, Herefordshire and Worcestershire) rose by 50 per cent in 1987/8 whilst population in rural areas of the counties is likely to grow by up to 17 per cent between 1988 and 1992. The NACRHT, under the slogan 'Village Homes for Village People', has established 14 charitable associations, including the one for the three counties, aiming to build small groups of houses, flats and cottages. They will not be sold, but will be retained for rent by particular groups in rural society. Two groups have been singled out. They are young adults from rural areas, who wish to remain in the countryside, and the elderly who wish to move from larger accommodation but remain within the countryside.

This use of housing associations and *village housing trusts* to provide low-cost accommodation throughout rural areas has received some central government encouragement in Britain (Conservative Political Centre, 1989). This solution sees the promotion of trusts nationwide to promote small-scale development to meet local needs as determined by local surveys. Rural low income families should have first claim to the housing which would be available at low rents, though, with schemes for easy purchase too, local landowners would be offered a stake in the resale value of homes built by trusts, thereby encouraging them to provide land. This scheme echoes sentiments expressed in previous policies which have never been translated into workable practicalities. For example, only 3 per cent of the budget of the Housing Corporation, the quango which funnels public money to housing associations, is devoted to small-scale housing schemes in the countryside. It is also clear that a more determined nationwide policy encouraging low-cost rural housing is required, perhaps building on some of the local initiatives of Community Trusts, but without limiting constraints currently imposed by central government which affect rentals and the degree of control the housing trusts can exert. This would then follow some of the recent developments in France, brought in since 1983 under legislation promoting decentralisation. This has led to a leading role being played by communes, whilst départements and regions provide support and co-ordination, i.e. initiatives come 'from below' rather than being imposed 'from above' (Coulmin, 1986).

12.6 Planning new villages

In Western Europe a common image of the village is of a settlement associated with slow, organic change and a sense of continuity with the past. Yet in most European countries there are examples of villages artificially introduced during the last two centuries. These villages were invented for a variety of reasons: aesthetic, philanthropic, political, for convenience and for ideals. Many are testaments to particular arcadian concepts of aristocrats, industrialists and idealists; others reflect particular planning schemes designed to meet the perceived needs of changing times, e.g. new retirement communities, and special tourist settlements (Darley, 1978b). Compared with the numerous schemes of wealthy individuals in the 19th century and new developments following the 'garden cities' concept in the early years of this century, new initiatives post-1945 have been more limited both in number and scope. The era of greater planning restrictions and financial vicissitudes has reduced opportunities for individual, corporate and public sector schemes.

In Britain the creation of new villages post-1945 has not formed a part of general planning policy. The new town, the extended town and the 'urban village' have all been important features in the attempt to draw off population from metropolitan areas, but the idea of the newly built publicly-funded village in open countryside has not been pursued. New villages have been proposed, but reliant on private funding and initiative, e.g. often associated with retirement or recreation schemes as in the £35 million project at Cardrona Mains, near Peebles, for 200 houses, a hotel and a golf course. Few schemes for such entities have been proposed and fewer carried out to their designated conclusion. Perhaps the experience of the early post-war forestry villages on the remote borders of England and Scotland contributed to this omission of the new village from the list of acceptable planning alternatives:

These villages . . . illustrate effectively the shortfall between the ideals of the most enlightened planners and the numerous intervening factors which can so easily invalidate the best of schemes (Darley, 1978a: 250).

One major exception to this, though, has been in the Netherlands on the Ijsselmeer Polders where new villages have been located on land reclaimed from the sea since the 1920s. Some settlement schemes have also accompanied land reform programmes in the Mediterranean.

In the Mezzogiorno several different plans were tried as part of schemes for the provision of services to nearly created farming settlements. The need for services was most critical in new dispersed settlements established on coastal plains previously devoid of service centres because of flooding and the presence of malaria. A common approach adopted was to locate facilities within a radius of 1.75 to 3.25 miles (3 to 5 km) from dispersed farming, serving in the order of 350 families, e.g. Gramola in the Sele River plain of Campania. Some of these new

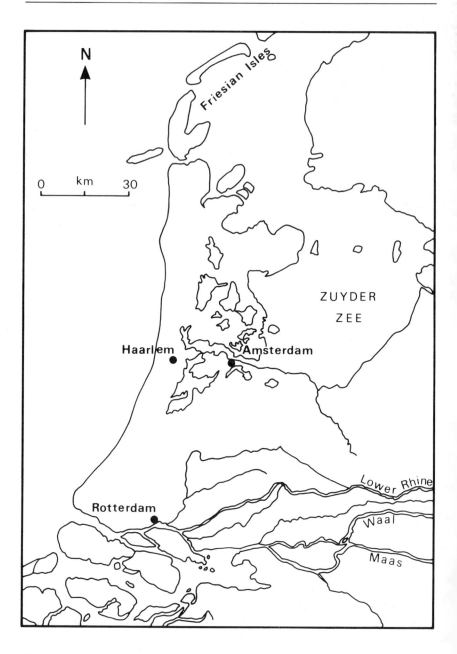

Figure 12.4 The Netherlands' coastline prior to the draining of the Zuyder Zee

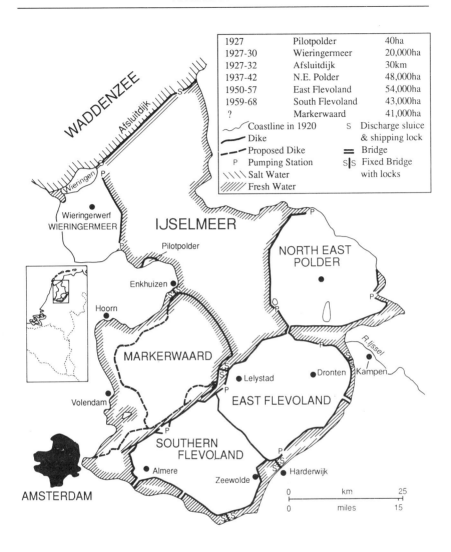

Figure 12.5 The Ijsselmeer Polders, the Netherlands (*Source:* Information and Documentation Centre for the Geography of The Netherlands, 1985)

service centres (*Borgo di Servicio*) were essentially non-residential and, except for service personnel, consisted of schools, a church, assembly hall, offices, restaurants, shops and workshops. In other cases they also included private residences. A similar scheme, but with a distinct hierarchy of service centres was established around Arborea in Sardinia. This gave a network of central places, each serving a 2.5 to 3.25 miles (4 to 5 km) radius. Schools were established at more regular intervals to reduce walking distances for pupils, but the resulting fragmentation of the school system did not help the quality of education (Maos, 1981: 385–6). Subsequently, several of the new settlement schemes involving the creation of dispersed farmsteads suffered from the abandonment of the farmsteads by younger farming families who preferred to live in the service centres.

12.6.1 New villages on the Ijsselmeer Polders

Since Roman times when settlement was confined to small islands, the *terpen* or *wierden*, the Dutch have reclaimed land from the sea. Dykes were first introduced around 1000 AD in the form of dams across tidal rivers and regulated by sluice gates in order to protect Amsterdam and Rotterdam. Dykes permitted the formation of *polders*, pieces of low-lying land reclaimed from the sea, though, with dykes frequently breached or flooded, there was a constant battle to retain land won from the sea. In 1334 the coast north of Den Helder was pierced to form the Friesian Islands with the Wadden Zee to the east. To the south the former Lake Flevo was enlarged to become the Zuider Zee to the north-east of Amsterdam (Figure 12.4). Further periodic serious flooding occurred before the invention of windmills in the 17th century which enabled the draining of large areas of low-lying land. Since this time the progress of reclamation has continued to make gradual progress, though often at high cost.

A major boost to reclamation was the application of steam power to pumping in the 19th century. This led to larger scale schemes being proposed and some implementation of such plans, e.g. the Harlemmermeer, to the south-west of Amsterdam, in 1884. The success of this prompted thoughts of reclaiming the Zuider Zee, and proposals were made which were the forerunners of the eventual *Ijsselmeer Polders scheme* first started in the 1920s (Figure 12.5). The aim of these early plans was to create a new area for agricultural settlement in a country in which, even in the late 19th century, there was great population pressure upon available land. However, from the outset it has been apparent that attempts to create small villages in the newly created lands would involve insurmountable problems related to the viability of such settlements. Consequently, numerous planning changes have had to be made to produce viable settlement schemes on the reclaimed land which, when established, would add 10 per cent to the land area of the Netherlands.

In 1927 a trail 40 ha (100 acres) was reclaimed in the north-west of the Zuider Zee as the precursor of the *Wieringermeer* project, begun in 1930 and covering 20,400 ha (50,410 acres). This compared with an average service area of between 500 and 1,200 ha (1,235 to 2,965 acres) per village in the rest of the Netherlands. As at this time 23 per cent of the Dutch population lived on farms, there was little questioning of the allocation of the reclaimed land to farming. Indeed, once problems associated with high salinity were overcome, the land proved to be of very high quality and was used for arable farming, but served only three new villages rather than the 14 proposed initially. This has meant that the three villages have service areas in excess of 6,500 ha (16,050 acres). One reason for not pursuing the original plan was that once farms had been established, service centres on the mainland were utilised by the farm population. So when a fourth service centre was proposed in the 1950s, it was apparent that the mainland centres plus the three new villages were perfectly adequate for what, by then, was a dwindling rural population on the new polder. This revision of initial settlement planning has characterised the other new polders.

Accompanying the Wieringermeer project was the enclosure of the Zuider Zee in 1932 by the Afsluitdijk, permitting greater water control and removal of salt. This control of salinity was slowed by war-time reflooding of some recently reclaimed land on the next polder created, the *North-East Polder*. However, it was redrained in 1946 to yield 49,000 ha (121,075 acres). This was occupied by farms of between 12 to 48 ha (30 to 120 acres) with a mean of 25 ha (60 acres) as compared with the Dutch national average of under 14 ha (35 acres). This distribution was very similar to the Wieringermeer where farms ranged from 21 ha to 45 ha (52 to 110 acres) and reflected the desire to create economic and efficient units on the new land so that the nation could benefit fully from the investment being placed in the creation of the new land. Farms were often arranged in clusters of four to help promote social links. Each farm consisted of one single owner-occupier holding containing the farmstead and farm buildings as well as the agricultural land. Labour was drawn from new villages built equidistant from the newly created service centre, Emmeloord. The initial plan for the North-East Polder stated a target population for each village of 4,000 with villages 4.5 to 5 miles (7 to 8 km) apart and serving 5,000 to 7,000 ha (12,350 to 17,300 acres) (Figure 12.6). Emmeloord was to reach at least 10,000 population. In fact, it achieved this very rapidly, a population of 25,000 being recorded in the mid-1980s. However, the villages have failed because of the use of less farm labour and the willingness of workers to commute longer distances from settlements off the Polder. The density of farm settlement is indicated by the fact that 43 per cent of the 44,000 population on the Polder live outside Emmeloord.

The next Polder to be drained, *East Flevoland*, had a rather different type of plan because this was to be the location for the new polderland capital, Lelystad. This was targeted to grow to 100,000 inhabitants and act as a fully functional regional centre. When the polder was drained between 1950 and 1957 the intention was to create ten new villages each with a 3,600 ha (8,900 acres) service area and nobody living more than 3 miles (5 km) from a village. But the reduction in the farm labour force prompted a revision from ten to four new villages and, on the opening of the 54,000 ha (133,435 acres) polder for settlement in the mid-1960s, only one, Dronten, was actually created. This was to service an area of 9,000 ha (22,240 acres) with farms up to 7.5 miles (12 km). In effect,

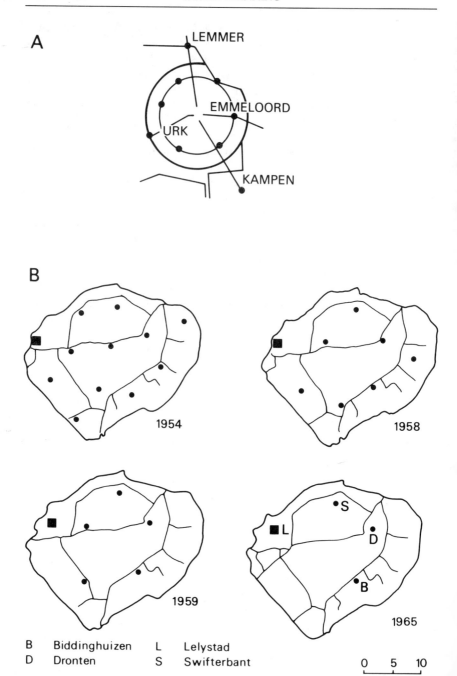

Figure 12.6 The planning of service centres on the Ijsselmeer Polders: North-East Polder

Plate 24 Creating a New Landscape – This modern farmstead and its new outbuildings was established on land reclaimed as part of the Ijsselmeer Polders in the Netherlands. Located close to the Polder capital, Lelystad, this arable farm is highly mechanised, relying on family labour plus some contract workers for certain activities

it has become a secondary centre to Lelystad, with two other new villages established as well. These have a minimum target population of 5,000, the same figure as planners in Norfolk, England, settled on around this time as the viable minimum population for a rural community. The mean holding size on East Flevoland is 31 ha (75 acres), but with holdings up to 125 ha (310 acres), very large by Dutch Standards (Plate 24).

The plans for the other two of the Ijsselmeer Polders, *South Flevoland* and *Markerwaard*, reflect the changing views on viable sizes for farms and villages and the proximity of Amsterdam to these two polders. For South Flevoland the principle of *deconcentrated concentration* has been applied, focusing settlement along the coastal strip which lies on the route from Amsterdam to Lelystad. Around 18 per cent of the polder has been designated for urban overspill and expansion in the form of residential housing. This is a marked departure from the earlier schemes. Also 25 per cent has been zoned as woods and for nature and recreational reserves. This compares with figures of 11 per cent for East Flevoland, 5 per cent for North-East Polder and 3 per cent for Wieringermeer (Table 12.4). Again, this reflects a desire to cater for the needs of the urban population in Amsterdam whilst also showing a lessened desire to

Table 12.4 Land use (%) in the Ijsselmeer Polders

	Wieringer-meer	North East Polder	Eastern Flevoland	Southern Flevoland
Agriculture	87	87	75	50
Woodland and nature areas	3	5	11	25
Residential areas	1	1	8	18
Dikes, roads, water courses	9	7	6	7

maximise agricultural output from the reclaimed land. Instead, the concern over how to cope with the rapid growth of population in Randstad Holland has tended to take precedence and the polders have been required to act as reception areas for urban overspill. Thus Lelystad has attempted to attract secondary and tertiary employment on a large-scale whilst Almere on South Flevoland has been designated for additional growth.

Lelystad is designated to reach a population of 100,000 and by 1987 had reached 60,000, its growth slowing in the early 1980s through economic stagnation and the fact that Almere, closer to Amsterdam, has acted as an intervening opportunity. Almere may reach a population of 130,000 early next century, on a poly-nuclear design meshing settlement with provision for recreation, agriculture and forestry.

Drainage of the largest of the new polders, Markerwaard, covering 62,000 ha (153,200 acres), was proposed for the late 1960s. Three new villages were included in one of the suggestions for this area, but its proximity to Amsterdam also led to suggestions that it should contain a second national airport as its principal function. However, the decision on when to commence reclamation has been postponed and the possible area to reclaim reduced to 41,000 ha (101,300 acres). The question mark over the future of the Markerwaard indicates the changed demand for agricultural land (Van Lier and Steiner, 1982). It is now recognised that reclamation is not required in order to promote increased agricultural production. Taking land out of agriculture to reduce output has become the requirement whilst settlement of urban commuters has replaced the concern for creating new farming communities.

One of the main problems with the new polders has been the lack of local non-agricultural employment for the maturing families of the polder communities. In the 1970s this became a major problem on North-East Polder where, by 1972, for a population of 35,000 only 16 firms had been attracted giving employment for less than 1,000 people. So commuting to mainland towns like Kampen and Zwolle had developed and there was a high rate of out-migration.

Farming on the new polders has been free-market oriented and geared to high output production. Only since the Land Reallocation Act of 1985 has greater consideration been given to ecological needs and sustainable development on newly reclaimed land. The new approach is illustrated in a small 9,000 ha (22,240 acres) polder in the north of the country, the Lauwerszeegebied, reclaimed in the late 1960s (Van Lier, 1988). Here farms are being created that will have both intensive and extensive systems of production, with a gradation so that some land is left as non-agricultural to form a nature reserve. Farms will be between 25 and 40 ha (62 to 100 acres), with up to 5 ha (12.5 acres) of 'transition zone' or more extensive activities. The latter will allow only certain activities; a similar situation to management agreements now being concluded with Dutch farmers in order to promote 'desirable practices' on over 1,000 farms in the country, e.g. mowing once a year, not mowing before June 15th, no grazing, no herbicides/pesticides, no fertilisers, no changes in the groundwater table, no changes in the soil structure.

Conclusion

In Britain the growing clamour for the countryside to be viewed as a sustainable, multi-purpose asset and not dominated by agricultural economics, has been joined by policy statements from the Countryside Commission. Increasingly, the Commission has pressed for policies that will ensure the retention of attractive, diverse and quality landscapes rather than ones produced simply by agricultural support policies giving rise to overproduction and significant environmental destruction (see Lichfield and Marinov, 1977; Nature Conservancy Council, 1986; O'Riordan, 1979). A wide-ranging statement by the Commission in June 1989 listed four key principles that should guide future management of the countryside:

(a) New housing must be more sympathetic to local styles and should normally be restricted to small towns and villages.
(b) Green Belts should be given a wider purpose than merely checking urban sprawl.
(c) New rural enterprises, including light industry and high technology, should be welcomed, particularly if they provide a new use for redundant farm buildings.
(d) Large-scale developments must be kept out of Areas of Outstanding Natural Beauty unless there is an over-riding national need.

This type of thinking is symbolic of the extent to which attitudes towards rural areas have changed post-1945. Although many in the urban population may still hold an idyllic view of the rustic inhabitants of pretty villages or of the 'great outdoors', policy-makers have increasingly been forced to take note of a range of rural problems, some of which have involved both a variety of conflicting interests and rapidly changing situations. No longer can policy simply be geared to preserving the agricultural interest and attempting to foster increased farm output.

Instead, competing social, economic and environmental considerations
need to be weighed against specific local needs. In short, the task of rural
planning has been made harder by the growing awareness of the need for
rural areas to serve multiple functions. Unfortunately in many countries
this multiplicity has tended to spawn a spaghetti-like tangle of interrelated agencies and interest groups that extends up to central government ministries which have responsibility for the countryside but are
often in conflict with one another.

A brief review of Chapters 9 to 12 conveys some of the range of rural
planning policies that have been put into practice post-1945. These
policies also demonstrate the extent to which planning has become more
heavily involved in conflict resolution: deciding between the merits of a
new tourist development and a fragile environment; persuading farmers
to de-intensify their production methods whilst the farmers continue to
respond to favourable market forces; preserving one area of countryside
from urban sprawl whilst condemning another to be changed irrevocably
by an influx of urban commuters. Perhaps the most over-riding impres-
sion of this planning activity is of its piecemeal nature. Throughout the
Developed World a clear goal for rural planning has been all too
frequently lacking. Hence, the Countryside Commission's statements
listed above represent one of relatively few attempts to set a clear goal
to work towards. There is a great difficulty in setting simple and
attainable goals for the entire rural population of the Developed World
given the variations in socio-cultural background, political characteristics,
economy and balance between pressured and remote countryside. Yet
there are remarkable similarities extending across the groups of countries
distinguished in Chapter 1, giving rise not only to some common
problems, but also common policies.

One continuing central problem is the reaction to the further development of agrarian capitalism as it continues to shed labour, concentrates
on larger production units and harnesses modern technology to raise
output. Governments seem to be caught between a continuing desire to
make land yield as much as possible, preventing surplus production,
reducing environmental diseconomies of modern farming and slowing
down the march of scale economies in order to retain viable communities
in rural areas. In the EEC a satisfactory balance between these conflicting aims has not been attained, primarily because the driving force of the
CAP remains the underwriting of large arable producers, coupled with
some preservation of smaller uneconomic farmers via price supports and
a range of 'green' and community-support policies, though the latter
receives much less finance. A similar policy conflict exists in North
America and Australia whilst New Zealand has removed farm income
supports and subsidies with a resulting dramatic decline in land prices
and farm incomes. The extent to which other governments will 'bite the
bullet' of scaling down farm price supports is one of the most intriguing
questions for rural development over the next two or three decades.

If evolving agrarian capitalism has brought new policy problems post-

1945 so too has the changing perception of the role of the countryside by the mass of the population. No longer are rural areas simply the 'town's larder' and supplier of in-migrants to the towns. Instead, rural areas have to fulfil a range of demands, therefore bringing about the planners' need to develop policies recognising multi-functioning rural areas and reconciliation of conflicting roles via integrated development schemes. Perhaps the most fundamental question here is the extent to which planning should be allowed to divert the needs of economic interests in order to provide both environmental protection and recreational provision. During the 1980s an environmental lobby has grown throughout the Developed World so that some of the tenets of traditional rural resource economics have been questioned. In those countries with substantial wilderness areas there is greater potential for designating large areas purely for environmental protection. Elsewhere, though, the 'green' lobby that sees a clear distinction between zoning of land for protection and zoning for recreational use has a harder task in eliminating a significant utilitarian element from the underlying aims of the creation of designated areas. Recreationists can easily damage the scenic landscape to which they are attracted, but with substantial pressure of demand on certain areas it has become increasingly difficult to separate protection from recreational provision.

Designation of protected areas and selection of villages as either growth poles or for conservation are decisions that are generally handed down by local or central government. There may be widespread consultation with rural inhabitants but such decisions are generally seen as 'top-down' responses to planning needs (e.g. Cloke, 1988c). It has been argued that planners at local and regional level merely serve to reinforce the basic characteristics of the economic and political objectives guided by central government under top-down responses. Therefore social and economic imbalances generated by the existing system cannot be removed by planners working within that system. However, there seems little likelihood that the various forms of capitalism within the Developed World are going to be replaced in the foreseeable future by new modes of production. Thus planners have to develop strategies that will work within the existing system or a close variant (Flynn and Lowe, 1987).

One set of strategies that has been applied to rural planning is to replace top-down by 'bottom-up' strategies. This implies that policies or solutions to particular problems are generated at a local level by rural communities themselves or organisations divorced in some way from central and regional government. 'Self-help' has been a characteristic of the frontier in North America and Australasia, and perhaps the growth of voluntary activity supported by official policies has been greater in some of the rural areas in these countries than in Western Europe. However, there have been various measures to encourage local initiatives in different parts of Europe, perhaps the decentralising measures in France and the new reverence for the decision-making capacity of the commune being the most sweeping expression of 'bottom-up'. But, of

course, decisions at a local level may not be allowed to conflict with
prevailing wisdom from central or regional government so that local decisions are often only allowed to affect particular issues, e.g. maintenance
of certain rural services, which do not impinge on broader development
strategies and power bases (Warner, 1974). They may also be unable to
overturn political considerations dictated by 'pork-barreling' in which
politicians steer funds to their own electorate (e.g. Johnston, 1988:
98–103).

   In terms of academic study of the countryside, there has been a growing recognition that to understand the changing social and economic
structure of the countryside, not to mention rural planning policies, it is
necessary not only to study spatial patterns, economic relationships and
social networks, but to seek out their driving forces. These forces
invariably can be traced to government, its prevailing ideology and the
way this is expressed in terms of constraints exerted upon individuals and
groups in society. It is not surprising therefore that there have been
attempts to focus upon the relationship between the evolution of
capitalism and rural economy and society, through the use of 'critical'
social theory (e.g. Bailey, 1975) and/or approaches using some form of
political economy framework.

   It is worth noting how this work has generated its own language and
'in'-terms in similar fashion to the wave of positivistic statistics-ridden
work of the 1960s and early 1970s. Theories from this earlier period
have proved the false gods of many academic studies of rural areas, and
so it is somewhat surprising to find the search for new theory back at
the top of the list of research agenda (Cloke, 1989b), especially given
that so much 'critical' social theory is remarkably uncritical of the
inherent weaknesses in its own determination and presents views of
society that often smack more of political dogma than a concerted
attempt to help understand social and economic change. In the past,
planners placed far too much store upon certain theories, notably central
place and growth pole theories, only to find that they did not necessarily
form a suitable basis for helping to improve the lot of rural inhabitants.
If a new wave of theories are to be utilised in practical planning applications, it is to be hoped that either the theories are better or that planners
take more note of the different scale of rural problems as compared with
urban ones, and also of significant economic, social and environmental
variation within rural areas. For the academics, chief tasks must be to
utilise their theories more effectively than in the past or perhaps reach
beyond the seemingly eternal search for new paradigms giving greater
explanatory power. Such searches seem to be full of self-justification and
lists of research agenda leading into blind alleys. The selection of theory
compatible only with preconceived dogma has consistently been a central
weakness of much so-called 'objective' research. A clearer focus upon key
decision-makers (individual, institutional and governmental) would seem
to be required to really generate greater understanding of rural change.

   For rural areas the decades on either side of 2000 AD promise numerous examples of both conflict and change. The growing awareness of the significance of the environmental dimension to rural resource developments means that certain projects in rural areas may face increased opposition. Yet, if demands for more rurally-based recreation are to be met, there may have to be more compromises between environmental protection and certain types of development. Attractions of rural locations for industrial development will continue to compete with centralising tendencies tying industry and services to higher-order central places. On the one hand, transport and energy costs may increase, thereby favouring concentration, but, on the other, land prices in rural areas may prove attractively low. The latter is likely to be closely related to changes in farm support policies: there have been signs that these are weakening, but how much and in what form remains to be seen. So too do prospects for radical alterations of rural power structures and the role of industrial capital in rural locations.

For rural inhabitants the dramatic changes post-1945 seem to have brought a deadly beauty: the beauty of spreading the urban 'good life' of the late 20th century throughout the countryside and the flip-side of that beauty via environmental degradation, new forms of impoverishment for some rural dwellers, and a destruction of the isolation and peace that for many was the chief quality of rural areas. It is a moot point to wonder whether rural inhabitants in the Developed World in 2020 AD will look back at the 1970s and 1980s and refer to them as a 'golden age' in the countryside in the way that previous generations have reminisced about their own rural past.

References

Abercrombie, N. and Urry, J. 1983, *Capital, labour and the Middle Classes* (Allen and Unwin, London).

Aceves, J.B. and Douglass, W.A. (eds) 1976, *The changing faces of rural Spain* (Schenkman Publishing Co., Cambridge, Mass.).

Adams, I.H. 1976, 'Agrarian landscape terms: a glossary for historical geography', *Special Publications, Institute of British Geographers*, No. 9.

Adams, W.H. 1984a, 'Sites of Special Scientific Interest and habitat protection: implications of the Wildlife and Countryside Act 1981', *Area*, 16: 273–80.

Adams, W.M. 1984b, *Implementing the Act: a study of habitat protection under Part II of the Wildlife and Countryside Act 1981* (British Association of Nature Conservationists and World Wildlife Fund, Oxford).

Adams, W.M. 1986, *Nature's place: conservation sites and countryside change* (Allen and Unwin, London).

Aitchison, J.W. 1984, 'Coefficients of specialisation and diversification: employment in rural France', *Area*, 16: 121–9.

Aitchison, J.W. and Aubrey, P. 1982, 'Part-time farming in Wales: a typological study', *Transactions of the Institute of British Geographers*, new series, 7: 88–97.

Aitchison, J.W. and Bontron, J.C. 1984, 'Les zones rurales fragiles en France', *Bulletin de la Société Neuchateloise de Géographie*, 28: 23–53.

Alam, S.M. and Pokshishevsky, V.V. 1976, *Urbanization in developing countries* (Osmania University, Hyderabad).

Albarre, G. 1977, 'The impact of second homes: second homes and conservation in southern Belgium', pp. 139–46 in Coppock, J.T. (ed.), *Second homes: curse or blessing?* (Pergamon Press, Oxford).

Albrecht, D.E. and Murdock, S.H. 1984, 'Toward a human ecological perspective on part-time farming', *Rural Sociology*, 49: 389–411.

Alden, J.D. and Sacha, S.K. 1978, 'An analysis of second job holding in the EEC', *Regional Studies*, 12: 639–50.

Almedal, S. 1983, 'Changes in rural settlement in northern Norway in the 1970s', *Nordia*, 17: 7–10.

Almond, G.A. and Verba, S. 1963, *The civic culture: political attitudes and democracy in five nations* (Princeton University Press, Princeton, NJ).

Alonso, W. 1960, 'A theory of the urban land market', *Papers and Proceedings of the Regional Science Association*, 6: 149–57.

Ambrose, P. 1974, *The quiet revolution* (Chatto and Windus, London).

Ambrose, P. 1986, *Whatever happened to planning?* (Methuen, London).

Ambrose, P. and Colenutt, B. 1975, *The property machine* (Penguin, Harmondsworth.

Anderson, C.B. 1961, 'The metamorphosis of American agrarian idealism in the 1920s and 1930s', *Agricultural History*, 35: 182–8.

Anderson, M.A. 1981, 'Planning policies and development control in the Sussex Downs AONB', *Town Planning Review*, 52: 5–25.

Anderson, P. and Yalden, D.W. 1981, 'Increased sheep numbers and loss of heather moorland in the Peak District', *Biological Conservation*, 20: 195–213.

Andrews, R.N.L. (ed.), 1979, *Land in America – community or natural resource?* (Lexington Books and D.C. Heath & Co., London and Toronto).

Andrian, J. 1981, 'Etude dur marche foncier et de la consummation de l'espace nonagricole en Picardie. Reurbanisation autour de quelques villes picarde: Amiens, Beauvais, Saint-Quentin, *Hommes et Terres du Nord*, 1–33.

Archer, B.H. 1974, 'The impact of recreation on local economies', *Planning Outlook*, 14: 16–27.

Arensberg, C.M. and Kimball, S.T. 1940, *Family and community in Ireland* (Peter Smith, Gloucester, Mass.).

Arkleton Trust 1982, *Schemes of assistance to farmers in Less Favoured Areas of the EEC* (The Trust, Langholm).

Armstrong, A. 1988, *Farmworkers: a social and economic history, 1770–1980* (Batsford, London).

Arnold, J.L. 1971, *The new deal in the suburbs: a history of the Greenbelt Town Program 1935–54* (Ohio State University Press, Columbus).

Ash, M. 1976, 'Time for a change in rural settlement policy', *Town and Country Planning*, 44: 528–31.

Attwood, E.A. 1978, 'Growth in Irish agriculture: fact or fiction?', *Irish Banking Review*, September: 14–26.

Ayton, J. 1980, 'Settlement policies can bring stabilisation', *The Planner*, 66: 98–9.

Bagnasco, A. 1981, 'Labour market, class structure and regional formations in Italy', *International Journal of Urban and Regional Research*, 5: 40–4.

Bailey, A.J. 1986, 'Regulatory reform and rural bus services: evidence from South Dakota', *Socio-Economic Planning Sciences*, 20: 291–8.

Bailey, F.G. 1957, *Caste and the economic frontier* (University of Manchester Press, Manchester).

Bailey, J. 1975, *Social theory for planning* (Routledge and Kegan Paul, London).

Baker, A.H.R. and Harley, J.B. (eds), 1973, *Man made the land: essays in English historical geography* (David and Charles, Newton Abbot).

Ball, R.M. 1987, 'Intermittent labour forms in UK agriculture: some implications for rural areas', *Journal of Rural Studies*, 3: 133–50.

Banaji, J. 1976, 'Summary of selected parts of Kantsky's "The agrarian question"', *Economy and Society*, 5: 2–49.

Banister, D. 1980, 'Transport mobility and deprivation in inter-urban areas:

research findings and policy perspectives', *Geographical Papers, Department of Geography, University of Reading*, No. 71.

Banister, D. 1983, 'Transport and accessibility', pp. 130–48 in Pacione, M. (ed.), *Progress in rural geography* (Croom Helm, London).

Banister, D. and Norton, F. 1988, 'The role of the voluntary sector in the provision of rural services – the case of transport', *Journal of Rural Studies*, 4: 57–71.

Barberis, C. 1968, 'The agricultural exodus in Italy', *Sociologia Ruralis*, 8: 179–88.

Barclay, H.B. 1964, *Buuri al Lamaab, a suburban village in the Sudan* (Cornell University Press, Ithaca, New York).

Barke, M. and France, L.A. 1988, 'Second homes in the Balearic Islands', *Geography*, 73: 143–6.

Barlett, P. 1986, 'Profile of full-time farm workers in a Georgia County', *Rural Sociology*, 51: 78–96.

Barlow, J. 1986a, 'Agribusiness in Britain: the appropriation of the food chain by capital', *Urban and Regional Studies Working Paper, University of Swansea*.

Barlow, J. 1986b, 'Landowners, property-ownership and the rural locality', *International Journal of Urban and Regional Research*, 10: 309–29.

Barnett, J.R. and Sheerin, I.G. 1978, 'Inefficiency and inequality: an evaluation of selected policy responses to medical maldistribution problems in New Zealand, *Community Health Studies*, 2: 65–72.

Barr, J. 1969, 'Durham's murdered villages', *New Society*, 340: 523–5.

Barrows, R. and Bromley, D. 1975, 'Employment impacts of the European Development Administration's Public Works Programme', *American Journal of Agricultural Economics*, 57: 46–54.

Barsby, S.L. and Cox, D.R. 1975, *Interstate migration of the elderly* (D.C. Heath, Lexington, Mass.).

Barthelemey, D. and Barthez, A. 1978, 'Propriété foncière, exploitation agricole et aménagement de l'espace rural', *Economie Rurale*, 126: 6–16.

Bartoli, P. 1981, *La politique de reconversion viticole. Resultats de la prime d'arrachage en Languedoc-Roussillon 1976–1979* (INRA, Montpellier).

Bassett, K. and Short, J. 1980, *Housing and residential structure* (Routledge and Kegan Paul, London).

Baudin, P.H. 1979, 'The Common Agricultural Policy', paper presented to the Salzburg Seminar, No. 193, 'Nutrition, food and population', Salzburg, 5–24/8/79.

Beale, C. 1977, 'The recent shift of United States population to non-metropolitan areas, 1970–75', *International Regional Science Review*, 2: 113–22.

Beavon, K.S.O. 1977, *Central place theory: a reinterpretation* (Longman, London).

Beck, J.M. 1988, *The rise of a subsidized periphery in Spain: a geographical study of state and market relations in the eastern Montes Orientales of Granada 1930–1982* (Netherlands Geographical Studies, 69, Department of Human Geography, University of Amsterdam, Amsterdam, Netherlands).

Beeching Report, 1963, *The reshaping of British railways* (HMSO, London).

Beed, C.S. 1981, *Melbourne's development and planning* (Clewara Press, Parkville).

Behar, R. 1986, *Santa Marcia del Monte. The presence of the past in a Spanish village* (Princeton University Press, Princeton, NJ).

Belil, M. and Clos, I. 1985, 'La descentralizacion industrial a Catalunya: l'eix el Vendrell-Valls-Montblanc', *Documents d'Analisi Geografica*, 6: 3–44.

Bell, C. and Newby, H. 1971, *Community studies: an introduction to the sociology of the local community* (Allen and Unwin London).

Bell, C. and Newby, H. 1974, 'Capitalist farmers in the British class structure', *Sociologia Ruralis*, 14: 86–107.

Bell, C. and Newby, H. 1976, 'Community, communion, class and community action', pp. 189–207 in Herbert, D.T. and Johnston, R.J. (eds), *Social areas in cities* (John Wiley, Chichester).

Bell, D. 1973, *The coming of post-industrial society* (Basic Books, New York).

Bell, P. and Cloke, P.J. 1989, 'The changing relationship between the private and public sectors: privatisation and rural Britain', *Journal of Rural Studies*, 5: 1–16.

Benelbas, F. 1981, *Economia agraria de Cataluna* (Ketres, Barcelona).

Bennett, R.J. 1986, 'Social and economic transition: a case study in Portugal's Western Algarve', *Journal of Regional Studies*, 2: 91–102.

Bentham, C.G. 1984, 'Mortality rates in the more rural areas of England and Wales', *Area*, 16: 219–26.

Bentham, C.G. and Haynes, R.M. 1985a, 'Geographical disparities in health needs and health service use in rural Norfolk', pp. 64–74 in Cloke, P.J. (ed.), *Rural accessibility and mobility* (Centre for rural Transport, Lampeter).

Bentham, C.G. and Haynes, R.M. 1985b, 'Health, personal mobility and the use of health services in rural Norfolk', *Journal of Rural Studies*, 1: 231–41.

Bentham, C.G. and Haynes, R.M. 1986, 'A raw deal in remoter rural areas', *Family Practitioner Services*, 13: 84–7.

Berger, B. 1960, *Working class suburb* (University of California Press, Berkeley).

Berger, M., Fruit, J-P., Plet, F. and Robic, M-C. 1980, 'Reurbanisation et analyse des espaces ruraux periurbain', *Espace Géographie*, 4: 303–13.

Berry, B.J.L. 1973, *Growth centers in the American urban system, Volume 1* (Ballinger Publishing Co., Cambridge, Mass.).

Berry, B.J.L. 1976, 'The counter-urbanization process: urban America since 1970', in Berry, B.J.L. (ed.), *Urbanization and counter-urbanization* (Urban Affairs Annual Review, Volume 11, Sage Publications, Beverley Hills, Ca.).

Best, R.H. 1977, 'Agricultural land loss – myth or reality?', *The Planner*, 63: 15–16.

Best, R.H. 1979, 'Land use structure and change in the EEC', *Town Planning Review*, 50: 395–411.

Best, R.H. 1981, *Land use and living space* (Methuen, London and New York).

Best, R.H. and Rogers, A.W. 1973, *The urban countryside* (Faber, London).

Beyers, W.B. 1979, 'Contemporary trends in the regional economic development of the United States', *Professional Geographer*, 31: 34–44.

Bhaskar, R. 1978, *A realist theory of science* (Harvester Press, Hassocks).

Bickerstaff, A., Wallace, W.L. and Evert, F. 1981, *Growth of forests in Canada,*

Part 2: A quantitative description of the land base and the mean annual increment (Information Report PX-X-I, Canadian Forestry Service, Environment Canada, Ottawa).

Bielkus, C.L. 1977, 'Second homes in Scandinavia', pp. 35–46 in Coppock, J.T. (ed.), *Second homes; curse or blessing?* (Pergamon, Oxford).

Biggar, J. 1984, *The greying of the Sunbelt: a look at the impact of the U.S. elderly migration* (Population Reference Bureau, Washington, DC).

Binns, J.A. and Funnell, D.C. 1983, 'Geography and integrated rural development', *Geografiska Annaler*, 65B: 57–67.

Birch, F. 1979, 'Leisure patterns 1973 and 1977', *Population Trends*, 19: 2–8.

Blacksell, M. 1979, 'Landscape protection and development control: an appraisal of planning in rural areas in England and Wales', *Geoforum*, 10: 267–74.

Blacksell, M. 1982, 'Leisure, recreation and environment', pp. 309–26 in Johnston, R.J. and Doornkamp, J.C. (eds), *The changing geography of the United Kingdom* (Methuen, London and New York).

Blacksell, M., Clark, A., Economides, K. and Watkins, C. 1988, 'Legal services in rural areas: problems of access and local need', *Progress in Human Geography*, 12: 47–65.

Blacksell, M. and Gilg, A.W. 1975, 'Landscape evaluation in practice: the case of south-east Devon', *Transactions of the Institute of British Geographers*, 66: 135–40.

Blacksell, M. and Gilg, A.W. 1981, *The countryside: planning and change* (George Allen and Unwin, London).

Blackwood, L. and Carpenter, E. 1978, 'The importance of anti-urbanism in determining residential preferences and migration patterns', *Rural Sociology*, 43: 31–47.

Blair, A.M. 1980, 'Urban influences on farming in Essex', *Geoforum*, 11: 371–84.

Blake, R.L. and Guild, P.Q. 1978, 'Mid-level practitioners in rural health care: a three year experience in Appalachia', *Journal of Community Health*, 4: 15.

Blowers, A. 1972, 'The declining villages of County Durham', in Open University, *Social Geography* (Open University Press, Bletchley).

Blowers, A. 1987, 'Transition or transformation? environmental policy under Thatcher', *Public Administration*, 65: 277–94.

Blunden, J. and Curry, N. (eds), 1985, *The changing countryside* (Croom Helm, London).

Blunden, J. and Curry, N. 1988, *A future for our countryside* (Basil Blackwell, Oxford).

Blunden, J. and Turner, G. 1985, *Critical countryside* (BBC, London).

Blythe, R. 1969, *Akenfield* (Penguin, Harmondsworth).

Body, R. 1982, *Agriculture: the triumph and the shame* (Maurice Temple Smith, London).

Body, R. 1984, *Farming in the clouds* (Maurice Temple Smith, London.

Bohland, J.R. 1988, 'Rural America', pp. 151–88 in Knox, P.L., Bartels, E.H., Bohland, J.R., Holcomb, B. and Johnston, R.J., *The United States, a contemporary geography* (Longman, London).

Bohland, J.R. and Treps, L. 1981, 'County pattern of elderly migration in the

United States', pp. 35–53 in Warnes, A. (ed.), *Geographical perspectives on the elderly* (John Wiley, Chichester).

Bokemeier, J.L. and Tait, J.L. 1980, 'Women as power actors: a comparative study of rural communities', *Rural Sociology*, 45: 238–55.

Bokemeier, J.L. and Tickamyer, A.R. 1985, 'Labor force experiences of nonmetropolitan women', *Rural Sociology*, 50: 51–73.

Bollman, R.D. 1982, 'Part-time farming in Canada: issues and non-issues', *Geojournal*, 6: 313–22.

Bollom, C. 1978, *Attitudes and second homes in rural Wales* (University of Wales Press, Cardiff).

Boserup, E. 1981, *Population and technology* (Oxford University Press, Oxford).

Bouquet, M. 1982, 'Production and reproduction of family farms in South-West England', *Sociologia Ruralis*, 22: 227–44.

Bouquet, M. and Winter, M. (eds), 1987, *Who from their labours rest? Conflict and practice in rural tourism* (Gower, Aldershot).

Bowers, J.K. 1985, 'The economics of agribusiness', pp. 29–44 in Healey, M.J. and Ilbery, B.W. (eds), *The industrialization of the countryside* (Geo Books, Norwich).

Bowers, J.K. and Cheshire, P. 1983, *Agriculture, the countryside and land use: an economic critique* (Methuen, London and New York).

Bowler, I.R. (ed.) 1975, *A register of research in rural geography* (Rural Geography Study Group, Institute of British Geographers).

Bowler, I.R. 1976a, 'The adoption of grant aid in agriculture', *Transactions of the Institute of British Geographers*, new series, 1: 143–58.

Bowler, I.R. 1976b, 'Regional agricultural policies: experience in the United Kingdom', *Economic Geography*, 52: 267–80.

Bowler, I.R. 1976c, 'Spatial responses to agricultural subsidies in England and Wales', *Area*, 8: 225–9.

Bowler, I.R. 1979, *Government and agriculture: a spatial perspective* (Longman, London).

Bowler, I.R. 1985, *Agriculture under the Common Agricultural Policy, a geography* (Manchester University Press, Manchester).

Bowler, I.R. 1986a, 'Direct supply control in agriculture: experience in Western Europe and North America', *Journal of Rural Studies*, 2: 19–30.

Bowler, I.R. (ed.), 1986b, 'Agriculture and the Common Agricultural Policy', *Department of Geography, University of Leicester, Occasional Papers*, No 15.

Bowler, I.R. 1987, 'The geography of agriculture under the CAP', *Progress in Human Geography*, 11: 24–40.

Bowler, I.R. 1988, 'The provision of job training for employment in the rural South Midlands', *East Midland Geographer*, 11: 22–30.

Bowler, I.R. and Ilbery, B.W. 1987, 'Redefining agricultural geography', *Area*, 19: 327–32.

Bozeman, B., Thornton, S. and McKinney, M. 1977, 'Continuity and change in opinion about sex roles', pp. 38–65 in Githens, M. and Prestage, J.L. (eds), *A portrait of marginality: the political behavior of the American woman* (David McKay and Co., New York).

Bracey, H.E. 1953, 'Towns as rural service centres', *Transactions of the Institute of British Geographers*, 19: 95–103.

Bradbeer, J.B. 1987, 'The future of mineral extraction in the rural environment', pp. 97–114 in Lockhart, D.G. and Ilbery, B.W. (eds), *The future of the British rural landscape* (Geo Books, Norwich).

Bradley, T. 1981, 'Capitalism and countryside: rural sociology as political economy', *International Journal of Urban and Regional Research*, 5: 581–7.

Bradley, T. and Lowe, P. 1984, 'Introduction: locality, rurality and social theory', pp. 1–23 in Bradley, T. and Lowe, P. (eds), *Locality and rurality* (Geo Books, Norwich).

Bradshaw, M. 1984, 'TVA at fifty', *Geography*, 69: 209–20.

Bradshaw, T.K. and Blakeley, E.J. 1979, *Rural communities in advanced industrial society* (Praeger, New York).

Brancher, D.M. 1969, 'Critique of K.D. Fines' landscape evaluation', *Regional Studies*, 3: 91–2.

Brandes, S.H. 1975, *Migration, kinship and community: tradition and transition in a Spanish village* (Academic Press, New York).

Brasch, E. 1979, 'Peri-urban agriculture in the areas of Kristianstad and Elsor, pp. 451–81 in OECD, *Agriculture in the planning and management of peri-urban areas – Volume II* (OECD, Paris).

Bray, C.E. 1980, 'Agricultural land regulation in several Canadian provinces', *Canadian Public Policy*, 6: 591–604.

Breathnach, P. 1984, 'Co-operation and community development: an aspect of rural development in the West of Ireland', pp. 155–78 in Jess, P.M., Greer, J.V., Buchanan, R.H. and Armstrong, W.J. (eds), *Planning and development in rural area* (Queen's University of Belfast, Belfast).

Breathnach, P. 1985, 'Rural industrialization in the west of Ireland', pp. 173–96 in Healey, M.J. and Ilbery, B.W. (eds), *The industrialization of the countryside* (Geo Books, Norwich).

Briggs, D. and Wyatt, B. 1988, 'Rural land-use change in Europe', pp. 7–25 in Whitby, M. and Ollerenshaw, J. (eds), *Land use and the European Environment* (Belhaven Press, London and New York).

Briggs, D. and Yurman, E. 1980, 'Disappearing farmland: a national concern', *Soil Conservation*, 45: 4–7.

Briggs, R. 1981, 'Federal policy in the United States and the transportation problems of low-density areas', pp. 238–61 in Lonsdale, R.E. and Holmes, J.H. (eds), *Settlement systems in sparsely populated regions: the United States and Australia* (Pergamon, New York).

Briggs, R. and Rees, J. 1982, 'Control factors in the economic development of non-metropolitan America', *Environment and Planning A*, 14: 1645–66.

British Columbia Select Standing Committee on Agriculture, 1978, *Inventory of Agricultural Land Reserves in British Colombia, Phase 1* (Legislative Assembly, Victoria, BC).

Britton, D.K. and Hill, B. 1975, *Size and efficiency in farming* (Saxon House, Farnborough).

Broads Authority, 1982, *What future for Broadland?* (The Authority, Norwich).

Brooks, D. 1985, 'Public policy and long-term timber supply in the South',

Forestry Service, 31: 342–57.

Broschen, E. and Himminghofen, W. 1983, 'The aged in the countryside; implications for social planning', *Sociologia Ruralis*, 23: 261–75.

Brotherton, D.I. 1982, 'Development pressures and control in the National Parks, 1966–1981', *Town Planning Review*, 53: 439–59.

Brotherton, D.I. 1985, 'Issues in National Park administration', *Environment and Planning A*, 17: 47–58.

Brown, D.A.H. and Taylor, K. 1988, 'The future of Britain's rural land', *Geographic Journal*, 154: 406–11.

Brown, D.L., Brewer, M.F., Boxley, R.F. and Beale, C.L. 1982, 'Assessing prospects for the adequacy of agricultural land in the United States', *International Regional Science Review*, 7: 273–84.

Brown, J.H., Phillips, R.S. and Roberts, N.A. 1981, 'Land markets at the urban fringe', *Journal of the American Planning Association*, 47: 131–4.

Brownrigg, M. 1982, 'The rural co-operative: an economist's view', *The Planner*, 68: 86–8.

Brunt, B.M. 1988, *The republic of Ireland* (Paul Chapman Publishing, London).

Brush, J.E. and Bracey, H.E. 1966, 'Rural service centres in south-western Wisconsin and southern England', *Geographical Review*, 45: 559–69.

Bruton, M. and Nicholson, D. 1985, 'Strategic land use planning and the British development plan system', *Town Planning Review*, 56: 21–41.

Bryant, C.R. 1980, 'Manufacturing in rural development', pp. 99–128 in Walker, R. (ed.), *Planning industrial development* (Wiley, Chichester).

Bryant, C.R. 1982, 'The rural real estate market: an analysis of geographic patterns of structure and change within an urban fringe environment', *Publication Series, Department of Geography, University of Waterloo*, No. 18.

Bryant, C.R. and Greaves, S.M. 1978, 'The importance of regional variation in the analysis of urbanisation-agriculture interactions', *Cahiers de Géographie du Quebec*, 22: 329–48.

Bryant, C.R. and Russwurm, L.H. 1982, 'North American farmland protection strategies in retrospect', *Geojournal*, 6: 501–11.

Bryant, C.R. Russwurm, L.H. and McLellan, A.G. 1982, *The city's countryside: land and its management in the rural-urban fringe* (Longman, London and New York).

Bryden, J.M. 1987, 'Crofting in the European context', *Scottish Geographical Magazine*, 103: 100–4.

Bunce, M. 1982, *Rural settlement in an urban world* (Croom Helm, London).

Burie, J.B. 1967, 'Prolegomena to a theoretical model of intercommunity variation', *Sociologia Ruralis*, 7: 347–64.

Burnley, I.H. 1988, 'Population turnaround and the peopling of the countryside? Migration from Sydney to country districts of New South Wales', *Australian Geographer*, 19: 268–83.

Burns, J.A., McInerny, J.P. and Swinbank, A. (eds), 1983, *The food industry: economics and policies* (Heinemann, London).

Burrell, A., Hill, B. and Medland, J. 1985, *A statistical handbook of UK agriculture* (Macmillan, London).

Burtin, J. 1987, *The Common Agricultural Policy and its reform* (Office for

Official Publications of the European Communities, Luxembourg), 4th edition.

Busch, L. Bonanno, A. and Lacy, W.B. 1989, 'Science, technology and the restructuring of agriculture', *Sociologia Ruralis*, 29: 118–30.

Buttel, F.H.L. 1980, 'Agricultural structure and rural ecology: toward a political economy of rural development', *Sociologia Ruralis*, 20: 44–62.

Buttel, F.H.L. 1982a, 'The political economy of agriculture in advanced industrial societies', *Current Perspectives in Social Theory*, 3: 27–55.

Buttel, F.H.L. 1982b, 'The political economy of part-time farming', *Geojournal*, 6: 293–300.

Buttel, F. and Newby, H. (eds), 1980, *The rural sociology of the advanced societies: critical perspectives* (Allanheld, Osmun, Montclair, NJ and Croom Helm, London).

Bylund, E. and Weissglass, G. 1970, 'The service structure of marginal areas. A preliminary report', *Urbaniseringen; Sverige*, 14(2) (Stockholm).

Cabeza, M.T. 1984, 'Un exemple de fragmentacio de l'estructura latifundia? Intent d'explicacio de les causes de l'augment del nombre d'exploitacions a la provincia de Jean: 1962–1972', *Documents d'Analisi Geografica*, 4: 3–30.

Caird, J.B. and Moisley, H.A. 1964, 'The Outer Hebrides', pp. 374–90 in Steers, J.A. (ed.), *Field studies in the British Isles* (Nelson, London).

Calvin, J.S., Dearinger, J. and Curtin, M. 1972, 'An attempt at assessing preferences for natural landscapes', *Environment and Behaviour*, 4: 447–70.

Campbell, J. 1984, 'Outlook for UK institutional forest management', *Journal of World Forest Resource Management*, 1: 105–16.

Carlson, J.E., Lassey, M.L. and Lassey, W.R. 1981, *Rural society and environment in America* (McGraw Hill, New York).

Carlson, J.E. and McLeod, M.E. 1978, 'A comparison of agrarianism in Washington, Idaho and Wisconsin', *Rural Sociology*, 43: 17–30.

Carlyle, W.J. 1975, 'Livestock markets in Scotland', *Annals of the Association of American Geographers*, 65: 449–60.

Carlyle, W.J. 1988, 'Rural change in the Prairies', pp. 243–67 in Robinson, G.M. (ed.), *A social geography of Canada: essays in honour of J. Wreford Watson* (North British Publishing, Edinburgh).

Carter, H. 1981, *The study of urban geography* (Edward Arnold, London), 3rd edition.

Carter, I.R. 1975, 'A socialist strategy for the Highlands', pp. 279–311 in de Kadt, E. and Williams, G. (eds), *Sociology and development* (Tavistock Publications, London).

Cartwright, A. and Anderson, R. 1981, *General practice revisited* (Tavistock, London).

Castells, M. 1976, 'Is there an urban sociology?', pp. 60–84 in Pickvance, C.G. (ed.), *Urban sociology: critical essays* (Tavistock, London).

Caulfield, C. 1984, 'Ministers admit Countryside Act has failed', *New Scientist*, 101: 10.

Cavazzani, A. and Fuller, A.M. 1982, 'International perspectives on part-time farming', *Geojournal*, 6: 383–9.

Cawley, M.E. 1979, 'Rural industrialisation and social change in western Ireland', *Sociologia Ruralis*, 19: 43–59.

Cawley, M.E. 1983, 'Part-time farming in rural development', *Sociologia Ruralis*, 23: 63–75.

Cawley, M.E. 1986, 'Disadvantaged groups and areas: problems of rural service provision', in Breathnach, P. and Cawley, M. (eds), 'Change and development in rural Ireland', *Special Publications, Geographical Society of Ireland*, No. 1 (Dublin).

Central Advisory Council 1967, *Children and their primary schools* (The Plowden Report, HMSO, London).

Centre for International Economics 1988, *The effects of farming support policies* (Trade Policy Research Centre, Canberra).

Centre for Policy Studies 1988, *Britain's biggest enterprise* (Centre for Policy Studies, London).

Cesaretti, P., De Benedictis, M. De Filippis, F., Giannola, A. and Perone-Pacifico, C. 1980, *Regional impact of the Common Agricultural Policy: Italian report* (Centro di Specializzazione e Richerche Economico-Agraria per il Mezzogiorno, Naples).

Cesarini, G. 1979, *Rural production co-operatives in Southern Italy* (Akleton Trust, Oxford).

Champion, A.G., Fielding, A.J. and Keeble, D.E. 1989, 'Counterurbanization in Europe', *Geographical Journal*, 155: 52–80.

Champion, A.G., Green, A.E., Owen, D.W., Ellin, D.J. and Coombes, M.G. 1987, *Changing places: Britain's demographic, economic and social complexion* (Edward Arnold, London).

Chayanov, A.V. 1966, *The theory of peasant economy* (R.D. Irwin, Homewood, Illinois).

Checkoway, B. 1980, 'Large builders, federal housing programmes and post-war suburbanization', *International Journal of Urban and Regional Research*, 4: 21–45.

Cherry, G.E. 1975, *Peacetime history: environmental planning – National Parks and recreation in the countryside* (HMSO, London).

Cherry, G.E. 1978, 'Comment on "Needed a new Scott Inquiry"', *Town Planning Review*, 49: 364–5.

Chisholm, M. 1985, 'The Development Commission's employment programmes in rural England', pp. 279–92 in Healey, M.J. and Ilbery, B.W. (eds), *The industrialisation of the countryside* (Geo Books, Norwich).

Christodoulou, D. 1976, 'Portugal's agrarian reform: a process of change with unique features', *Land Reform, Land Settlement and the Co-operatives*, 2: 1–21.

Chubb, M. and Bauman, E.H. 1977, 'Assessing the recreation potential of rivers', *Journal of Soil and Water Conservation*, 32: 97–102.

Cicchetti, C. J. and Smith, K.V. 1973, 'Congestion, quality deterioration and optimal use: wilderness recreation in the Spanish peaks primitive area;, *Social Science Research*, 2: 15–30.

Clark, C. 1958, 'Transport – maker and breaker of cities', *Town Planning Review*, 28: 237–50.

Clark, D.H. 1979, 'The influence of organisational structure on the spatial pattern of commodity linkages: an example from the New Zealand dairy industry', *New Zealand Geographer*, 35: 51–63.

Clark, G. 1982a, 'Housing policy in the Lake District', *Transactions of the Institute of British Geographers*, 7: 59–70.

Clark, G. 1982b, *Housing and planning in the countryside* (Wiley, Chichester).

Clark, S.D. 1968, *The suburban society* (University of Toronto Press, Toronto).

Clark, T.D. 1984, *The greening of the South: the recovery of land and forest* (University Press of Kentucky, Lexington).

Clawson, M. 1963, *Land and water for recreation: opportunities, problems and policies* (Rand McNally, New York).

Clawson, M. and Knestch, J.L. 1966, *Economics of outdoor recreation* (Johns Hopkins Press, Baltimore).

Clepper, H. 1971, *Professional forestry in the United States* (Johns Hopkins Press, Baltimore).

Cloke, P.J. 1977, 'An index of rurality for England and Wales', *Regional Studies*, 11: 31–46.

Cloke, P.J. 1979, *Key settlements in rural areas* (Methuen, London).

Cloke, P.J. 1980a, 'New emphases for applied rural geography', *Progress in Human Geography*, 4: 182–217.

Cloke, P.J. 1980b, 'Key settlements', *Town and Country Planning*, 49: 187–9.

Cloke, P.J. 1981, 'Key settlement policies at the local level', *The Village*, 35: 28–30.

Cloke, P.J. 1983, *An introduction to rural settlement planning* (Methuen, London and New York).

Cloke, P.J. 1985, 'Whither rural studies?' *Journal of Rural Studies*, 1: 1–9.

Cloke, P.J. 1986, 'Implementation, intergovernmental relations and rural studies: a review', *Journal of Rural Studies*, 2: 245–53.

Cloke, P.J. 1987, 'Rurality and change: some cautionary notes', *Journal of Rural Studies*, 3: 71–6.

Cloke, P.J. 1988a, review of Hoggart and Buller, 'Rural development: a geographical perspective', p. 20 in *Times Higher Educational Supplement*, 26.2.88.

Cloke, P.J. (ed.), 1988b, *Policies and plans for rural people* (Unwin Hyman, London).

Cloke, P.J. 1988c, 'Conclusions: rural policies – responses to problems or problematic responses?, pp. 240–58 in Cloke, P.J. (ed.), *Policies and plans for rural people* (Unwin Hyman, London).

Cloke, P.J. (ed.), 1989a, *Rural land-use planning in Developed Nations* (Unwin Hyman, London).

Cloke, P.J. 1989b, 'Rural geography and political economy' in Peet, R. and Thrift, N. (eds), *New models of geography: the political economy perspective, Volume 1* (Unwin Hyman, London).

Cloke, P.J. 1989c, 'State deregulation and New Zealand's agricultural sector', *Sociologia Ruralis*, 29: 34–48.

Cloke, P.J. and Edwards, G. 1984, 'Changing needs and provision: what future?' pp. 131–45 in Cloke, P.J. (ed.), *Wheels within Wales* (Centre for Rural Transport, Lampeter).

Cloke, P.J. and Edwards, G. 1986, 'Rurality in England and Wales 1981: a replication of the 1971 index', *Regional Studies*, 20: 289–306.

Cloke, P.J. and Hanrahan, P. 1984, 'Policy and implementation – in rural planning', *Geoforum*, 15: 261–70.

Cloke, P.J. and Little, J.K. 1986, 'The implementation of rural policies – a survey of county planning authorities', *Town Planning Review*, 57: 265–84.

Cloke, P.J. and Little, J.K. 1987, 'Policy, planning and the state in rural localities', *Journal of Rural Studies*, 3: 343–51.

Cloke, P.J. and McLaughlin, B. 1989, 'Politics of the alternative land use and rural economy (ALURE) proposals in the UK: crossroads or blind alley?', *Land Use Policy*, 6: 235–48.

Cloke, P.J. and Park, C.C. 1985, *Rural resource management* (Croom Helm, London).

Cloke, P.J. and Shaw, D.P. 1983, 'Rural settlement policies in structure plans', *Town Planning Review*, 54: 338–54.

Cloke, P.J. and Thrift, N. 1987, 'Intra-class conflict in rural areas', *Journal of Rural Studies*, 3: 321–33.

Clout, H.D. 1969, 'Second homes in France', *Journal of the Town Planning Institute*, 55: 440–3.

Clout, H.D. 1970, 'Social aspects of second-home occupation in the Auvergne', *Planning Outlook*, 9: 33–49.

Clout, H.D. 1971, 'Second homes in the Auvergne', *Geographical Review*, 61: 530–53.

Clout, H.D. 1972, *Rural geography: an introductory survey* (Pergamon, Oxford).

Clout, H.D. 1975, 'Structural change in French farming: the case of the Puy-de-Dome', *Tijdschrift Voor Economische en Sociale Geographie*, 66: 234–45.

Clout, H.D. 1984, *A rural policy for the EEC?* (Methuen, London and New York).

Clout, H.D. 1987a, 'Rural space', pp. 19–38 in Clout, H.D. (ed.), *Regional development in Western Europe* (David Fulton Publishers, London), 3rd edition.

Clout, H.D. 1987b, 'France', pp. 165–94 in Clout, H.D. (ed.), *Regional development in Western Europe* (David Fulton Publishers, London), 3rd edition.

Clout, H.D. 1988, 'France', pp. 98–119 in Cloke, P.J. (ed.), *Policies and plans for rural people: an international perspective* (Unwin Hyman, London).

Cocklin, C., Smith, B. and Johnston, T. (eds), 1987, *Demands on rural lands: planning for resource use* (Westview Press, Boulder and London).

Coleman, A. 1978, 'Last bid for land use sanity', *Geographical Magazine*, 50: 820–4.

Comby, J. 1973, 'Un nouvel aspect de la politique de DATAR: les villes moyennes, poles de développement et d'aménagement', *Norois*, 80: 647–60.

Commins, P. 1978, 'Socio-economic adjustments to rural depopulation', *Regional Studies*, 12: 79–94.

Commins, P. 1980, 'Imbalances in agricultural modernization; with illustrations from Ireland', *Sociologia Ruralis*, 20: 63–81.

Commission of the European Communities, 1985, *Perspectives for the Common Agricultural Policy. The green paper of the Commission* (Commission of the European Communities, Brussels).

Connell, J. 1978, *The end of tradition: country life in central Surrey* (Routledge and Kegan Paul, London).

Conservative Political Centre, 1989, *Low-cost housing in rural areas: a Conservative strategy* (CPC, London).

Constandse, A.K., Cepede, M., Bracey, H.E. and Gerl, F. (eds), 1964, *Sociologia Ruralis, Volume IV*, Special Issue, Nos 3–4, pp. 211–467.

Cook, A.K. 1987, 'Nonmetropolitan migration: the influence of neglected variables', *Rural Sociology*, 52: 409–18.

Cooke, P. and Pires, A. de R. 1985, 'Productive decentralisation in three European regions', *Environment and Planning A*, 17: 527–54.

Coombes, M.G. and Dallalonga, R. 1987, 'Counterurbanisation in Britain and Italy: a comparative critique of the concept, causation and evidence', *Centre for Urban and Regional Development Studies, Occasional Papers*, No. 84.

Cooper, A., Murray, R. and Warnock, S. 1989, 'Agriculture and environment in the Mourne AONB', *Applied Geography*, 9: 35–56.

Coppock, J.T. (ed.), 1977, *Second homes: curse or blessing?* (Pergamon Press, Oxford).

Coppock, J.T. 1986, 'Planning for countryside recreation in Great Britain', pp. 295–316 in Shafi, M. and Raza, M. (eds), *Spectrum of modern geography* (Concept Publishing Co., New Delhi).

Coppock, J.T. and Duffield, B.S. 1975, *Outdoor recreation: a spatial analysis* (Macmillan, London).

Cortz, D. 1978, 'Corporate farming: a tough row to hoe', pp. 144–51 in Rodefield, R.D. *et al.* (eds), *Change in rural America* (C.V. Mosby, St Louis).

Coughenour, C.M. and Christenson, J.A. 1983, 'Farm structure, social class and farmers' policy perspectives', pp. 67–86 in Brewster, D.E. *et al.* (eds), *Farms in transition* (Iowa Sate University Press, Ames).

Coulmin, P. 1986, *La decentralisation: la dynamique du developpement local* (Syros Adels, Paris).

Countryside Commission 1974, *New agricultural landscapes* (Countryside Commission, Cheltenham).

Countryside Commission 1976, *The Lake District Upland Management Experiment* (Countryside Commission, Cheltenham).

Countryside Commission 1980, *Areas of Outstanding Natural Beauty* (Countryside Commission, Cheltenham).

Countryside Commission 1981a, *The future of the uplands* (Countryside Commission, Cheltenham).

Countryside Commission 1981b, *Countryside management in the urban fringe* (Countryside Commission, Cheltenham).

Countryside Commission 1983, *The changing uplands* (Countryside Commission, Cheltenham).

Countryside Commission 1987a, *Enjoying the countryside: priorities for action* (Countryside Commission, Cheltenham).

Countryside Commission 1987b, *Policies for enjoying the countryside* (Countryside Commission, Cheltenham).

Countryside Commission 1987c, *Forestry in the countryside* (Countryside Commission, Cheltenham).

Countryside Commission 1987d, *Recreation 2000* (Countryside Commission, Cheltenham).

Countryside Commission 1988, *Changing the rights-of-way network* (Countryside Commission, Manchester).

Countryside Policy Review Panel 1987, *New opportunities for the countryside* (Countryside Commission, Cheltenham).

Countryside Review Committee 1977, 'Leisure and the countryside', *Topic Papers*, No. 2 (HMSO, London).

Coventry, Solihull and Warwickshire Councils 1971, *A strategy for the sub-region. Supplementary report 5: Countryside* (The Councils, Coventry).

Cox, G. and Lowe, P.D. 1983, 'A battle not the war: the politics of the Wildlife and Countryside Act', pp. 48–76 in Gilg, A.W. (ed.), *Countryside Planning Yearbook 1983* (Geo Books, Norwich).

Cox, G. and Winter, M. 1986, 'Agriculture and conservation in Britain: a policy community under siege', pp. 181–215 in Cox, G., Lowe, P. and Winter, M. (eds), *Agriculture: people and policies* (Allen and Unwin, London).

Cox, G. Lowe, P. and Winter, M. 1985, 'Land use conflict after the Wildlife and Countryside Act 1981: the role of the Farming and Wildlife Advisory Group', *Journal of Rural Studies*, 1: 173–84.

Cox, G., Lowe, P. and Winter, M. (eds) 1986, *Agriculture: people and policies* (Allen and Unwin, London).

Cox, H. 1965, *The secular city* (SCM Press, New York).

Crabb, P. 1985, 'Rural change in Prince Edward Island, Canada: an interactive or directive economy?', *Journal of Rural Studies*, 1:241–52.

Craig, G.M., Jollans, J.L. and Korbey, A. 1982, *The case for agriculture: an independent assessment* (Centre for Agricultural Strategy, Reading).

Craik, K.H. 1972, 'Appraising the objectivity of landscape dimensions', pp. 292–346 in Krutilla, J.V. (ed.), *Natural environments: studies in theoretical and applied analysis* (Johns Hopkins Press, Baltimore).

Cromley, R.G. and Leinbach, T.R. 1981, 'The pattern and impact of the filter down process in non-metropolitan Kentucky', *Economic Geography*, 57: 208–24.

Curry, J. 1980, 'Rural poverty' in Kennedy, S. (ed.), *One million poor* (Turoe Press, Dublin).

Curry, N.R. 1985, 'Countryside recreation sites policy, a review', *Town Planning Review*, 56: 70–89.

Curtis, L. 1983, 'Reflections on management agreements for conservation of Exmoor moorland', *Journal of Agricultural Economics*, 34: 397–406.

Curtis, L.F. and Walker, A.J. 1982, 'Conservation and protection', pp. 381–402 in Johnston, R.J. and Doornkamp, J.C. (eds), *The changing geography of the United Kingdom* (Methuen, London).

D'Aeth, R. 1981, *A positive approach to rural primary schools* (University Institute of Education, Cambridge).

Dahms, F.A. 1977, 'How Ontario's Guelph District developed', *Canadian Geographical Journal*, 94: 48–55.

Dahms, F.A. 1980, 'The evolving spatial organisation of small settlements in the countryside – An Ontario example', *Tijdschrift voor Economische en Sociale Geografie*, 71: 295–306.

Daniel, T.C. and Boster, R.S. 1976, 'Measuring landscape esthetics: the Scenic Beauty Estimation Method', *USDA Forest Service Research Papers*, RM–167 (Rocky Mountain Forest and Range Experiment Station, Fort Collins, Colorado).

Daniels, T.L. 1986, 'Hobby farming in America: rural development or threat to commercial agriculture?', *Journal of Rural Studies*, 2: 31–40.

Daniels, T.L. and Lapping, M.B. 1987, 'Small town triage: a rural settlement policy for the American Midwest', *Journal of Rural Studies*, 3: 273–80.

Darley, G. 1978a, *Villages of vision* (Granada Publishing, St Albans).

Darley, G. 1978b, 'Rural settlement – rural resettlement: the future', *Built Environment*, 4: 299–310.

Davidson, J. and Wibberley, G.P. 1977, *Planning and the rural environment* (Pergamon Press, Oxford).

Davies, E.T. 1983, *The role of farm tourism in the Less Favoured Areas of England and Wales, 1981* (Agricultural Economics Unit, University of Exeter).

Davies, H.W.E. 1988a, 'Development control in England', *Town Planning Review*, 59: 127–36.

Davies, H.W.E. 1988b, 'The control of development in the Netherlands', *Town Planning Review*, 59: 207–26.

Davies, R.B. and O'Farrell, P.N. 1981, 'A spatial and temporal analysis of second home ownership in Wales', *Geoforum*, 12: 161–78.

Davis, J.C. 1973, *Statistics and data analysis in Geology* (John Wiley and Sons, New York).

Dawson, A.H. 1980, 'The great increase in barley growing in Scotland', *Geography*, 65: 213–17.

Day, G., Rees, G. and Murdoch, J. 1989, 'Social change, rural localities and the state: the restructuring of rural Wales', *Journal of Rural Studies*, 5: 227–44.

Dean, C. 1985, 'The impact of oil developments on rural southern England', pp. 263–78 in Healey, M.J. and Ilbery, B.W. (eds), *The Industrialisation of the Countryside* (Geo Books, Norwich).

Dean, K.G. 1987, 'The disaggregation of migration flows: the case of Brittany, 1975–1982', *Regional Studies*, 21: 313–26.

Dean, K.G., Brown, B.J.H., Perry, R.W. and Shaw, D.P. 1984a, 'The conceptualisation of counterurbanisation', *Area*, 16: 9–16.

Dean, K.G., Brown, B.J.H., Perry, R.W. and Shaw, D.P. 1984b, 'Counterurbanisation and the characteristics of persons migrating to West Cornwall', *Geoforum*, 15: 177–90.

Dear, M. and Scott, A. 1981, 'Towards a framework for analysis', pp. 13–18 in Dear, M. and Scott, A. (eds), *Urbanization and urban planning in capitalist society* (Methuen, London and New York).

Dearden, P. 1980a, 'Landscape assessment: the last decade', *Canadian Geographer*, 24: 316–25.

Dearden, P. 1980b, 'Aesthetic encounters of the statistical kind', *Area*, 12: 171–3.

Dearden, P. 1985, 'Philosophy, theory and method in landscape evaluation', *Canadian Geographer*, 29: 263–5.

Dearden, P. 1987, 'Consensus and a theoretical framework for landscape evaluation', *Journal of Environmental Management*, 24: 267–87.

Dearden, P. and Rosenblood, L. 1980, 'Some observations on multivariate techniques in landscape evaluation', *Regional Studies*, 14: 99–110.

De Barros, A. 1980, 'Portuguese agrarian reform and economic and social development', *Sociologia Ruralis*, 20: 82–96.

De Bendedictis, M. 1981, 'Agricultural development in Italy: national problems in community framework', *Journal of Agricultural Economics*, 32: 275–86.

Deere, C.D. and De Janvry, A. 1979, 'A conceptual analysis for the analysis of peasants', *American Journal of Agricultural Economics*, 61: 601–11.

De Jong, G.F. and Sell, R.R. 1977, 'Population redistribution, migration and residential preferences', *Annals of the American Academy of Political and Social Sciences*, 429: 130–44.

Demangeon, A. 1935, *Problèmes de géographie humaine* (A. Colin, Paris).

Dematteis, G. 1987, 'Urbanisation and counter-urbanisation in Italy', *Ekistiks*, 53: 26–33.

Dennis, N., Henriques, F.M. and Slaughter, C. 1957, *Coal is our life* (Eyre and Spottiswoode, London).

Department of the Environment (DoE) 1971, 'The nature of rural areas of England and Wales', *Dept. of the Environment, Internal Working Paper*.

Department of the Environment (DoE) 1975, *Sport and recreation* (White Paper, Cmnd. 6200, HMSO, London).

Department of the Environment (DoE) 1987, *Rural enterprise and development* (HMSO, London).

De Smidt, M. 1987, 'In pursuit of deconcentration: the evolution of the Dutch urban system from an organizational perspective', *Geografiska Annaler*, 69B: 133–44.

Development Board for Rural Wales 1984, *The impact of regional industrial policy on Wales: Minutes of evidence of the Development Board for Rural Wales to the Committee on Welsh Affairs* (HMSO, London).

Development Commission 1980, *Encouraging enterprise in the countryside: thirty-seventh report of the Development Commissioners 1978-9* (HMSO, London).

Development Commission 1984, *Guidelines for development of RDPs* (Development Commission, London).

Dexter, K. 1977, 'The impact of technology on the political economy of agriculture', *Journal of Agricultural Economics*, 28: 211–19.

Dhillon, H.S., Doermann, A.C. and Walcoff, P. 1978, 'Telemedicine and rural primary health care: an analysis of the impact of telecommunications technology', *Socio-Economic Planning Sciences*, 12: 37–48.

Dicken, P. 1982, 'The industrial structure and the geography of manufacturing', pp. 171–202 in Johnston, R.J. and Doornkamp, J.C. (eds). *The changing geography of the United Kingdom* (Methuen, London and New York).

Dickens, P., Duncan, S., Goodwin, M. and Gray, F. 1985, *Housing, states and localities* (Methuen, London).

Diem, A. 1980, 'Valley renaissance in the High Alps', *Geographical Magazine*, 52: 492–7.

Dillman, B.L. and Cousins, C.F. 1982, 'Urban encroachment on prime agricultural land: a micro-analysis of one SMSA', *International Regional Science Review*, 7: 285–92.

Dillman, D.A. 1979, 'Residential preferences, quality of life, and the population turnaround', *American Journal of Agricultural Economics*, 61: 960–6.

Dillman, D.A. and Hobbs, D.J. (eds) 1982, *Rural society in the US. Issues for the 1980s* (Westview Press, Boulder, Col.).

Dobson, S.M. 1987a, 'Manufacturing establishments, linkage patterns and the implications for peripheral area development: the case of Devon and Cornwall', *Geoforum*, 18: 37–54.

Dobson, S.M. 1987b, 'Public assistance and employment growth in the rural periphery: some issues in policy design and implementation', *Geoforum*, 18: 55–64.

Dodgson, J.S. 1984, 'Railway costs and closures', *Journal of Transport Economics and Policy*, 219–35.

Donaldson, J.G.S. and F. 1969, *Farming in Britain today* (Penguin, Harmondsworth).

Douglass, W.A. 1975, *Echalar and Murelaga: opportunity and rural exodus in two Spanish Basque villages* (C. Hurst and Co., London).

Dourand-Droughin, J.-L., Szwengrub, L.-M. and Mihailescu, I. (eds), 1981, *Rural community studies in Europe: trends, selected and annotated bibliographies, analyses – Volume 1* (Pergamon, Oxford).

Dourand-Droughin, J.-L., Szwengrub, L.-M. and Mihailescu, I. (eds), 1982, *Rural community studies in Europe: trends, selected and annotated bibliographies, analyses – Volume 2* (Pergamon, Oxford).

Dourand-Droughin, J.-L., Szwengrub, L.-M. and Mihailescu, I. (eds), 1985, *Rural community studies in Europe: trends, selected and annotated bibliographies, analyses – Volume 3* (Pergamon, Oxford).

Dower, M. 1965, 'Fourth wave: the challenge of leisure', *Architects' Journal*, 122–90.

Dower, M. 1980, *Jobs in the countryside* (National Council for Voluntary Organisations, London).

Drudy, P.J. 1978, 'Depopulation in a prosperous agricultural sub-region', *Regional Studies*, 12: 49–60.

Drudy, P.J. and S.M. 1979, 'Population mobility and labour supply in rural regions: North Norfolk and the Galway Gaeltacht', *Regional Studies*, 13: 91–9.

Duffield, B.S. and Long, J.A. 1981, 'Tourism in the Highlands and Islands of Scotland: rewards and conflicts', *Annals of Tourism Research*, 8: 403–31.

Duffield, B.S. and Owen, M. 1970, *The Pentland Hills: some aspects of outdoor recreation, a research study* (Tourism and Recreation Research Unit, Department of Geography, University of Edinburgh).

Duffy, P.J. 1986, 'Planning problems in the countryside' in Breathnach, P. and Cawley, M.E. (eds), 'Change and development in rural Ireland', *Special Publications, Geographical Society of Ireland*, No. 1 (Dublin).

Dunkle, R.E., Powell, J.H. and Goode, R. 1983, 'Service implications for a community based geriatric assessment unit', *Journal of Gerontology*, 23: 64–5.

Dunleavy, P. 1981, 'Perspectives on urban studies', pp. 1–16 in Blowers, A.T.,

Brook, C., Dunleavy, P. and McDowell, L. (eds), *Urban change and conflict: an interdisciplinary reader* (Harper and Row, London).

Dunn, M.C. 1974, 'Landscape evaluation: a further perspective', *The Planner*, 60: 935–6.

Dunn, M.C. 1976, 'Landscape with photographs: testing the preference approach to landscape evaluation', *Journal of Environmental Management*, 4: 15–26.

Dunn, M. Rawson, M. and Rogers, A. 1981, *Rural housing: competition and choice* (Allen and Unwin, London).

Durant, R. 1939, *Watling: a survey of social life on a new housing estate* (P.S. King, London).

Durkheim, E. 1933, *The division of labour in society* (Macmillan, Toronto).

Dutch Ministry of Agriculture and Fisheries (MAF) 1967, *Selected agricultural figures* (Dutch MAF, The Hague).

Eberle, N. 1982, *Return to Main Street: a journey to another America* (W.W. Norton and Co., London and New York).

Economic Development Committee (EDC) for Agriculture, Land Utilisation Group, 1987, *Directions for change: land use in the 1990s* (National Economic Development Office, London).

Edel, M. 1981, 'Land policy, economic cycles and social conflict', pp. 19–44 in De Neufville, J.I. (ed.), *The land use policy debate in the United States* (Plenum Press, London).

Edwards, C. 1986, 'The role of natural resources in regional agricultural growth', pp. 692–702 in Maunder, A. and Renborg, U. (eds), *Agriculture in a turbulent world economy: Proceedings of the 19th International Conference of Agricultural Economists, Malaga, 1986* (Gower, Aldershot).

Edwards, D. 1988, 'The planning system and the control of development in Denmark', *Town Planning Review*, 59: 137–58.

Edwards, G.W. 1985, 'Rural public transport alternatives in Central Powys', pp. 95–120 in Cloke, P.J. (ed.), *Rural accessibility and mobility* (Centre for Rural Transport, Lampeter).

Edwards, G.W. 1987, 'Transport and highways policy – a dead end?', pp. 137–63 in Cloke, P.J. (ed.), *Rural planning: policy into action?* (Harper and Row, London).

Edwards, J.A. 1971, 'The viability of lower size-order settlements in rural areas: the case of North-East England', *Sociologia Ruralis*, 11: 247–76.

Eire Department of Education 1965, *Investment in education* (Stationery Office, Dublin).

Eire Department of Education 1986, *Health statistics* (Stationery Office, Dublin).

Ekhaugen, K., Gronmo, S. and Kirby, D. 1980, 'State support to small stores: a Nordic form of consumer policy', *Journal of Consumer Policy*, 4: 195–211.

Elgie, R.A. 1984, 'The changing structure of non-metropolitan migration in the American Deep South', *Tijdschrift Voor Economische en Sociale Geografie*, 75: 14–21.

Elson, M.J. 1981, 'Structure Plan policies for pressured rural areas', *Countryside Planning Yearbook*, 2: 49–71.

Elson, M.J. 1986, *Green Belts: conflict mediation in the urban fringe* (Heinemann, London).

Emery, F.V. 1974, *The making of the English landscape – Oxfordshire* (Hodder and Stoughton, London).

Ennew, J., Hirst P., and Tribe, K. 1977, '"Peasantry" as an economic category', *Journal of Peasant Studies*, 4: 295–322.

Engels, R. and Forstall, R. 1985, 'Metropolitan areas are growing again', *American Demographics*, 7: 23–5.

Epstein, T.S. 1962, *Economic development and social change in South India* (Manchester University Press, Manchester).

Erasmus, C. 1967, 'The upper limits of peasantry and agrarian reform: Bolivia, Venezuela and Mexico compared', *Ethnology*, 6.

Erickson, R.A. 1976, 'The filtering-down process: industrial location in a non-metropolitan area', *Professional Geographer*, 28: 254–60.

Erickson, R.A. 1980, 'Corporate organisation and manufacturing branch plant closures in nonmetropolitan areas', *Regional Studies*, 14: 491–501.

Ervin, D.E. 1988, 'Set Aside programmes: using United States experience to evaluate United Kingdom proposals', *Journal of Rural Studies*, 4: 181–91.

Estall, R. 1983, 'The decentralization of manufacturing industry: recent American experience in perspective, *Geoforum*, 14: 133–48.

Evans, N.J. and Ilbery, B.W. 1989, 'A conceptual framework for investigating farm-based accommodation and tourism in Britain', *Journal of Rural Studies*, 5: 257–66.

Evans, R.J. and Lee, W.R. (eds) 1986, *The German peasantry: conflict and community in rural society from the 18th to the 20th centuries* (Croom Helm, London and Sydney).

Fairweather, J.R. and Gillies, J.L. 1982, 'A content analysis of "Rural Sociology" and "Sociologia Ruralis": preliminary results', *Sociologia Ruralis*, 22: 172–9.

Falk, W.W. and Pinhey, T.K. 1978, 'Making sense of the concept 'rural' and doing rural sociology – interpretive perspective', *Rural Sociology*, 43: 547–58.

Farmer, R.S.J. 1979, 'International migration', pp. 31–59 in Neville, R.J.W. and O'Neill, C.J. (eds), *The population of New Zealand: interdisciplinary perspectives* (Longman Paul, Auckland).

Farrington, J.H. and Harrison, R.J. 1985, 'SCOTMAP, rural bus services and deregulation', *Scottish Geographical Magazine*, 101: 111–23.

Fedden, R. 1974, *The National Trust: past and present* (Fontana, London).

Fennell, R. 1979, *The Common Agricultural Policy of the European Community: its institutional and administrative organisation* (Granada, London).

Ferguson, M.J. and Munton, R.J.C. 1979, 'Informal recreation sites in London's Green Belt', *Area*, 11: 196–205.

Fernandez, R.R. and Dillman, D.A. 1979, 'The influence of community attachment on geographic mobility', *Rural Sociology*, 44: 345–60.

Ferrando, M.G. 1975, 'Social stratification in the agricultural sector of Spain: a sociological study of census data', *Sociologia Ruralis*, 15: 107–18.

Ferrao, J. and Jensen-Butler, C. 1986, 'Industrial development in Portuguese regions during the 1970s', *Tijdschrift Voor Economische en Sociale Geografie*, 77: 132–48.

Fielder, J.L. 1981, 'A review of the literature on access and utilization of medical care with special emphasis on rural primary care', *Social Science and Medicine*, 15c: 129–42.

Fielding, A.J. 1982, 'Counterurbanisation in Western Europe', *Progress in Planning*, 17: 3–52.

Fielding, A.J. 1986, 'Counterurbanisation', pp. 224–56 in Pacione, M. (ed.), *Population geography: progress and prospect* (Croom Helm, Beckenham).

Fincher, R.M. 1983, 'Guest editorial: Socialist planning in capitalist countries', *Environment and Planning A*, 15: 867–8.

Fines, K.D. 1968, 'Landscape evaluation: a research project in East Sussex', *Regional Studies*, 2: 41–55.

Fischel, W. 1982, 'The urbanisation of agricultural land: a review of the National Agricultural Lands Study', *Land Economics*, 58: 236–59.

Fischer, C.S. 1973, 'On urban alienations and anomie: powerlessness and social isolation', *American Sociological Review*, 38: 311–26.

Fischer, C.S. 1982, 'Introduction: public policy and the urbanisation of farmland', *International Regional Science Review*, 7: 249–56.

Fitchen, J.M. 1981, *Poverty in rural America: a case study* (Westview Press, Boulder, Col.).

Fitzsimmons, M. 1986, 'The new industrial agriculture: the regional integration of speciality crop production', *Economic Geography*, 62: 334–53.

Flinn, W.L. 1982, 'Communities and their relationships to agrarian values', pp. 19–32 in Browne, W.P. and Hadwiger, D.F. (eds), *Rural policy problems* (D.C. Heath, Lexington).

Flinn, W.L. and Johnson, D.E. 1974, 'Agrarianism amongst Wisconsin farmers', *Rural Sociology*, 39: 187–204.

Flynn, A. and Lowe, P. 1987, 'The problems of analysing party politics: Labour and Conservative approaches to rural conservation', *Environment and Planning A*, 19: 409–14.

Ford, T.R. 1978, *Contemporary rural America: persistence and change* (Iowa State University Press, Iowa).

Foresta, R.A. 1985, 'Natural regions for national parks: the Canadian experience', *Applied Geography*, 5: 179–94.

Forestry Commission 1987, *67th Annual Report* (Forestry Commission) (HMSO, London).

Forrest, J. and Johnston, R.J. 1973, 'Migration and mobility', pp. 132–49 in Johnston, R.J. (ed.), *Urbanisation in New Zealand* (Reed, Wellington).

Forsythe, D.E. 1974, 'Escape to fulfilment: urban-rural migration and the future of a small island community', unpublished Ph.D. thesis, Cornell University.

Forsythe, D.E. 1980, 'Urban incomers and rural change: the impact of migrants from the city on life in an Orkney community', *Sociologia Ruralis*, 20: 287–307.

Forsythe, D.E. 1982a, 'Urban-rural migration, change and conflict in an Orkney island community', *SSRC North Sea Oil Panel Occasional Papers*, No. 14.

Forsythe, D.E. 1982b, 'Gross migration and social change: an Orkney case study', pp. 90–104 in Jones, J.R. (ed.), 'Recent migration in northern Scotland: pattern, process, impact', *SSRC North Sea Oil Panel Occasional Papers*, No. 13.

Forsythe, D.E. *et al.*, 1983, *The rural community and the small school* (Aberdeen University Press, Aberdeen).

Fothergill, S. 1986, 'Industrial employment and planning restraint in the London green belt', pp. 27–40 in Towse, R.J. (ed.), *Industrial/office development in areas of planning restraint – the London green belt* (Proceedings of a conference held at Kingston Polytechnic, 11 April 1986, School of Geography).

Fothergill, S., Gudgin, G., Kitson, M. and Monk, S. 1985, 'Rural industrialization: trends and causes', pp. 147–60 in Healey, M.J. and Ilbery, B.W. (eds), *The industrialization of the countryside* (Geo Books, Norwich).

Fothergill, S. and Gudgin, G. 1979, 'Regional employment change: sub-regional explanation', *Progress in Planning*, 12: 155–219.

Fothergill, S., Kitson, M. and Monk, S. 1985, 'Urban industrial change: the causes of the urban-rural contrast in manufacturing employment trends', *Inner Cities Research Programme Research Reports*, No. 11 (HMSO, London).

Frank, W. 1985, 'Part-time farming, underemployment and double activity of farmers in the EEC', *Sociologia Ruralis*, 23: 20–7.

Frankena, M.W. and Scheffman, D.T. 1980, *Economic analysis of provincial land use policies in Ontario* (University of Toronto Press, Toronto).

Frankenberg, R. 1957, *Village on the border* (Cohen and West, London).

Frankenberg, R. 1966a, *Communities in Britain* (Penguin, Harmondsworth).

Frankenberg, R. 1966b, 'British community studies: problems of synthesis', pp. 123–54 in Banton, M. (ed.), *The social anthropology of complex societies* (Tavistock, London).

Franklin, S.H. 1964, 'Gosheim, Baden-Wurttemberg: a Mercedes dorf', *Pacific Viewpoint*, 5: 127–58.

Franklin, S.H. 1969, *The European peasantry: the final phase* (Methuen, London).

Frater, J.M. 1983, 'Farm tourism in England: planning, funding, promotion and some lessons from Europe', *Tourism Management*, 4: 167–79.

Frazer, R.M. 1971, 'Patterns of Maori migration: an East Coast example', *Proceedings, Sixth New Zealand Geography Conference*, pp. 209–51.

Freudenberg, W.R. 1982, 'Social impact assessment', in Dillman, D.A. and Hobbs, D.J. (eds), *Rural society in the United States: issues for the 1980s* (Westview Press, Boulder, Col.).

Fricke, W. 1971, 'Socialgeographische untersuchung zur bevolkerungs und siedlungsentwicklung in Frankfurter Raum', *Rhein-Mainsche Forshungen*, 71: 7–15.

Fricke, W. 1976, 'Bevolkerung und raum eines ballungsgebietes seit der industrialisierung. Eine geographische analyze des modellgebiets Rhein-Neckar', *Veroff. Akad. Raumforsch. Landesplan*, 11: 1–68.

Friedl, E. 1964, 'Lagging emulation in post-peasant society', *American Anthropologist*, 66: 569–86.

Freidmann, H. and McMichael, P. 1989, 'Agriculture and the state system. The rise and decline of national agricultures, 1870 to the present', *Sociologia Ruralis*, 29: 93–117.

Frost, S. 1986, 'Socioeconomic development policies for rural revival', *Land Use Policy*, 3: 122–6.

Frundt, H.J. 1975, 'American agribusiness and US foreign agricultural policy', unpublished Ph.D. thesis, Rutgers University, New Brunswick, New York.

Frutos, L.M. 1984, 'La accion estatal en el desarrollo industrial de Extremadura', Documents d'Analisi Geografica, 4: 69–102.

Fuller, A.M. (ed.) 1985, Farming and the rural community in Ontario: an introduction (University of Toronto Press, Toronto).

Fullerton, B. and Williams, A.F. 1972, Scandinavia (Chatto and Windus, London).

Furuseth, O.J. 1985, 'Local farmland conservation programmes in the US: a study of California counties', Applied Geography, 5: 211–28.

Furuseth, O.J. and Pierce, J.T. 1982, Agricultural land in an urban society (Resource Publications in Geography, Association of American Geographers, Washington, DC).

Gagliardo, J.G. 1969, From pariah to patriot. The changing image of the German peasant (Kentucky).

Galeski, B. 1968, 'Social organisation and rural social change', Sociologia Ruralis, 8: 256–88.

Galeski, B. 1972, Basic concepts of rural sociology (Manchester University Press, Manchester).

Gans, H.J. 1962, The urban villages (Free Press, New York).

Gans, H.J. 1967, The Levittowners (Vintage Books, New York).

Gans, H.J. 1968, 'Urbanism and suburbanism as ways of life', pp. 95–118 in Pahl, R.E. (ed.) Readings in urban sociology (Pergamon, Oxford).

Garcia-Ramon, M.D. 1985a, 'Old and new in Spanish farming', Geographical Magazine, 113: 128–33.

Garcia-Ramon, M.D. 1985b, 'Agricultural change in an industrializing area, 1955–71', pp. 140–54 in Hudson, R. and Lewis, J. (eds), Uneven development in southern Europe (Methuen, London).

Garcia-Ramon, M.D. 1987, 'Part-time peasants in Catalonia, Spain: between tourism and industrialization', Ekistiks, 53: 82–8.

Garrison, C.B. 1968, 'The impact of new industry: an application of the economic base multiplier to small rural areas', Land Economics, 48: 329–37.

Garrison, C.B. 1970, The impact of rural industry on local government finances in five small towns in Kentucky (Economic Research Service, USDA, AER-191).

Garst, R.D. 1974, 'Innovation diffusion among the Gusic of Kenya', Economic Geography, 50: 300–12.

Gasson, R.M. 1966, 'The influence of urbanization on farm ownership and practice, Studies in Rural Land Use, Department of Agricultural Economics, Wye College, University of London, No. 7.

Gasson, R.M. 1967, 'Some economic characteristics of part-time farming in Britain', Journal of Agricultural Economics, 18: 111–20.

Gasson, R.M. 1971, 'Use of sociology in agricultural economics', Journal of Agricultural Economics, 22: 29–38.

Gasson, R.M. 1980, 'Roles of farm women in England', Sociologia Ruralis, 20: 165–80.

Gasson, R.M. 1981, 'Roles of women on farms: a pilot study', *Journal of Agricultural Economics*, 32: 11–20.

Gasson, R.M. 1984, 'Farm women in Europe: their needs for off farm employment', *Sociologia Ruralis*, 24: 216–28.

Gasson, R. M. 1986, 'Part-time farming in England and Wales', *Journal of the Royal Agricultural Society*, 147: 34–41.

Gasson, R.M. 1988, *The economics of part-time farming* (Longman, London).

Gault, I. 1981, 'Green Belt policies in development plans' *Working Papers, Department of Town Planning, Oxford Polytechnic*, No. 41.

Gayler, H.J. 1982, 'The problems of adjusting to slow growth in the Niagara Region of Ontario', *Canadian Geographer*, 26: 191–206.

Geertz, C. 1963, *Agricultural involution* (University of California Press, Berkeley).

Gelfand, M.J. 1975, *A nation of cities: the federal government and urban America, 1933–65* (Oxford University Press, New York).

Geoffroy, B. 1982, 'Quelques éléments de réflexion sur les communes periurbains Yonnaises', *Norois*, 114: 303–11.

Gertler, L.O. and Crowley, R. 1982, *Changing Canadian cities: the next 25 years* (McClelland and Stewart, Toronto), 2nd edition.

Giddens, A. 1979, *Central problems in social theory* (Macmillan, London).

Giddens, A. 1981, *A contemporary critique of historical materialism, Volume 1* (Macmillan, London).

Giddens, A. 1984, *The constitution of society* (Polity Press, London).

Gilder, I.M. 1979, 'Rural planning policies: an economic appraisal', *Progress in Planning*, 11: 213–71.

Gilder, I.M. 1980, 'Do we need key settlement policies?' *The Planner*, 66: 99–112.

Gilg, A.W. 1974, 'A critique of Linton's method of assessing scenery as a natural resource', *Scottish Geographical Magazine*, 90: 125–9.

Gilg, A.W. 1975, 'The objectivity of Linton type methods of assessing scenery as a natural resource', *Regional Studies*, 9: 181–9.

Gilg, A.W. 1976a, 'Rural employment', pp. 125–72 in Cherry G.E. (ed.), *Rural planning problems* (Leonard Hill, London).

Gilg, A.W. 1976b, 'Assessing scenery as a natural resource', *Scottish Geographical Magazine*, 92: 41–9.

Gilg, A.W. 1978a, *Countryside planning: the first three decades 1945–1976* (David and Charles, Newton Abbot).

Gilg, A.W. 1978b, 'Policy forum: needed: a new 'Scott' inquiry', *Town Planning Review*, 49: 353–71.

Gilg, A.W. 1983, 'Population and employment', pp. 74–105 in Pacione, M. (ed.), *Progress in rural geography* (Croom Helm, Beckenham).

Gilg, A.W. 1985, *An introduction to rural geography* (Edward Arnold, London).

Gillmor, D.A. 1987, 'Concentration of enterprises and spatial change in the agriculture of the Republic of Ireland', *Transactions of the Institute of British Geographers*, new series, 12: 204–16.

Gilmore, H.W. 1953, *Transportation and the growth of cities* (Free Press of Glencoe, Illinois).

Ginatempo, N. 1985, 'Social reproduction and the structure of marginal areas in southern Italy: some remarks on the role of the family in the present crisis', *International Journal of Urban and Regional Research*, 9: 99–112.

Giner, S. and Sevilla, E. 1984, 'Spain: from corporatism to corporation', pp. 113–44 in Williams, A. (ed.), *Southern Europe transformed: political and economic change in Greece, Italy, Portugal and Spain* (Harper and Row, London).

Glass, R. 1959, 'The evaluation of planning: some sociological considerations', *International Social Science Journal*, 11: 393–409.

Glendining, D. 1978, *Why did they leave Eketahuna?* (Wairarapa Education and Rural Services Committee, Masterton).

Glenn, N.D. and Hill, L. 1977, 'Rural-urban differences in attitudes and behavior in the United States', *Annals of the American Academy of Political and Social Science*, 429: 36–50.

Gold, R.L. 1985, *Ranching, mining and the human impact of natural resource development* (Transaction Press, New Brunswick, NJ).

Goldenweiser, A. 1936, *Anthropology: an introduction to primitive culture* (F.S. Crofts, New York).

Goldschmidt, W. 1978a, *As you sow* (Allanheld, Osmun and Co., Montclair, NJ).

Goldschmidt, W. 1978b, 'Large-scale farming and the rural social structure', *Rural Sociology*, 43: 362–6.

Goldstein, S. 1976, 'Facets of redistribution: research challenges and opportunities', *Demography*, 13: 423–34.

Goodchild, R. and Munton, R.J.C. 1985, *Development and the landowner: an analysis of the British experience* (George Allen and Unwin, London).

Goodenough, R. 1984, 'The great American crop surplus: 1983 solution', *Geography*, 69: 351–3.

Goodman, D. and Redclift, M. 1985, 'Capitalism, petty commodity production and the farm enterprise', *Sociologia Ruralis*, 25: 231–47.

Gould, A. and Keeble, D.E. 1984, 'New firms and rural industrialization in East Anglia', *Regional Studies*, 18: 189–201.

Government of Canada, 1981, *A forest sector strategy for Canada* (Minister of Supply and Services Canada, Ottawa).

Gould, P.R. 1969, *Spatial diffusion* (Commission on College Geography, Association of American Geographers, Washington, DC).

Grafton, D.J. 1982, 'Net migration, outmigration and remote rural areas: a cautionary note', *Area*, 14: 313–18.

Grafton, D.J. 1984, 'Small-scale growth centres in remote rural regions: the case of Alpine Switzerland', *Applied Geography*, 4: 29–46.

Granath, O. 1978, 'Swedish agricultural policy', *Swedish Institute Papers*, 181.

Graziani, A. 1978, 'The Mezzogiorno and the Italian economy', *Cambridge Journal of Economics*, 2: 355–72.

Graziani, A. 1983, 'The subsidized South', *Mezzogiorno d'Europa*, 3: 79–85.

Green, B. 1981, *Countryside conservation* (Allen and Unwin, London).

Green, B. 1986, 'Agriculture and the environment: a review of major issues in the UK', *Land Use Policy*, 3: 193–204.

Green, G.P. 1985, 'Large-scale farming and the quality of life in rural communities: further specification of the Goldsmith hypothesis', *Rural Sociology*, 50: 262–74.

Green, R.J. 1971, *Country planning: the future of rural regions* (Manchester University Press, Manchester).

Greenwood, D.J. 1972, 'Tourism as an agent of change: a Spanish Basque case', *Ethnology*, 11: 80–91.

Gregor, H. 1982, *Industrialization of US agriculture: an interpretive atlas* (Westview Press, Boulder, Col.).

Gregory, D. 1982, 'A realist construction of the social', *Transactions of the Institute of British Geographers*, new series, 7: 254–6.

Gregory, D. 1985, 'People, places and practices: the future of human geography', pp. 56–76 in King, R.L. (ed.), *Geographical futures* (Geographical Association, Sheffield).

Griffin, K. 1973, 'Policy options for rural development', *Oxford Bulletin of Economics and Statistics*, 35: 239–74.

Grigg, D.B. 1977, 'E.G. Ravenstein and the laws of migration', *Journal of Historical Geography*, 3: 41–54.

Grigg, D.B. 1980, 'Migration and overpopulation', pp. 60–83 in White, P.E. and Woods, R.I. (eds), *The geographical impact of migration* (Longman, London).

Grigg, D.B. 1989, *English agriculture: an historical perspective* (Blackwell, Oxford).

Groenendijk, J. 1983, 'A key to settlement growth in rural areas: local administrators, their powers and the size of their territories', pp. 283–96 in Clark, G., Groenendijk, J. and Thissen, F. (eds), *The changing countryside: proceedings of the First British-Dutch Symposium on Rural Geography* (Geo Books, Norwich).

Groome, D. and Tarrant, C. 1985, 'Countryside recreation: achieving access for all?', *Countryside Planning Yearbook* 6: 72–100.

Grove, R. 1983, *The future of forestry* (British Association of Conservationists, Stanton St John, Oxford).

Gullickson, G.L. 1986, *Spinners and weavers of Auffay: rural industry and the sexual division of labour in a French village, 1750–1850* (Cambridge University Press, Cambridge).

Gunter, W.P. and Ellis, C.M. 1977, 'Income inequality in a depressed area: A Principal Components Analysis', *Review of Regional Studies*, 5: 42–51.

Hall, P. (ed.), 1973, *The containment of urban England* 2 vols (Allen and Unwin/Sage, London and Beverly Hills).

Hall, P. 1985, 'The people: where will they go?', *The Planner*, 71: 3–12.

Hall, P. 1986, 'The New Zealand urban system: deurbanisation at the southern periphery of the urban world', *New Zealand Geographer*, 42: 65–9.

Hall, P. and Hay, D. 1980, *Growth centres in the European urban system* (Heinemann Education, London).

Hamill, L. 1975, 'Analysis of Leopold's quantitative comparisons of landscape aesthetics', *Journal of Leisure Research*, 7: 16–28.

Hamill, L. 1985, 'On the presence of error in scholarly communication: the case of landscape aesthetics', *Canadian Geographer*, 29: 270–3.

Hansen, J.C. 1975, 'Population trends and prospects in marginal areas of Norway', pp. 255–75 in Kosinski, L. and Pothero, R.M. (eds), *People on the move – studies in internal migration* (Methuen, London).

Haren, C.C. and Holling, R.W. 1979, 'Industrial development in non-metropolitan America: a locational perspective', pp. 13–45 in Lonsdale, R.E. and Seyler, H.L. (eds), *Non-metropolitan industrialisation* (Winston, Washington, DC).

Hareven, T.K. 1977, 'Family time and historical time', *Daedalus*, 108.

Harper, S. 1987, 'A humanistic approach to the study of rural populations', *Sociologia Ruralis*, 3: 309–19.

Harper, S. 1989, 'The British rural community: an overview of perspectives', *Journal of Rural Studies*, 5: 161–84.

Harper, S. and Donnelly, P. 1987, 'British rural settlements in the hinterland of conurbations: a classification', *Geografiska Annaler*, 69B: 55–64.

Harris, R.C. 1977, 'The simplification of Europe overseas', *Annals of the Association of American Geographers*, 67: 469–83.

Harrison, C. 1981, 'A playground for whom? Informal recreation in London's Green Belt', *Area*, 13: 109–14.

Harriss, J. (ed.), 1982, *Rural development: theories of peasant economy and agrarian change* (Hutchinson University Library, London).

Hart, J.F. 1978, 'Cropland concentrations in the South', *Annals of the Association of American Geographers*, 68: 505–17.

Hart, J.F. 1980, 'Land use change in a Piedmont Country', *Annals of the Association of American Geographers*, 70: 492–527.

Hart, J.F. 1984, 'Population change in the upper lake states', *Annals of the Association of American Geographers*, 74: 221–43.

Hart, J.F. 1986, 'Change in the Corn Belt', *Geographical Review*, 76: 51–72.

Hartke, W. 1956, 'Die sozialbrache also phanomen der geographischen diffenzierung der landschaft', *Erdkunde*, 34: 257–69.

Harvey, D.W. 1973, *Social justice and the city* (Edward Arnold, London).

Harvey, D.W. 1975, 'The political economy of urbanization in advanced capitalist societies – the case of the United States', pp. 119–64 in Gappert, G. and Rose, H. (eds), *The political economy of cities* (Sage, Beverly Hills).

Harvey, D.W. 1982, *The limits to capital* (Basil Blackwell, Oxford).

Haushofer, H. 1978, 'Die idealvorstellung vom deutschen bauern', *Z.A.A.*, 26: 147–60.

Hausler, R. 1974, 'The emergence of area development', pp. 15–29 in Brinkman, G. (ed.), *The development of rural America* (University of Kansas Press, Lawrence).

Hayami, Y. and Ruttan, V.W. 1971, *Agricultural development: an international perspective* (Johns Hopkins University Press, Baltimore).

Haynes, R.M. and Bentham, C.G. 1979a, *Community hospitals and rural accessibility* (Saxon House, Farnborough).

Haynes, R.M. and Bentham, C.G. 1979b, 'Accessibility and the use of hospitals in rural areas', *Area*, 11: 186–91.

Haynes, R.M. and Bentham, C.G. 1982, 'The effects of accessibility on general practitioner consultations, out-patient attendances and in-patient admissions in

Norfolk, England', *Social Science and Medicine*, 16: 561–9.

Healey, P. 1980, 'The implementation of selective restraint policies', *Working Papers, Department of Town Planning, Oxford Polytechnic*, No. 45.

Healy, R.G. and Short, J.L. 1981, *The market for rural land: trends, issues and policies* (The Conservation Foundation, Washington DC).

Hebbert, M. 1982, 'Regional policy in Spain', *Geoforum*, 13: 1178–1202.

Hechter, M. 1975, *Internal colonialism: the Celtic fringe in British national development 1536–1966* (Routledge and Kegan Paul, London).

Heeley, J. 1988, 'The rural attractions base of Scotland', *Scottish Association of Geography Teachers*, 17: 61–4.

Heenan, L.D.B. 1968a, 'Internal migration in the South Island', *New Zealand Geographer*, 24: 84–90.

Heenan, L.D.B. 1968b, 'Internal migration in the South Island', *Publications, Department of Geography, University of Otago*, No. 109.

Heenan, L.D.B. 1979, 'Internal migration: inventory and appraisal', pp. 60–88 in Neville, R.J.W. and O'Neill, C.J. (eds), *The population of New Zealand: interdisciplinary perspectives* (Longman Paul, Auckland).

Heenan, L.D.B. 1988, 'Population studies', *Progress in Human Geography*, 12: 282–92.

Heffernan, W.D. 1982, 'The structure of agriculture and quality of life in rural communities', pp. 337–46 in Dillman, D. and Hobbs, D. (eds), *Rural Sociology: research issues for the 1980s* (Westview Press, Boulder, Col.).

Henderson, L.J. 1976, 'A critical look at thematic map design', *Cartography*, 9: 175–80.

Hendrix, W.G. and Fabos, J.G. 1975, 'Visual land-use compatibility as a significant contributor to visual resource quality', *International Journal of Environmental Studies*, 8: 21–8.

Henwood, W.D. 1982, 'The national parks system of New Zealand: its evolution and prospects', *Park News*, 18 (1): 3–11 and 19 (3): 3–9.

Herden, W. 1983, 'Die rezente bevolkerungs- und bausubstanzentwicklung des weslichen Rhein-Neckar-Raumes', *Heidelberger Georgraphische Arbeiten*, 60: 1–229.

Herdt, R.W. and Cochrane, W.W. 1966, 'Farm land prices and farm technological advance', *Journal of Farm Economics*, 48: 243–71.

Herington, J.M. 1982, 'Circular 22/80 – the demise of settlement planning?', *Area*, 14: 157–66.

Herington, J.M. 1984, *The outer city* (Harper and Row, London).

Herington, J.M. 1986, 'Exurban housing mobility: the implications for future study', *Tijdschrift Voor Economische en Sociale Geographie*, 77: 178–86.

Higgins, J. 1983, *A study of part-time farmers in the Republic of Ireland* (Economics and Rural Welfare Centre, An Foras Taluntais, Dublin).

Highlands and Islands Development Board (HIDB) 1967, *First Annual Report* (HIDB, Inverness).

HIDB 1983, *Seventeenth Annual Report* (HIDB, Inverness).

Hill, B. and Gasson, R.M. 1985, 'Farm tenure and farming practice', *Journal of Agricultural Economics*, 35: 187–99.

Hill, B. and Ray, D. 1987, *Economics for agriculture: food, farming and the rural economy* (Macmillan Education, Basingstoke).

Hill, B.E. 1984, *The Common Agricultural Policy: past, present and future* (Methuen, London and New York).

Hillery, G.A. 1955, 'Definitions of community: areas of agreement', *Rural Sociology*, 20: 111–23.

Hillman, M. and Whalley, A. 1980, *The social consequences of rail closure* (Policy Studies Institute, London).

Hinton, W.L. 1986, 'Enlargement of the EEC in the fruit and vegetable sector', *Food Marketing*, 2: 3–13.

Hirsch, G.P. and Maunder, A. H. 1978, *Farm amalgamation in Western Europe* (Saxon House, Farnborough).

Hirschman, A.O. 1958, *The strategy of economic development* (Yale University Press, New Haven, Ct.).

Hobson, P.M. 1949, 'The parish of Barra', *Scottish Geographical Magazine*, 65: 71–81.

Hockin, R., Goodall, B. and Whittow, J. 1978, 'The site requirements of outdoor recreation activities', *Geographical Papers, Department of Geography, University of Reading*, No. 54.

Hodge, G.A. and Quadeer, M.A. 1983, *Towns and villages in Canada: the importance of being unimportant* (Butterworths, Toronto).

Hodge, I.D. 1978, 'On the local environmental impact of livestock production', *Journal of Agricultural Economics*, 29: 279–90.

Hodge, I.D. 1986, 'Rural development and the environment: a review', *Town Planning Review*, 57: 175–86.

Hodge, I.D. and Monk, S. 1987, 'Manufacturing employment change within rural areas', *Journal of Rural Studies*, 3: 65–9.

Hodge, I.D. and Whitby, M. 1979, 'New jobs in the Eastern Borders: an economic evaluation of the Development Commission factory programme', *Monographs, Agricultural Adjustment Unit, University of Newcastle-upon-Tyne*, No. 8.

Hodge, I.D. and Whitby, M. 1981, *Rural employment* (Methuen, London).

Hodges, M.W. and Smith, C.S. 1954, 'The Sheffield estates', pp. 79–134 in Department of Social Science, University of Liverpool, *Neighbourhood and community: an enquiry into social relationships on housing estates in Liverpool and Sheffield* (Liverpool University Press, Liverpool).

Hoffman, D.W. 1982, 'Saving farmland, a Canadian program', *Geojournal*, 6: 539–46.

Hoggart, K. and Buller, H. 1987, *Rural development: a geographical perspective* (Croom Helm, Beckenham).

Holderness, B.A. 1972, 'Open and closed parishes in England in the eighteen and nineteenth centuries', *Agricultural History Review*, 20: 126–39.

Holderness, B.A. 1985, *British agriculture since 1945* (Manchester University Press, Manchester).

Holland, S. 1976, *The regional problem* (Macmillan, London).

Hollingham, M.A. and Howarth, R.W. 1989, *British milk marketing and the Common Agricultural Policy: the origins of confusion and crisis* (Avebury, London).

Holman, R. 1978, *Poverty: explanations of social deprivation* (Martin Robertson, Oxford).

Holmes, J.H. 1977, 'Population', pp. 331–53 in Jeans, D.N. (ed.), *Australia: a geography* (Sydney University Press, Sydney).

Holmes, J.H. 1981, 'Lands of distant promise', pp. 1–13 in Lonsdale, R.E. and Holmes, J.H. (eds), *Settlement systems in sparsely populated regions: the United States and Australia* (Pergamon Press, New York).

Holmes, J.H. 1988, 'Australia', pp. 192–217 in Cloke, P.J. (ed.), *Policies and plans for rural people: an international perspective* (Unwin Hyman, London).

Hooper, A.J. 1988, 'Planning and the control of development in the federal Republic of Germany', *Town Planning Review*, 59: 183–206.

Hoover, E.M. 1968, *Location theory and the shoe leather industries* (Johnson Reprint Corp., New York).

Horner, A.A. 1986, 'Rural population change in Ireland', in Breathnach, P. and Cawley, M.E. (eds), 'Change and development in rural Ireland', *Special Publications, Geographical Society of Ireland*, No. 1.

Horner, A.A. and Daultrey, S.G. 1980, 'Recent population changes in the Republic of Ireland', *Area*, 12: 129–35.

Houck, J.P. 1986, 'Views on agricultural economics' role in economic thought', *American Journal of Agricultural Economics*, 68: 375–80.

Houston, G. 1987, 'Assessing the IDP for the Western Isles', *Scottish Geographical Magazine*, 103: 163–5.

Howes, R. and Law, D. 1972, *Mid-Wales: an assessment of the impact of the Development Commission Factory Programme* (HMSO, London).

Hughes, J. 1984, 'Policies and practice: the Scottish experience', pp. 179–95 in Jess, P.M., Greer, J.V., Buchanan, R.H. and Armstrong, W.J. (eds), *Planning and development in rural areas* (Institute of Irish Studies, Queen's University, Belfast).

Hugo, G.J. 1983, 'Population change in Australian urban and rural areas', *National Institute of Labour Studies Working Papers*, No. 51.

Hugo, G.J. and Smailes, P.J. 1985, 'Urban-rural migration in Australia: a process view of the turnaround', *Journal of Rural Studies*, 1: 11–30.

Huigen, P. 1983, 'Access in a remote area', pp. 87–98 in Clark, G., Groenendijk, J. and Thissen, F. (eds), *The changing countryside: Proceedings of the First British-Dutch Symposium on Rural Geography* (Geo Books, Norwich).

Hunt, M.E. *et al.*, 1984, *Retirement communities: an American original* (The Haworth Press, New York).

Hunter, J. 1976, *The making of the crofting community* (John Donald, Edinburgh).

Ilbery, B.W. 1978, 'Agricultural decision-making: a behavioural perspective', *Progress in Human Geography*, 2: 448–66.

Ilbery, B.W. 1983, 'A behavioural analysis of hop farming in Hereford and Worcestershire', *Geoforum*, 14: 447–59.

Ilbery, B.W. 1985, 'Behavioural interpretation of horticulture in the Vale of Evesham', *Journal of Rural Studies*, 1: 121–33.

Industrial Development Authority (IDA), Eire, 1972, *Jobs for the people: summary of regional industrial plans* (IDA, Dublin).

Industry Department for Scotland 1987, *Review of the Highlands and Islands Development Board: Report of Review Group to the Secretary of State for Scotland* (HMSO, Edinburgh).

Information and Documentation Centre for the Geography of the Netherlands, 1985, *Compact geography of the Netherlands* (Ministry of Foreign Affairs, The Hague).

Institute of Manpower Studies, 1986, *Changing working patterns: how companies achieve flexibility to meet new needs* (Institute of Manpower Studies/National Economic Development Office, Department of the Environment, London).

International Union for Conservation of Nature and Natural Resources (IUCN), 1969, *Red data book*, 2 vols (IUCN, Morges, Switzerland).

IUCN, 1985, *1985 United Nations' list of National Parks and protected areas* (IUCN, Gland, Switzerland).

Jackson, H.T. 1985, *Crabgrass frontier: the suburbanization of the United States* (Oxford University Press, New York).

Jackson, J.N. 1982, 'The Niagara Fruit Belt: The Ontario Municipal Board decision of 1981', *Canadian Geographer*, 26: 172–6.

Jackson, J.N. (ed.) 1986, 'The Niagara region: trends and prospects – the challenge of change', *Occasional Publications, Department of Geography, Brock University*, No. 4.

James, N.D.G. 1981, *A history of English forestry* (Blackwell, Oxford).

Jarvie, W.K. 1981, 'Internal migration and structural change in Australia 1966–71: some preliminary observations', pp. 25–55 in *Papers of the Australian and New Zealand Section of the Regional Science Association*, sixth meeting (Surfers' Paradise, Queensland).

Jobes, P.C. 1987, 'The disintegration of gemeinschaft social structure from energy development: observations from ranch communities in the western United States', *Journal of Rural Studies*, 3: 219–29.

Johansen, H.E. and Fuguitt, G.V. 1979, 'Population growth and retail decline: conflicting effects of urban accessibility in American villages', *Rural Sociology*, 44: 24–38.

Johansen, H.E. and Fuguitt, G.V. 1984, *The changing rural village in America: demographic trends since 1950* (Ballinger Publishing Co., Cambridge, Mass.).

John, B.S. 1984, *Scandinavia: a new geography* (Longman, London and New York).

Johnsen, J.T. 1978, 'Rechtshilfe in Norwegen', *Jahrbuch fur Rechtssociologie und Rechtstheorie*, 5: 185–205.

Johnson, D.G. 1986, 'Agricultural economics: contributions; discussion', *American Journal of Agricultural Economics, 68: 395–6.

Johnson, J.A. and Price, C. 1987, 'Afforestation, employment and depopulation in the Snowdonia National Park', *Journal of Rural Studies*, 3: 195–205.

Johnston, B.F. and Kilby, P. 1975, *Agriculture and structural transformation* (Oxford University Press, London and New York).

Johnston, R.J. 1966a, 'Components of rural population change', *Town Planning Review*, 37: 279–93.

Johnston, R.J. 1966b, 'An index of accessibility and its use in the study of bus services and settlement patterns', *Tijdschrift Voor Economische en Sociale Geografie*, 57: 33-8.

Johnston, R.J. 1971, 'Resistance to migration and the mover/stayer dichotomy: aspects of kinship and population stability in an English rural area', *Geografiska Annaler*, 53B: 16-27.

Johnston, R.J. 1979, *Geography and geographers: Anglo-American human geography since 1945* (Edward Arnold, London).

Johnston, R.J. 1983, 'Urbanization', pp. 363-4 in Johnston, R.J. (ed.), *The dictionary of human geography* (Blackwell, Oxford), 2nd edition.

Johnston, R.J. 1988, 'The political organization of US space', pp. 81-110 in Knox, P.L., Bartels, E.H., Holcomb, B., Bohland, J.R. and Johnston, R.J., *The United States: a contemporary geography* (Longman, London).

Johnston, T. and Smit, B. 1985, 'An evaluation of the rationale for farmland policy in Ontario', *Land Use Policy*, 2: 225-37.

Jones, A.R. 1989a, 'The role of the SAFER in agricultural restructuring: the case of Languedoc-Roussillon, France', *Land Use Policy*, 6: 249-61.

Jones, A.R. 1989b, 'The reform of the EEC's table wine sector: agricultural despecialisation in the Languedoc', *Geography*, 74: 29-37.

Jones, A.R. 1984, 'Agriculture: organization, reform and the EEC', pp. 236-67 in Williams, A. (ed.), *Southern Europe transformed: political and economic change in Greece, Italy, Portugal and Spain* (Harper and Row, London).

Jones, G.E. 1973, *Rural life: patterns and processes* (Longman, London).

Jones, H.R. 1973, 'Modern emigration from Malta, *Transactions of the Institute of British Geographers*, 60: 101-20.

Jones, H.R. 1976, 'The structure of the migration process: findings from a growth point in mid-Wales', *Transactions of the Institute of British Geographers*, new series, 1: 421-32.

Jones, H.R. (ed.) 1982, 'Recent migration in Northern Scotland: pattern, process, impact', *SSRC North Sea Oil Panel, Occasional Papers*, No. 13.

Jones, H.R., Caird, J.B., Berry, W. and Dewhurst, J. 1986, 'Peripheral counter-urbanisation: findings from an integration of census and survey data in northern Scotland', *Regional Studies*, 20: 15-26.

Jones, H.R., Ford, N., Caird, J.B. and Berry, W. 1984, 'Counter-urbanization in societal context: long-distance migration to the Highlands and Islands of Scotland', *Professional Geographer*, 36: 437-44.

Jones, P. 1980, 'Primary school provision in rural areas', *The Planner*, 66: 4-6.

Jones, P.E. 1985, 'Recent changes in rural Wales: some implications for schools', *Education for Development*, 9: 8-16.

Joseph, A.E., Keddie, P.D. and Smit, B. 1988, 'Unravelling the population turnaround in rural Canada', *Canadian Geographer*, 32: 17-30.

Joseph, A.E. and Poyner, A. 1982, 'Interpreting patterns of public service utilization in rural areas', *Economic Geography*, 58: 262-73.

Joseph, A.E. and Smit, B. 1983, 'Preferences for public service provision in rural areas undergoing exurban residential development: a Canadian view', *Tijdschrift voor en economische sociale geografie*, 74: 41-52.

Joseph, A.E. and Smit, B. 1985, 'Rural residential development and municipal service provision: a Canadian case study', *Journal of Rural Studies*, 1: 321-37.

Joseph, A.E., Smit, B. and McIlravey, G.P. 1989, 'Consumer preferences for rural residences: a conjoint analysis in Ontario, Canada', Environment and Planning A, 21: 47–64.

Josling, T.E. and Hamway, D. 1976, 'Income transfer effects of the CAP', pp. 180–205 in Davey, B. et al. (eds), Agriculture and the state, (Macmillan, London).

Kada, P. 1980, Part-time family farming. Off-farm employment and farm adjustments in the United States and Japan (Center for Academic Publications, Tokyo).

Kale, S.R. and Lonsdale, R.E. 1979, 'Factors encouraging and discouraging plant location in nonmetropolitan areas', pp. 47–56 in Lonsdale, R.E. and Seyler, M.R. (eds), Nonmetropolitan industrialization (John Wiley, New York).

Kale, S.R. and Lonsdale, R.E. 1987, 'Recent trends in US and Canadian nonmetropolitan manufacturing', Journal of Rural Studies, 3: 1–13.

Kariel, H.G. and Kariel, P.E. 1982, 'Socio-cultural impacts of tourism: an example from the Austrian Alps', Geografiska Annaler, 64B: 1–16.

Keane, M.J. and O'Cinneide, M.S. 1983, 'Promoting economic development amongst rural communities', Journal of Rural Studies, 2: 281–9.

Keane, M.J., Cawley, M. and O'Cinneide, M.S. 1983, 'Industrial development in Gaeltacht areas: the work of Udaras na Gaeltachta', Cambria, 10: 47–60.

Keeble, D.E. 1980, 'Industrial decline, regional policy and the urban-rural manufacturing shift in the European Community', Environment and Planning A, 12: 945–62.

Keeble, D.E. 1984, 'The rural-urban manufacturing shift', Geography, 69: 163–6.

Keeble, D.E. and Gould, A. 1985, 'Entrepreneurship and manufacturing firm formation in rural regions: the East Anglian case', pp. 197–220 in Healey, M.J. and Ilbery, B.W. (eds), The industrialization of the countryside (Geo Books, Norwich).

Keeble, D.E., Owens, P.L. and Thompson, C. 1983, 'The urban-rural manufacturing shift in the European Community', Urban Studies, 20: 405–18.

Keen, P.A. 1978, 'Rural rail services', pp. 139–46 in Cresswell, R. (ed.), Rural transport and country planning (Leonard Hill, London).

Keinath, W.F. 1982, 'The decentralization of American economic life: an economic evaluation', Economic Geography, 58: 343–57.

Keinath, W.F. 1985, 'The spatial component of the post-industrial society', Economic Geography, 61: 223–40.

Kenney, M., Lobao, L.M., Curry, J. and Goe, W.R. 1989, 'Mid-western agriculture in US Fordism. From the New Deal to economic restructuring', Sociologia Ruralis, 29: 131–48.

Keown, P.A. 1971, 'The career cycle and the stepwise migration process', New Zealand Geographer, 27: 175–84.

King, R.L. 1973, Land reform: the Italian experience (Butterworth, London).

King, R.L. 1976, 'Long-range migration patterns in the EEC: an Italian case study', pp. 108–25 in Lee, R. and Ogden, P.E. (eds), Economy and society in the EEC (Saxon House, Westmead).

King, R.L. 1979, 'The Maltese migration cycle: an archival survey', Area, 11: 245–9.

King, R.L. 1987a, *Italy* (Harper and Row, London).

King, R.L. 1987b, 'Italy', pp. 129–64 in Clout, H.D. (ed.), *Regional development in Western Europe* (David Fulton, London), 3rd edition.

King, R.L. and Burton, S. 1982, 'Land fragmentation: notes on fundamental rural spatial problems', *Progress in Human Geography*, 6: 475–94.

King, R.L. and Strachan, A.J. 1978, 'Sicilian agro-towns', *Erdkunde*, 32: 110–23.

King, R.L. and Strachan, A.J. 1980, 'Spatial variation in Sicilian migration: a stepwise multiple regression analysis', *Mediterranean Studies*, 2: 62–87.

King, R.L., Strachan, A.J. and Mortimer, J. 1983, 'Return migration: a review of the literature', *Discussion Papers, Department of Geography, Oxford Polytechnic*, 20.

King, R.L. and Took, L.J. 1983, 'Land tenure and rural social change: the Italian case', *Erdkunde*, 37: 186–98.

Kinsey, B.H. 1987, *Agribusiness and rural enterprise* (Croom Helm, London).

Klein, R. 1983, *The politics of the National Health Service* (Longman, London).

Knowles, D.J. 1987, 'Agricultural policy and change in the rural economy: the case of the dairy sector (UK)', pp. 45–54 in Lockhart, D.G. and Ilbery, B.W. (eds), *The future of the British rural landscape* (Geo Books, Norwich).

Knox, P.L. 1987, *Urban social geography: an introduction* (Longman, London).

Knox, P.L. 1988, 'The economic organization of US space', pp. 111–50 in Knox, P.L., Bartels, E.H., Holcomb, B., Bohland, J.R. and Johnston, R.J., *The United States: a contemporary geography* (Longman, London).

Knox, P.L. and Cullen, J.D. 1981, 'Town planning and the internal survival mechanisms of urbanised capital', *Area*, 13: 183–8.

Knox, P.L. and Cottam, B. 1981, 'Rural deprivation in Scotland: a preliminary assessment', *Tijdschrift Voor Economische en Sociale Geografie*, 72: 162–75.

Knox, P.L. and Pacione, M. 1980, 'Locational behaviour, place preferences and the inverse care law in the distribution of primary medical care', *Geoforum*, 11: 43–55.

Knudson, D.M. 1976, 'A system for evaluating scenic rivers', *Water Resources Bulletin*, 12: 281–90.

Kohl, R.L. and Uhl, J.N. 1985, *Marketing of agricultural products* (Macmillan, New York).

Kontuly, T. and Vogelsang, R. 1988, 'Explanations for the intensification of counterurbanisation in the Federal Republic of Germany', *Professional Geographer*, 40: 42–53.

Kontuly, T., Wiord, S. and Vogelsang, R. 1986, 'Counterurbanization in the Federal Republic of Germany', *Professional Geographer*, 38: 170–81.

Kreuger, R.R. 1977, 'The destruction of a unique renewable resource: the case of the Niagara Fruit Belt', pp. 132–48 in Kreuger, R.R. and Mitchell, B. (eds), *Mapping Canada's renewable resources,* (Methuen, London).

Kreuger, R.R. 1978, 'Urbanization of the Niagara Fruit Belt', *Canadian Geographer*, 22: 179–94.

Kreuger, R.R. 1984, 'The struggle to preserve speciality crop land in the rural-urban fringe of the Niagara peninsula of Ontario', pp. 292–313 in Bunce, M.F. and Troughton, M.J. (eds), 'The pressures of change in rural Canada',

Geographical Monographs, Department of Geography, Atkinson College, York University, No. 14.

Krocher, U. 1953, 'Die sozialgeographische entwicklung der funf feldbergdorfer im Taunus in den letzen 150 jahren', *Rhein-Mainische Forschungen,* 37.

Kroeber, A.L. 1948, *Anthropology* (Harrap, London).

Krout, J.A. 1988, 'The elderly in rural environments', *Journal of Rural Studies,* 4: 103–14.

Krutilla, J.V. and Fisher, A.C. (eds) 1975, *The economics of natural environments* (Johns Hopkins Press, Baltimore).

Kunnecke, B.H. 1974, 'Sozialbrache – a phenomenon of the rural landscape of Germany', *Professional Geographer,* 26: 412–15.

Kunst, F. 1985, 'Siedlungstruktur, distanz und lebensraum', *Erdkunde,* 39: 307–16.

Labasse, J. 1961, 'Structures et paysages nouveaux en Allemagne du Sud', *Revue de Geographie Lyonnaise,* 2: 93–116.

Lands Directorate, 1980, 'CLUMP', *Discussion Papers, Land Use Monitoring Division* (Environment Canada, Ottawa).

Lands Directorate, 1985a, 'Urbanization of rural land in Canada', *Land use change in Canada, Fact Sheets,* No. 85–4 (Environment Canada, Ottawa).

Lands Directorate, 1985b, 'Victoria urban-centred region 1976–1980', *Land use change in Canada, Fact Sheets,* No. 85–6 (Environment Canada, Ottawa).

Lands Directorate, 1986, *Urbanization of rural land in Canada: Supplementary tables for fact sheet 85–4* (Environment Canada, Ottawa).

Lang, M. 1986, 'New urban and rural definitions for the United States Census of Population', *Regional Studies,* 20: 77–83.

Lapping, M.B. and Clemenson, H.A. 1984, 'Recent developments in North American rural planning', *Countryside Planning Yearbook,* 5: 42–61.

Lapping, M.B. and Forster, V.D. 1982, 'Farmland and agricultural policy in Sweden: an integrated approach', *International Regional Science Review,* 7: 293–302.

Larsen, O.F. 1968, 'Rural society', pp. 580–8 in Sills, D.L. (ed.), *International encyclopaedia of the social sciences* (Macmillan and Co. and the Free Press, USA).

Lassey, W.R. 1977, *Planning in rural environments* (McGraw-Hill, New York).

Lassey, W.R., Lapping, M.B. and Carlso, J.E. 1988, 'The USA', pp. 142–65 in Cloke, P.J. (ed.), *Policies and plans for rural people: an international perspective* (Unwin Hyman, London).

Law, C.M. and Warnes, A.M. 1975, 'Life begins at sixty: the increase in retirement migration', *Town and Country Planning,* 43: 531–4.

Law, C.M. and Warnes, A.M. 1976, 'The changing geography of the elderly in England and Wales', *Transactions of the Institute of British Geographers,* new series, 1: 453–71.

Law, C.M. and Warnes, A.M. 1982, 'The destination decision in retirement migration', pp. 53–82 in Warnes, A.M. (ed.), *Geographical perspectives on the elderly* (John Wiley, Chichester).

Lawrence, G. 1987, *Capitalism and the countryside: the rural crisis in Australia* (Pluto Press, Sydney).

Lawton, R. 1964, 'Historical geography: the Industrial Revolution', pp. 221–44 in Watson, J.W. and Sissons, J.B. (eds), *The British Isles: a systematic geography* (Nelson, London).

Lawton, R. 1986, 'Population', pp. 10–29 in Langton J. and Morris, R.J. (eds), *Atlas of industrializing Britain* (Methuen, London).

Layton, R.L. 1978, 'The operational structure of the hobby farm', *Area*, 10: 242–6.

Layton, R.L. 1979, 'Hobby Farming', *Geography*, 65: 220–3.

Layton, R.L. 1981, 'Attitudes of hobby and commercial farmers in the rural-urban fringe of London, Ontario', *Cambria*, 8: 33–44.

Leavis, F.R. and Thompson, D. 1932, *Culture and environment. The training of critical awareness* (Chatto and Windus, London).

Le Coz, J. 1984, 'Niveaux de décision et d'organisation dans l'espace rural français', *Cahiers de Fontenay*, 35: 41–52.

Lee, C. 1986, 'Regional structure and change', pp. 30–3 in Langton J. and Morris, R.J. (eds), *Atlas of industrializing Britain* (London and New York).

Lee, L. 1959, *Cider with Rosie* (The Hogarth Press, London).

Lefebvre, H. 1976, *The survival of capitalism* (Allison and Busby, London).

Lefebvre, H. 1977, 'Reflections on the politics of space', pp. 339–52 in Peet, R. (ed.), *Radical geography* (Maaroufa Press, Chicago).

Le Heron, R.B. 'Food and fibre production under capitalism: a conceptual agenda', *Progress in Human Geography*, 12: 409–30.

Le Heron, R.B. 1989, 'A political economy perspective on the expansion of New Zealand livestock farming, 1960–1984. Part 1: Agricultural policy; Part 2: Aggregate farmer responses – evidence and policy implications', *Journal of Rural Studies*, 5: 17–44.

Leonard, P. 1982, 'Management agreements: a tool for conservation', *Journal of Agricultural Economics*, 33: 351–60.

Leopold, L.B. 1969, 'Landscape aesthetics', *Natural History*, 78: 37–44.

Leopold, L.B. and Marchand, M.O. 1968, 'On the quantitative inventory of the riverscape', *Water Resources Research*, 4: 709–17.

Lewis, D. and Wallace, H. (eds) 1984, *Policies into practice – national and international case studies in implementation* (Heinemann, London).

Lewis, G.J. 1967, 'Commuting and the village in mid-Wales', *Geography*, 52: 294–304.

Lewis, G.J. 1979, *Rural communities – a social geography* (David and Charles, Newton Abbot).

Lewis, G.J. 1988, 'Counterurbanisation and social change in the rural South Midlands', *East Midlands Geographer*, 11: 3–12.

Lewis, G.J. and Maund, D.J. 1976, 'The urbanisation of the countryside: a framework for analysis', *Geografiska Annaler*, 58B: 17–27.

Lewis, J.R. 1989, *The village school* (Robert Hale, London).

Lewis, J.R. and Williams, A.M. 1981, 'Regional uneven development in the European periphery: the case of Portugal 1950–1978', *Tijdschrift Voor Economische en Sociale Geografie*, 72: 81–98.

Lewis, J.R. and Williams, A.M. 1985, 'The Sines project: Portugal's growth centre or white elephant?', *Town Planning Review*, 56: 339–66.

Lewis, J.R. and Williams, A.M. 1987, 'Productive decentralization or indigenous growth? Small manufacturing enterprises and regional development in Central Portugal', *Regional Studies,* 21: 343–61.

Ley, D.F. 1984, 'Inner city revitalization in Canada: a Vancouver case study', pp. 186–204 in Palen, J.J. and London, B. (eds), *Gentrification, displacement and neighbourhood revitalization* (State of New York Press, Albany, NY).

Ley, D.F. 1986, 'Alternative explanations for inner-city gentrification: a Canadian assessment', *Annals of the Association of American Geographers,* 76: 521–35.

Li, P.S. and MacLean, B.D. 1989, 'Changes in the rural elderly population and their effects on the small town economy: the case of Saskatchewan 1971–1986', *Rural Sociology,* 54: 213–26.

Lichfield, N. and Marinov, U. 1977, 'Land-use planning and environmental protection: convergence or divergence?', *Environment and Planning A,* 9: 985–1002.

Lichter, D.T. 1989, 'The underemployment of American rural women: prevalence, trends and spatial inequality', *Journal of Rural Studies,* 5: 199–208.

Lichter, D.T., Fuguitt, G.V. and Heaton, T.B. 1985, 'Components of non-metropolitan population change: the contribution of rural areas', *Rural Sociology,* 50: 88–98.

Liddle, M.J. and Scorgie, H.R.A. 1980, 'The effects of recreation on freshwater plants and animals. A review', *Biological Conservation,* 17: 183–206.

Linton, D. 1968, 'The assessment of scenery as a natural resource', *Scottish Geographical Magazine,* 84: 219–38.

Lipietz, A. 1980, 'The structuration of space: the problem of land and spatial policy', pp. 60–75 in Carney, J.G., Lewis, J.R. and Hudson, R. (eds), *Regions in crisis* (St Martin's Press, New York).

Little, J.K. 1986, 'Guest editorial: Feminist perspectives in rural geography – an introduction', *Journal of Rural Studies,* 2: 1–8.

Little, J.K. 1987, 'Gender relations in rural areas: the importance of women's domestic role', *Journal of Rural Studies,* 3: 335–42.

Littlejohn, G. 1977, 'Chayanov and the theory of peasant economy', in Hindess, B. (ed.), *Sociological theories of the economy* (Macmillan, London).

Litton, R.B. 1968, 'Forest landscape description and inventories: a basis for land planning and design', *USDA Forest Service Research Papers,* No. PSW-49 (Pacific Southwest Forest and Range Experiment Station, Berkeley, Ca.).

Litton, R.B. 1972, 'Aesthetic dimensions of the landscape', pp. 262–91 in Krutilla, J.V. (ed.), *Natural environments: studies in theoretical and applied analysis* (Johns Hopkins Press, Baltimore).

Litton, R.B. 1974, 'Visual vulnerability of forest landscapes', *Journal of Forestry,* 72: 392–7.

Lockwood, D. 1964, 'Social integration and system integration', pp. 244–57 in Zollschan, C.K. and Hirsch, W. (eds), *Exploration in social change* (Routledge and Kegan Paul, London).

London, B. 1980, 'Gentrification as urban reinvasion: some preliminary definitional and theoretical considerations', in Leska, S.B. and Spain, D. (eds), *Back to the city* (Pergamon Press, New York).

Long, L.H. and De Are, D. 1980, *Migration to nonmetropolitan areas: apprais-ing the trends and reasons for moving* (Special Demographic Analyses, US Bureau of the Census, Washington, DC).

Long, L.H. and De Are, D. 1983, 'The slowing of urbanisation in the United States', *Scientific American*, 249: 31–9.

Long, L. and Frey, W. 1982, *Migration and settlement: United States* (Inter-national Institute for Applied Systems Analysis, Laxenburg, Austria).

Lonsdale, R.E. 1981, 'Industry's role in nonmetropolitan economic development and population change', pp. 129–48 in Roseman, C.C. (ed.), *Population redistribution in the Midwest* (North Central Regional Center for Rural Development, Ames, Iowa).

Lonsdale, R.E. 1985, 'Industrialization of the countryside: the case of the United States', pp. 161–72 in Healey, M.J. and Ilbery, B.W. (eds), *The industrializa-tion of the countryside* (Geo Books, Norwich).

Lonsdale, R.E. and Enyedi, G. (eds) 1984, *Rural public services: international comparisons* (Westview Press, Boulder, Col.).

Lonsdale, R.E. and Holmes, J.H. (eds) 1981, *Settlement systems in sparsely populated regions: the United States and Australia* (Pergamon, New York).

Lonsdale, R.E. and Seyler, H.L. 1979, *Nonmetropolitan industrialization* (V.H. Winston and Sons, Washington, DC).

Loomis, C.P. 1957, *Community and society* (Michigan State University Press, Ann Arbor).

Lopreato, J. 1967, *Peasants no more: social class and social change in an underdeveloped society* (Chandler Publishing Co., San Francisco).

Louwes, S.L. 1977, 'Inhibitors to change in agriculture: is a pluridisciplinary approach needed?', *European Review of Agricultural Economics*, 4: 271–98.

Lowe, P., Cox, G., MacEwan, M., O'Riordan, T. and Winter, M. 1986, *Coun-tryside conflicts: the politics of farming, forestry and conservation* (Gower/Maurice Temple Smith, Aldershot).

Lowe, P. and Goyder, J. 1982, *Environmental groups in politics* (George Allen and Unwin, London).

Lowenthal, D. 1968, 'The American scene', *Geographical Review*, 58: 61–88.

Lowenthal, D. 1975, 'The place of the past in the American landscape', pp. 89–118 in Lowenthal, D. and Bowden M.J. (eds), *Geographies of the mind: essays in historical geography* (Oxford University Press, New York).

Lowenthal, D. 1978, 'Finding valued landscapes', *Progress in Human Geography*, 2: 373–418.

Lowenthal, D. and Prince, H.C. 1964, 'The English landscape', *Geographical Review*, 54: 309–46.

Lowenthal, D. and Prince, H.C. 1965, 'English landscape tastes', *Geographical Review*, 55: 186–222.

Lowenthal, D. and Prince, H.C. 1976, 'Transcendental experience', pp. 117–31 in Wapner, S., Cohen, S.B. and Kaplan, B. (eds), *Experiencing the environ-ment* (Plenum Press, New York).

Low Pay Unit, 1984, *The problems of low pay* (Evidence to the House of Commons Employment Committee on Low Pay, June 1984, HMSO, London).

Lucey, D.I.F. and Kaldor, D.R. 1969, *Rural industrialization: the impact of industrialization on two rural communities in Western Ireland* (Chapman, London).

Lumb, R.A. 1980, 'Migration in the Highlands and Islands of Scotland', *Research Papers, Institute of the Study of Sparsely Populated Areas,* No. 3, (University of Aberdeen).

Lumb, R.A. 1981, 'A community based approach to the analysis of migration in the Highlands and Islands of Scotland', *Sociological Review,* 28: 611–27.

Lupton, T. and Mitchell, D. 1954, 'The Liverpool Estate', in Department of Social Science, University of Liverpool, *Neighbourhood and Community: an enquiry into social relationships on housing estates* (Liverpool University Press, Liverpool).

Lyson, T.A. 1984, 'Pathways into production agriculture: the structuring of farm recruitment in the United States', pp. 79–103 in Schwarzweller, H.K. (ed.), *Research in rural sociology and development, volume 1* (JAI Press, Greenwich, Ct.).

McCallum, J.D. and Adams, J.G.L. 1980, 'Charging for countryside recreation: a review with implications for Scotland', *Transactions of the Institute of British Geographers,* new series, 5: 350–68.

McCarthy, K.F. and Morrison, P.A. 1977, 'The changing demographic and economic structure of non-metropolitan areas', *International Regional Science Review,* 2: 123–42.

McCarthy, K.F. and Morrison, P.A. 1979, *The changing demographic and economic structure of nonmetropolitan areas in the United States* (Rand Corporation, Santa Monica, Ca.).

McCaskill, M. 1964, 'Population changes by migration 1956–1961', *New Zealand Geographer,* 20: 74–87.

McCleery, A. 1988, 'The Highland Board reviewed: a note on the analysis of economic and social change', *Scottish Geographical Magazine,* 104: 171–5.

McDermott, D. and Horner, A. 1978, 'Aspects of rural renewal in western Connemara', *Irish Geography,* 11: 176–9.

McEwan, A. and M. 1987, *Greenprints for the countryside? The story of Britain's National Parks* (Allen and Unwin, London).

MacEwan, M. and Sinclair, G. 1983, *New life for the hills* (Council for National Parks, London).

McEwan, M. and A. 1981, *National Parks* (George Allen and Unwin, London).

McEwen, J. 1981, *Who owns Scotland?* (Polygon Books, Edinburgh).

MacFarlane, A. 1978, *The origins of English individualism: family, property and social transition* (Basil Blackwell, Oxford).

McGranahan, D. 1980, 'The spatial structure of income distribution in rural regions', *American Sociological Review,* 45: 313–24.

MacGregor, B.D. 1978, 'The Highland problem: the difficulties of development in a remote rural area', unpublished M.Sc. thesis, Heriot-Watt University.

MacGregor, B.D., Robertson, D.A. and Shucksmith, M. (eds) 1987, *Rural housing in Scotland: recent research and policy* (Aberdeen University Press, Aberdeen).

Mackay, G.A. and Moir, A.C. 1980, 'Norwegian policies for assisting rural shops', pp. 21–6 in Mackay, G.A. and Moir, A.C. (eds), *Rural Scotland price survey Autumn 1980* (Institute for the Study of Sparsely Populated Areas, University of Aberdeen, Aberdeen).

MacKenzie, J. 1983, *Rural planning in France: the evolution of sub-regional planning procedures* (Department of Town Planning, Polytechnic of the South Bank).

McLaughlin, B.P. 1976, 'Rural settlement planning: a new approach', *Town and Country Planning*, 44: 156–60.

McLaughlin, B.P. 1983, *Country crisis: the lid off the chocolate box* (Channel 4 TV, Plymouth).

McLaughlin, B.P. 1985, 'Assessing the extent of rural deprivation', *Journal of Agricultural Economics*, 36: 77–80.

McLaughlin, B.P. 1986a, 'Rural policy in the 1980s: the revival of the rural idyll', *Journal of Rural Studies*, 2: 81–90.

McLaughlin, B.P. 1986b, 'The rhetoric and the reality of rural deprivation', *Journal of Rural Studies*, 2: 292–307.

McLaughlin, B.P. 1987, 'Rural policy into the 1990s – self help or self deception?', *Journal of Rural Studies*, 3: 361–4.

MacLeary, A.R. 1981, 'Rural planning: problems and policies', *Journal of Agricultural Economics*, 32: 317–29.

Macmillan, J.A. and Graham, J.D. 1978, 'Rural development planning: a science?', *American Journal of Agricultural Economics*, 60: 945–9.

McNeill, W.H. 1978, *The metamorphosis of Greece since World War Two* (University of Chicago Press, Chicago).

Macpherson, I. 1988, 'Farming and forestry – a study in separate development', *Bulletin of the Scottish Association of Geography Teachers*, 17: 51–60.

McQuin, P. 1978, *Rural retreating: a review of an Australian case study* (Department of Geography, University of New England, Armidale).

Maas, J.H.M. 1983, 'The behaviour of landowners as an explanation of regional differences in agriculture: latifundists in Sevilla and Cordoba (Spain)', *Tijdschrift Voor Economische en Sociale Geografie*, 74: 87–95.

Maher, C.A. 1982, *Austrialian cities in transition* (Shillington House, Melbourne).

Maisel, S.J. 1953, *Housebuilding in transition* (University of California Press, Berkeley).

Majumdar, D.M. 1958, *Caste and communication in an Indian village* (Asia Publishing House, Bombay).

Malecki, E.J. 1986, 'Technical imperatives and modern corporate strategy', pp. 67–79 in Scott, A.J. and Storper, M. (eds), *Production, work and territory* (Allen and Unwin, Boston).

Maltby, E. 1986, *Waterlogged wealth: why waste the world's wet places?* (Earthscan, London).

Malzahn, M. 1979, 'B.C.'s green acres: a look at the future of farmland in B.C.', *Urban Reader*, 7: 14–19.

Mann, S.A. and Dickinson, J.M. 1978, 'Obstacles to the development of a capitalist agriculture', *Journal of Peasant Studies*, 5: 466–81.

Manners, I.R. 1974, 'The environmental impact of modern agricultural technologies', pp. 181–212 in Manners, I.R. and Mikesell, M.W. (eds),

'Perspectives on environment', *Publications, Association of American Geographers, Commission on College Geography,* No. 13.

Manning, E.W. and Eddy, S.S. 1978, 'The Agricultural Land Reserves of British Colombia: an impact analysis', *Land Use in Canada Series, Lands Directorate, Environment Canada,* No. 13.

Maos, J.A. 1981, 'Land settlement in the Mediterranean region: lessons gleaned from Italy, Spain and Israel', *Ekistiks,* 48: 382–93.

Maos, J.A. and Prior, I. 1988, 'The spatial organisation of rural services: an operational model for regional development planning', *Applied Geography,* 8: 65–80.

Marlow, J. 1971, *The Tolpuddle martyrs* (Deutsch, London).

Marsden, T.K. 1984, 'Capitalist farming and the farm family', *Sociology,* 18: 205–24.

Marsden, T.K. 1988, 'Exploring political economy approaches in agriculture', *Area,* 20: 315–22.

Marsden, T.K., Munton, R.J.C., Whatmore, S.J. and Little, J.K. 1989, 'Strategies for coping in capitalist agriculture: an examination of the responses of farm families in British agriculture', *Geoforum,* 20: 1–14.

Marsden, T.K., Whatmore, S.J., Munton, R.J.C. and Little, J.K. 1986, 'The restructuring process and economic centrality in capitalist agriculture', *Journal of Rural Studies,* 2: 271–80.

Marsden, T.K., Whatmore, S.J. and Munton, R.J.C. 1987, 'Uneven development and the restructuring process in British agriculture: a preliminary exploration', *Journal of Rural Studies,* 3: 297–308.

Martin, I. 1976, 'Rural communities', pp. 49–84 in Cherry, G.E. (ed.), *Rural planning problems* (Leonard Hill, London).

Marx, K. 1961, *Capital – Volume 1* (Foreign Languages Publishing House, Moscow).

Marx, K. 1964, *Pre-capitalist economic formations* (International Publishers, New York).

Marx, K. 1976, *Capital: a critique of political economy, Volume 1* (Penguin, Harmondsworth).

Masser, I. 1980, 'The limits to planning', *Town Planning Review,* 51: 39–49.

Massey, D.B. 1977, 'The analysis of capitalist landownership: an investigation of the case of Great Britain', *International Journal of Urban and Regional Research,* 1: 404–24.

Massey, D.B. 1978, 'The ideology of locational analysis', pp. 5–13 in Anderson, J., Leach, B. and Williams, P. (eds), *New approaches to geography and planning* (Social Geography Study Group, Institute of British Geographers).

Massey, D.B. 1983, 'Industrial restructuring as class restructuring: production decentralization and local uniqueness', *Regional Studies,* 17: 73–90.

Massey, D.B. 1984, *Spatial divisions of labour: social structures and the geography of production* (Macmillan, London).

Massey, D.B. and Catalano, A. 1977, *Capital and land: land ownership by capital in Great Britain* (Edward Arnold, London).

Mather, A.S. 1978, 'Patterns of afforestation in Britain since 1945', *Geography,* 63: 157–66.

Mather, A.S. 1979, 'Land use changes in the Highlands and Islands, 1946–75', *Scottish Geographical Magazine*, 95: 114–22.

Mather, A.S. 1985, 'The rise and fall of government-assisted land settlement in Scotland', *Land Use Policy*, 2: 217–24.

Mather, A.S. 1986, *Land use* (Longman, London and New York).

Mather, A.S. 1987, 'The structure of forest ownership in Scotland: a first approximation', *Journal of Rural Studies*, 3: 175–82.

Mather, A.S. 1988, 'New private forests in Scotland: characteristics and contrasts', *Area*, 20: 135–43.

Mather, A.S. and Murray, N.C. 1986, 'Disposal of Forestry Commission land in Scotland', *Area*, 18: 109–16.

Mather, A.S. and Murray, N.C. 1987, 'Employment and private-sector afforestation in Scotland', *Journal of Rural Studies*, 3: 207–18.

Mather, A.S. and Murray, N.C. 1988, 'The dynamics of rural land-use change: the case of private sector afforestation in Scotland', *Land Use Policy*, 5: 103–20.

Mather, P.M. 1976, *Computational methods of multivariate analysis in physical geography* (John Wiley and Sons, London).

Maude, A.J.S. and Van Rest, D.J. 1985, 'The social and economic effects of farm tourism in the United Kingdom', *Agricultural Administration*, 20: 85–99.

Mendels, F.F. 1981, *Industrialisation and population pressure in eighteenth century Flanders* (Arno Press, New York).

Meredith, S.A. 1987, 'Agricultural production in Less Favoured Areas', *Journal of the Royal Agricultural Society*, 148: 36–45.

Mignon, C. 1971, 'L'agriculture à temps partiel dans le département du Puy-de-Dome', *Revue d'Auvergne*, 85: 1–41.

Mills, D.R. 1972, 'Has historical geography changed?', pp. 41–78 in Open University, *Political, historical and regional geography* (Units 13–15 in New Trends in Geography, Open University, Open University Press, Milton Keynes).

Mills, D.R. 1980, *Lord and peasant in nineteenth century Britain* (Croom Helm, London).

Mills, D.R. and Short, B.M. 1983, 'Social change and social conflict in nineteenth century England – the use of the open-closed village model', *Journal of Peasant Studies*, 10: 253–62.

Milner, H. 1952, 'The folk-urban continuum', *American Sociological Review*, 17: 529–37.

Ministry of Agriculture, Fisheries and Food (MAFF), 1987, *Food from our own resources* (Cmnd. 6020, HMSO, London).

Ministry of Housing, 1944, *The control of land use* (HMSO, London).

Ministry of Housing, 1948, *The design of dwellings* (HMSO, London).

Mitchell, B.W. 1989, *Geography and resource analysis* (Longman, Harlow), 2nd edition.

Mitchell, C.B. 1976, 'Some social aspects of public passenger transport', in Transport and Road Research Laboratory (TRRL), *Transport and Road Research Laboratory Symposium on unconventional bus services* (TRRL, Crowthorne, Berks.).

Mitchell, G.D. 1950, 'Social disintegration in a rural community', *Human Relations*, 3: 279–306.

Mitchell, G.D. 1951, 'The relevance of group dynamics to rural planning problems', *Sociological Review*, 43: 1–16.

Mitchell, G.F.C. 1973, 'Rural-urban preferences with respect to agricultural policy objectives', *Occasional Publications, Department of Economics, University of Bristol*.

Mitchelson, R.L. and Fischer, J.S. 1987, 'Long distance commuting and income change in the towns of upstate New York', *Economic Geography*, 63: 48–65.

Moller, J. 1985, 'The landed estate and the landscape: landownership and changing landscape in Southern Sweden during the 19th and 20th centuries', *Geografiska Annaler*, 67B: 45–52.

Monteiro, J.C. and Malheiro, H. 1983, 'O ambiente economico e a sua repercursao na actividade empresarial: o caso Portugues', *Journadas de Gestao* (Faculdade de Economia, Coimbra).

Mooney, P.H. 1982, 'Labour time, production time, and capitalist development in agriculture', *Sociologia Ruralis*, 22: 279–92.

Moran, W. 1974, 'Systems of agriculture: regional patterns and locational influences', pp. 123–49 in Johnston, R.J. (ed.), *Society and environment in New Zealand* (Whitcombe and Tombs, Christchurch).

Moran, W. 1978, 'Land value, distance and productivity on the Auckland periphery', *New Zealand Geographer*, 34: 85–96.

Moran, W. 1979, 'Spatial patterns of agriculture on the urban periphery: the Auckland case', *Tijdschrift Voor Economische en Sociale Geografie*, 70: 164–76.

Morris, J. and Hess, T.M. 1986, 'Farmer uptake of agriculture land drainage benefits', *Environment and Planning A*, 18: 1649–64.

Morrison, P.A. and Abrahams, A. 1982, *Is population decentralization lengthening commuting distances?* (Rand Corporation, Santa Monica, Ca.).

Moseley, M.J. 1973a, 'The impact of growth centres in rural regions', *Regional Studies*, 7: 57–94.

Moseley, M.J. 1973b, 'Some problems of small expanding towns', *Town Planning Review*, 44: 263–78.

Moseley, M.J. 1974, *Growth centres in spatial planning* (Pergamon, Oxford).

Moseley, M.J. 1977, 'A look at rural transport and accessibility', *The Village*, 23: 33–5.

Moseley, M.J. (ed.) 1978, *Social issues in rural Norfolk* (University of East Anglia, Norwich).

Moseley, M.J. 1979, *Accessibility: the rural challenge* (Methuen, London).

Moseley, M.J. 1980, 'Is rural deprivation really rural?', *The Planner*, 66: 97.

Moseley, M.J. 1984, 'The revival of rural areas in advanced economies: a review of some causes and consequences', *Geoforum*, 15: 447–56.

Moseley, M.J. 1987, 'The Waveney Project: the role of the catalyst in rural community development', *Social Work Monographs, University of East Anglia*.

Moseley, M.J. and Darby, J. 1978, 'The determinants of female activity rates in rural areas. An analysis of Norfolk parishes', *Regional Studies*, 12: 297–309.

Moseley, M.J. Harman, R.O., Coles, O.B. and Spencer, M.B. 1977, *Rural transport and accessibility*, 2 vols (Centre of East Anglian Studies, University of East Anglia, Norwich).

Moseley, M.J. and Packman, J. 1983, 'Mobile services and the rural accessibility problem', pp. 79–86 in Clark, G., Groenendijk, J. and Thissen, F. (eds), *The changing countryside. Proceedings of the First Anglo-Dutch Symposium of Rural Geography* (Geo Books, Norwich).

Moseley, M.J. and Spencer, M.B. 1978, 'Access to shops: the situation in rural Norfolk', pp. 33–44 in Moseley, M.J. (ed.), *Social issues in rural Norfolk* (University of East Anglia, Norwich).

Mougenot, C. 1982, 'Les mécanismes sociaux de la "rurbanisation"', *Sociologia Ruralis*, 22: 264–78.

Mountjoy, A.B. 1966, 'Industrial development in Apulia', *Geography*, 51: 369–72.

Mountjoy, A.B. 1970, 'Industrial development in eastern Sicily', *Geography*, 55: 441–4.

Mountjoy, A.B. 1973, *Problem regions of Europe – The Mezzogiorno* (Oxford University Press, London).

Mrohs, E, 1983, 'Zur sozialen lage der nebenerwerbslandwirte in der Bundesrepublik Deutschland 1980', *Sociologia Ruralis*, 23: 28–49.

Muller, P.D. 1976, *The outer city: geographical consequences of the urbanization of suburbs* (American Association of Geographers, Washington, DC).

Mumford, L. 1973, *The fall of megalopolis* (Penguin, Harmondsworth).

Munton, R.J.C. 1977, 'Financial institutions: their ownership of agricultural land in Great Britain', *Area*, 9: 29–37.

Munton, R.J.C. 1981, 'Agricultural land use in the London Green Belt', *Town and Country Planning*, 50: 17–19.

Munton, R.J.C. 1983, *London's Green Belt: Containment in practice* (George Allen and Unwin, London).

Munton, R.J.C. 1985, 'Investment in British agriculture by the financial institutions', *Sociologia Ruralis*, 25: 155–73.

Munton, R.J.C. 1988, 'Agricultural change and technological development: tendencies apart?', paper delivered to the Institute of British Geographers Annual Conference, Loughborough University of Technology.

Munton, R.J.C., Whatmore, S.J. and Marsden, T.K. 1988, 'Reconsidering urban-fringe agriculture: a longitudinal analysis of capital restructuring on farms in the Metropolitan Green Belt', *Transactions of the Institute of British Geographers*, new series, 13: 324–36.

Munton, R.J.C., Whatmore, S.J. and Marsden, T.K. 1989, 'Part-time farming and its implications for the rural landscape: a preliminary analysis', *Environment and Planning A*, 21: 523–36.

Myrdal, G.M. 1957, *Economic theory and under-developed regions* (G. Duckworth, London).

Nash, R., Williams, H. and Evans, M. 1976, 'The one-teacher school', *British Journal of Educational Studies*, 24: 12–32.

Nature Conservancy Council (NCC), 1978, *The endless village* (NCC, Peterborough).

NCC, 1986, *Nature conservation in rural development. The need for new thinking about rural sector policies* (NCC, Peterborough).

NCC, 1987, *Birds, bogs and forestry – the peatlands of Caithness and Sutherland* (NCC, Peterborough).

NCC, 1988, *The Flow Country – the peatlands of Caithness and Sutherland* (NCC, Peterborough).

NCC, 1989, *The effects of agricultural land use change on the flora of three grazing marsh areas* (NCC, Peterborough).

Naylon, J. 1959, 'Land consolidation in Spain', *Annals of the Association of American Geographers*, 49: 361–73.

Naylon, J. 1961, 'Progress in land consolidation in Spain', *Annals of the Association of American Geographers*, 51: 335–8.

Naylon, J. 1966, 'The Badajoz Plan; an example of land settlement and regional development in Spain', *Erdkunde*, 20: 44–60.

Naylon, J. 1967, 'Irrigation and internal colonization in Spain', *Geographical Journal*, 133: 178–91.

Naylon, J. 1973, 'An appraisement of Spanish irrigation and land settlement policies since 1939', *Iberian Studies*, 2: 2–17.

Naylon, J. 1987, 'Iberia', pp. 383–418 in Clout, H.D. (ed.), *Regional development in Western Europe* (David Fulton Publishers, London).

Naylor, E.L. 1976, 'Les réformes sociales et structurales de l'agriculture dans le Finistere: l'activité de FASASA de 1962 à 1973', *Etudes Rurales*, 62: 89–111.

Naylor, E.L. 1981, 'Farm structure policy in north-east Scotland', *Scottish Journal of Politics*, 28: 266–72.

Naylor, E.L. 1982, 'Retirement policy in French agriculture', *Journal of Agricultural Economics*, 33: 25–36.

Naylor, E.L. 1983, 'Peripheral industrialisation – the case of Brittany', *Scottish Geographical Magazine*, 99: 101–10.

Naylor, E.L. 1985, 'Socio-structural policy in French agriculture', *O'Dell Memorial Monographs, Department of Geography, University of Aberdeen*, No. 18: 1–180.

Naylor, E.L. 1986, 'Milk quotas in France', *Journal of Rural Studies*, 2: 153–61.

Naylor, E.L. 1987, 'EEC dairy policy', *Geography*, 72: 239–41.

Neate, S. 1981, *Rural deprivation: an annotated bibliography* (Geo Books, Norwich), 2nd edition.

Neimanis, V.P. 1979, 'Canada's cities and their surrounding land resources', *Canada Land Inventory Reports*, No. 15 (Lands Directorate, Environment Canada, Ottawa).

Nelson, J.G. 1970, 'Man and landscape change in Banff National Park: a national park problem in perspective', pp. 63–98 in Nelson, J.G. (ed.), *Canadian parks in perspective* (Harvest House, Montreal).

Nelson, J.G. 1973, 'Canadian National Parks and related reserves: research needs and management', pp. 348–79 in Nelson, J.G. et al., (eds), *Canadian public land-use in perspective* (Social Science Research Council, Ottawa).

Nelson, J.G. 1987, 'National Parks and Protected Areas, National Conservation Strategies and sustainable development', *Geoforum*, 18: 291–319.

Nelson, J.G. and Butler, R.W. 1974, 'Recreation and the environment', pp. 290–

310 in Manners, I.R. and Mikesell, M.W. (eds), 'Perspectives on environment', *Association of American Geographers, Commission on College Geography*, No. 13.

Nelson, J.G., Needham, R.D. and Mann, D.L. (eds) 1978, 'International experience with National Parks and related reserves', *Occasional Publications, Department of Geography, University of Waterloo*, No. 12.

Neville-Rolfe, E. 1973, *Food and agriculture in the Common Market* (European Research Bureau, Oxford).

Newby, H. 1977, *The differential worker: a study of farm workers in East Anglia* (Allen Lane, London).

Newby, H. 1980a, 'Trend report: rural sociology', *Current Sociology*, 28: 1–141.

Newby, H. 1980b, *Green and pleasant land? Social change in rural England* (Penguin, Harmondsworth).

Newby, H. 1982, 'Rural sociology and its relevance to the agricultural economist – a review', *Journal of Agricultural Economics*, 33: 125–65.

Newby, H. 1983, 'The sociology of agriculture: towards a new rural society', *Annual Review of Sociology*, 9: 67–81.

Newby, H. 1986, 'Locality and rurality: the restructuring of rural social relations', *Regional Studies*, 20: 209–16.

Newby, H. 1987a, 'Emergent views in theories of agrarian development', *Arkleton Research Occasional Papers*, 2 (Arkleton Trust, Oxford).

Newby, H. 1987b, *Country life: a social history of rural England* (Wiedenfeld and Nicolson, London).

Newby, H. 1988a, review of Armstrong, 'Farmworkers, a social and economic history, 1770–1980', *Times Higher Education Supplement*.

Newby, H. 1988b, *The countryside in question* (Hutchinson, London).

Newby, H., Bell, C., Rose, D. and Saunders, P. 1978, *Property, paternalism and power: class and control in rural England* (Hutchinson, London).

Nicholls, J.R. 1978, *The impact of the EC on the UK food industry* (Wilton Publications, Farnborough, Hants.).

Nicholson, B. 1975, 'Return migration to a marginal rural area – an example from northern Norway', *Rural Sociology*, 24: 227–44.

Nicholson, M. 1987, *The new environmental age* (Cambridge University Press, Cambridge).

Nidenberg, S. 1978, 'French agriculture in the EEC', *Journal of Agricultural Economics*, 29: 325–34.

Nix, J.S. 1983, *Farm management pocketbook* (Wye College, University of London), 13th edition.

Northcott, P.L. 1981, 'Canada: a forest nation', pp. 3–8 in Mullins, E.J. and McKnight, T.S. (eds), *Canadian woods, their properties and uses* (University of Toronto Press, Toronto) 3rd edition.

Nurminen, E. 1981, 'The spatial structure and process of the primary school network in Kanta-Hame, Finland', *Acta Universitatis Tamperensis*, series A, 129.

Nurminen, E. and Robinson, G.M. 1985, 'Demographic changes and planning initiatives in Scotland's Northern and Western Isles', *Research Discussion*

Papers, Department of Geography, University of Edinburgh, No. 20.

Nutley, S.D. 1979, 'Patterns of regional accessibility in the north-west Highlands and Islands', *Scottish Geographical Magazine,* 95: 142–54.

Nutley, S.D. 1980, 'The concept of isolation: a method of evaluation and a West Highland example', *Regional Studies,* 14: 111–23.

Nutley, S.D. 1983, *Transport policy appraisal and personal accessibility in rural Wales* (Geo Books, Norwich).

Nutley, S.D. 1984, 'Planning for rural accessibility provision: welfare, economy and equity', *Environment and Planning A,* 16: 357–76.

Nutley, S.D. 1985, 'Planning options for the improvement of rural accessibility: uses of the time-space approach', *Regional Studies,* 19: 37–50.

Nutley, S.D. 1988, '"Unconventional modes" of transport in rural Britain: progress to 1985', *Journal of Rural Studies,* 4: 73–86.

Office of Population Censuses and Surveys (OPCS), 1981, *Area mortality: decennial supplement 1969–73, England and Wales* (HMSO, London).

O'Brien, L. 1989, 'Rural housing in Ireland: a description and classification for the 1980s', *Irish Geography,* 22: 1–12.

O'Cinneide, M.S. 1987, 'The role of development agencies in peripheral areas with special reference to Udaras na Gaeltachta', *Regional Studies,* 21: 65–9.

O'Cinneide, M.S. and Keane, M.J. 1983, 'Employment growth and population change in rural areas: an example from the West of Ireland', *Irish Geography,* 16: 108–12.

O'Farrell, P.N. 1980, 'Multi-national enterprises and regional development: Irish evidence', *Regional Studies,* 14: 141–51.

O'Farrell, P.N. 1984, 'Components of manufacturing change in Ireland 1973–1981', *Urban Studies,* 21: 155–76.

O'Farrell, P.N. and Markham, J. 1975, 'Commuting costs and residential location: a process of urban sprawl', *Tijdschrift Voor Economische en Sociale Geografie,* 66: 66–74.

O'Flanagan, T.P. 1980, 'Agrarian structures in North West Iberia: responses and their implications for development', *Geoforum,* 1: 157–69.

O'Flanagan, T.P. 1982, 'Land reform and rural modernization in Spain', *Erdkunde,* 36: 48–53.

Ogden, P.E. 1980, 'Migration, marriage and the collapse of traditional peasant society in France', pp. 52–79 in White, P.E. and Woods, R.I. (eds), *The geographical impact of migration* (Longman, London).

Ogden, P.E. 1984, *Migration and geographical change* (Cambridge University Press, Cambridge).

Ogden, P.E. 1985, 'Counterurbanization in France: the 1982 population census', *Geography,* 70: 24–35.

O'Hagan, J. 1978, *Growth and adjustment in national agriculture* (Allanheld, Osmun and Co., Montclair, NJ).

Oldfield, R. 1979, *The effect of car ownership on bus patronage* (Transport and Road Research Laboratory, Crowthorne, Berks.).

O'Neill, P. 1989, 'National economic change and the locality', *Environment and Planning A,* 21: 666–70.

Ontario Ministry of Agriculture and Food (MAF), 1977, *Food land guidelines* (The Queen's Printer, Toronto).

OECD, 1978, *Agricultural policy in Sweden* (OECD, Paris).

OECD, 1983, *Urban statistics in OECD countries* (OECD, Paris).

O'Riordan, T. 1976, *Environmentalism* (Pion, London).

O'Riordan, T. 1979, 'Ecological studies and political decisions', *Environment and Planning A,* 11: 805-13.

O'Riordan, T. 1985, 'Research policy and review 6: future directions for environmental policy', *Environment and Planning A,* 17: 1431-46.

O'Riordan, T. 1987, 'Agriculture and environmental protection', *Geography Review,* 1: 35-40.

Orkney Islands Council (OIC), 1983, *Orkney Economic Review 1982* (OIC, Kirkwall).

Orwin, C.S. and Whetham, E.H. 1971, *History of British agriculture 1846-1914* (David and Charles, Newton Abbot).

Oscarsson, G. and Oberg, S. 1987, 'Northern Europe', pp. 319-52 in Clout, H.D. (ed.), *Regional development in Western Europe* (David Fulton, London), 3rd edition.

Outdoor Recreation Resources Review Commission (ORRRC), 1962, *Outdoor recreation for America* (US Government Printing Office, Washington, DC).

Oxfordshire County Council (OCC), 1976, *Local transport in Oxfordshire* (OCC, Oxford).

Oxley, P.R. 1976, *Dial-a-Ride in the UK: a general study symposium on unconventional bus services* (Transport and Road Research Laboratory, Crowthorne, Berks.).

Pacione, M. 1979, 'Second homes on Arran', *Norsk Geografisk Tidsskrift,* 33: 33-8.

Pacione, M. 1980, 'Differential quality of life in a metropolitan village', *Transactions of the Institute of British Geographers,* new series, 5: 185-206.

Pacione, M. 1982, 'The viability of smaller rural settlements', *Tijdschrift Voor Economische en Sociale Geografie,* 73: 149-61.

Pacione, M. 1983, *Progress in rural geography* (Croom Helm, London).

Pacione, M. 1984, *Rural geography,* (Harper and Row, London).

Packman, J. and Wallace, D. 1982, 'Rural services in Norfolk and Suffolk', pp. 155-75 in Moseley, M.J. (ed.), *Power, planning and people in rural East Anglia,* (University of East Anglia, Norwich).

Pahl, R.E. 1965a, 'Urbs in rure: the metropolitan fringe in Hertfordshire', *Geographical Papers, London School of Economics,* No. 2.

Pahl, R.E. 1965b, 'Class and community in English commuter villages', *Sociological Review,* 6: 5-23.

Pahl, R.E. 1967, 'The rural-urban continuum: a reply to Eugen Lupri', *Sociologia Ruralis,* 7: 21-9.

Pahl, R.E. 1968, 'The rural-urban continuum', pp. 263-305 in Pahl, R.E. (ed.), *Readings in Urban Sociology* (Pergamon Press, London).

Pahl, R.E. 1975, *Whose city?* (Penguin, Harmondsworth), 2nd edition.

Pahl, R.E. 1977, 'Managers, technical experts and the state', in Harloe, M. (ed.), *Captive cities* (John Wiley, London).

Palmer, C.J., Robinson, M.E. and Thomas, R.W. 1977, 'The countryside image: an investigation of structure and meaning', *Environment and Planning A*, 9: 739–49.

Palmer, R. 1955, 'Realtors as social gatekeepers: a study in social control', unpublished Ph.D. thesis, Yale University.

Parenteau, R. 1980, 'Le milieu peri-urban: l'example Montrealai', *Cahiers de géographie du Quebec*, 24: 249–76.

Park, C.C. 1982, 'The supply of and demand for water', pp. 129–46 in Johnston, R.J. and Doornkamp, J.C. (eds), *The changing geography of the United Kingdom* (Methuen, London and New York).

Parker, K. 1984, *A tale of two villages* (Peak Park Joint Planning Board, Bakewell).

Parson, D.J. no date, 'Rural gentrification: the influence of rural settlement planning policies', *Research Papers in Geography, University of Sussex*, No. 3.

Pearce, D.G. and Richez, G. 1987, 'Antipodean contrasts: National Parks in New Zealand and Europe', *New Zealand Geographer*, 43: 53–9.

Pearce, J. 1981, 'The Common Agricultural Policy', *Chatham House Papers*, No. 13 (Royal Institute of International Affairs, London).

Pearse, P.H. 1980, *The forest industries of British Columbia* (B.C. Tel Series in Business Economics, No. 1, Vancouver, BC).

Peet, R. 1983, 'Relations of production and the relocation of US manufacturing industry since 1960', *Economic Geography*, 59: 112–43.

Pellenberg, H. and Kok, J.A.A.M. 1985, 'Small and medium-sized innovative firms in the Netherlands' urban and rural regions', *Tijdschrift Voor Economische en Sociale Geografie*, 76: 242–52.

Penn, B.H.E. 1988, 'Recreational access to land in Scotland and British Columbia', unpublished Ph.D. thesis, University of Edinburgh.

Penning-Rowsell, E.C. 1975, 'Constraints on the application of landscape evaluation', *Transactions of the Institute of British Geographers*, 66: 149–55.

Penning-Rowsell, E.C. 1981, 'Fluctuating fortunes in gauging landscape value', *Progress in Human Geography*, 5: 25–41.

Penning-Rowsell, E.C. and Searle, G.H. 1977, 'The Manchester landscape evaluation method: a critical appraisal', *Landscape Research*, 2 (3): 6–11.

Pepper, D. 1984, *The social roots of environmentalism* (Croom Helm, London).

Percy, M.B. 1986, *Forest management and economic growth in British Columbia* (Economic Council of Canada, Minister of Supply and Services Canada, Ottawa).

Perrons, D.C. 1981, 'The role of Ireland in the new international division of labour: a proposed framework for regional analysis', *Regional Studies*, 16: 81–100.

Perry, M. 1987, 'Eligibility and access to small factories: a case study in Cornwall', *Journal of Rural Studies*, 3: 15–22.

Perry, P.J. 1969, 'Structural reform in French agriculture: the role of the SAFER', *Revue Geographie Montreal*, 23: 137–51.

Perry, R., Dean, K. and Brown, B. (eds), 1986, *Counterurbanisation: international case studies of socio-economic change in rural areas* (Geo Books, Norwich).

Persson, L.O. 1983, 'Part-time farming – cornerstone or obstacle in rural development?', *Sociologia Ruralis,* 23: 50–62.

Peterson, G.L. and Neumann, E.S. 1969, 'Modelling and predicting human response to the visual recreation environment', *Journal of Leisure Research,* 1: 219–37.

Pfeffer, M.J. 1989, 'The feminization of production on part-time farms in the Federal Republic of Germany', *Rural Sociology,* 54: 60–73.

Phillips, A. and Roberts, M. 1973, 'The recreation and amenity value of the countryside', *Journal of Agricultural Economics,* 24: 85–102.

Phillips, D.R. and Williams, A.M. 1981, 'Council house sales and village life', *New Society,* 58: 367–8.

Phillips, D.R. and Williams, A.M. 1982, *Rural housing and the public sector* (Gower, Aldershot).

Phillips, D.R. and Williams, A.M. 1984, *Rural Britain: a social geography* (Basil Blackwell, Oxford).

Photiadis, J.D. 1965, 'The position of the coffee house in the social structure of the Greek village', *Sociologia Ruralis,* 5: 45–56.

Photiadis, J.D. 1974, 'Religion in an Appalachian state', *Appalachian Center, Research Report,* No. 6.

Photiadis, J.D. 1976, 'Changes in the social organisation of Greek village', *Sociologia Ruralis,* 25: 25–40.

Pickvance, C. (ed.) 1976, *Urban sociology: critical essays* (Tavistock, London).

Pierce, J.T. 1981a, 'The B.C. Agricultural Land Commission: a review and evaluation', *Plan Canada,* 21: 48–56.

Pierce, J.T. 1981b, 'Conversion of rural land to urban: a Canadian profile', *Professional Geographer,* 33: 163–73.

Pierson, G.W. 1973, *The moving American* (Alfred Knopf, New York).

Pirie, G.H. 1979, 'Measuring accessibility: a review and prospect', *Environment and Planning A,* 11: 199–312.

Pitt-Rivers, J. 1976, 'Preface', pp. vii–x in Aceves, J.B. and Douglass, W.A. (eds), *The changing faces of rural Spain* (Schenkman Publishing Co., Cambridge, Mass.).

Plaut, T. 1980, 'Urban expansion and the loss of farmland in the United States: implications for the future', *American Journal of Agricultural Economics,* 62: 537–42.

Pope, C. 1984, 'Ronald Reagan and the limits of responsibility', *Sierra Club Bulletin,* 69, 3: 51–4.

Popenoe, D. 1980, 'Urban form in advanced societies – a cross national enquiry', in Ungerson, C. and Karn, V. (eds), *The consumer experience of housing* (Gower, London).

Porter, E. 1978, *Water management in England and Wales* (Cambridge University Press, Cambridge).

Porto, M. 1984, 'Portugal: twenty years of change', pp. 84–112 in Williams, A.M. (ed.), *Southern Europe transformed: political and economic change in Greece, Italy, Portugal and Spain* (Harper and Row, London).

Potter, C.A. 1983, *Investing in rural harmony* (World Wildlife Fund, Godalming).

Potter, C.A. 1986a, 'Processes of countryside change in lowland England', *Journal of Rural Studies*, 2: 187–95.

Potter, C.A. 1986b, 'The environmental effects of CAP reform', *Countryside Planning Yearbook*, 7: 76–88.

Preece, R.A. 1980, *An evaluation by the general public of scenic quality in the Cotswolds Area of Outstanding Natural Beauty: a basis for monitoring future change* (Department of Town Planning, Oxford Polytechnic, Oxford).

Preece, R.A. 1981, 'Patterns of development control in the Cotswolds Area of Outstanding Natural Beauty', *Research Papers, School of Geography, University of Oxford*, No. 27.

Prince, H.C. 1985, 'Landscape through painting, *Geography*, 69: 3–18.

Pryke, R.W.S. and Dobson, J.S. 1975, *The rail problem* (Martin Robertson, Oxford).

Punter, J.V. 1974, *The impact of exurban development on land and landscape in the Toronto-centred region (1954–1971)* (Canada Mortgage and Housing Corporation, Ottawa).

Punter, J.V. 1988, 'Planning control in France', *Town Planning Review*, 59: 159–82.

Pye-Smith, C. and Hall, C. 1987, *The countryside we want: a manifesto for the year 2000* (Green Books, Hartland).

Pye-Smith, C. and North, R. 1984, *Working the land: a new plan for a healthy agriculture* (Maurice Temple Smith, London).

Queen, S.A. and Carpenter, D.B. 1953, *The American city* (McGraw Hill, New York).

Radford, E. 1970, *The new villagers: urban pressure on rural areas in Worcestershire* (Cass, London).

Ramblers' Association, 1980, *Afforestation: the case against expansion* (Brief for the Countryside, No. 7, Ramblers' Association, London).

Radetzki, M. 1982, 'Regional development benefits of mineral projects', *Regional Policy*, 8: 193–200.

Raven, G.P. 1988, 'The development of the Portuguese agrarian reform', *Journal of Rural Studies*, 4: 35–43.

Ravenstein, E.G. 1885, 'The laws of migration', *Journal of the Royal Statistical Society*, 48: 167–235.

Rawlyk, G.A. and McDonald, J. 1979, 'Prince Edward Island and the problems of Confederation, 1967–1978', pp. 171–96 in Rawlyk, G.A. (ed.), *The Atlantic provinces and the problems of Confederation* (Breakwater Press, St John's).

Raynor, I. 1980, *Canadian public properties: the Canadian Mortgage and Housing Corporation Survey* (Supply and Services Canada, Ottawa).

Redclift, M. 1973, 'The effects of socio-economic changes in a Spanish pueblo on community cohesion', *Sociologia Ruralis*, 13: 1–14.

Redfield, R. 1941, *The folk culture of Yucatan* (University of Chicago Press, Chicago).

Redfield, R. 1947, 'The folk society', *American Journal of Sociology*, 52: 293–308.

Redfield, R. 1956, *Peasant society and culture,* (University of Chicago Press, Chicago).

Redfield, R. 1973, *Peasant society and culture: an anthropological approach to civilization* (University of Chicago Press, Chicago), 3rd edition.

Rees, A.D. 1950, *Life in a Welsh countryside* (University of Wales Press, Cardiff).

Rees, R. 1973, 'Geography and landscape painting: an introduction to a neglected field, *Scottish Geographical Magazine*, 89: 147–57.

Rees, R. 1976, 'Images of the Prairie: landscape painting and perception in the Western Interior of Canada', *Canadian Geographer*, 20: 259–78.

Reid, R. and Wilson, G. 1986, *Agroforestry in Australia and New Zealand: the growing of productive trees on farms* (Goddard and Dobson, Box Hill, Vic.), 3rd edition.

Reid, R.A. 1975, 'Simulation and evaluation of an experimental rural medical care delivery system', *Socio-Economic Planning Services*, 9: 111–19.

Reissman, L. 1964, *The urban process* (Free Press of Glencoe, London and New York).

Revelle, R. 1966, 'Outdoor recreation in a hyper-productive society', *Daedalus*, 96: 1172–91.

Richardson, H.W. 1971, *Urban economics* (Penguin, Harmondsworth).

Rickard, T.J. 1986, 'Problems in implementing farmland preservation policies in Connecticut', *Journal of Rural Studies*, 2: 197–207.

Riddell, M.A. 1986, 'EEC milk quotas – the experience of four member states', *Journal of the Royal Agricultural Society*, 147: 90–9.

Ritchie, W. 1967, 'The machair of South Uist', *Scottish Geographical Magazine*, 83: 161–73.

Robert, P. and Randolph, W.G. 1983, 'Beyond decentralization: the evolution of population distribution in England and Wales 1961–81', *Geoforum*, 14: 75–102.

Roberts, B.K. 1977, *Rural settlement in Britain* (Dawson, Folkstone).

Robin, J. 1980, *Elmdon: continuity and change in a north-west Essex village 1861–1964* (Cambridge University Press, Cambridge).

Robinson, D.A. and Blackman, J.D. 1989, 'Soil erosion, soil conservation and agricultural policy for arable land in the United Kingdom', *Geoforum*, 20: 83–92.

Robinson, G.M. 1983a, 'The evolution of the horticultural industry in the Vale of Evesham', *Scottish Geographical Magazine*, 99: 89–100.

Robinson, G.M. 1983b, 'West Midlands farming 1840s to 1970s: agricultural change in the period between the Corn Laws and the Common Market', *Occasional Publications, Department of Land Economy, University of Cambridge*, No. 15.

Robinson, G.M. 1986a, 'Migration: aspirations and reality amongst school-leavers in a small Canterbury town', *New Zealand Population Review*, 12: 218–34.

Robinson, G.M. 1986b, 'The expansion of the Wytch Farm oilfield', *Geography*, 71: 355–8.

Robinson, G.M. (ed.) 1987, 'A register of research in rural geography', *Occasional Papers, Department of Geography, University of Edinburgh*, No. 6.

Robinson, G.M. 1988a, *Agricultural change: geographical studies of British agriculture* (North British Publishing, Edinburgh).

Robinson, G.M. 1988b, 'The city beyond the city', pp. 221–42 in Robinson, G.M. (ed.), *A social geography of Canada: Essays in honour of J. Wreford Watson* (North British Publishing, Edinburgh).

Robinson, G.M. 1988c, 'Spatial changes in New Zealand's food processing industry, 1973–84', *New Zealand Geographer,* 44: 69–79.

Roemer, J.M. 1976, *Rural health care* (Mosby, St Louis).

Rogers, A.W. 1983, 'Housing', pp. 106–29 in Pacione, M. (ed.), *Progress in rural geography* (Croom Helm, Beckenham).

Rogers, A.W. 1985, 'Local claims on rural housing: a review', *Town Planning Review,* 56: 36–80.

Rogers, A.W. 1987, 'Voluntarism, self-help and rural community development: some current approaches', *Journal of Rural Studies,* 3: 353–60.

Rogers, A.W., Blunden, J. and Curry, N. (eds) for the Open University, 1985, *The countryside handbook* (Croom Helm, Beckenham, Kent).

Rogers, R. 1979, *Schools under threat: a handbook on closures* (Advisory Centre for Education, London).

Rohr-Zanker, R. 1989, 'A review of the literature on elderly migration in the Federal Republic of Germany', *Progress in Human Geography,* 13: 209–22.

Rose, D., Saunders, P., Newby, H. and Bell, C. 1979, 'The economic and political basis of rural deprivation: a case study', pp. 11–20 in Shaw, J.M. (ed.), *Rural deprivation and planning* (Geo Books, Norwich).

Roseman, C.C. and Crothers, C. 1984, 'Changing inter-regional migration patterns in New Zealand, 1966 to 1981', *New Zealand Geographer,* 32: 160–76.

Roseman, C.C. and Williams, J. 1980, 'Metropolitan to nonmetropolitan migration: a decision-making perspective', *Urban Geography,* 1: 283–94.

Rosenfeld, A. 1989, 'A new theory of class locations in US family farm agriculture and non-farm corporations', *Journal of Rural Studies,* 5: 45–60.

Roweis, S. and Scott, A. 1978, 'The urban land question', pp. 38–75 in Cox, K.R. (ed.), *Urbanization and conflict in market societies* (Methuen, London).

Rucker, G. 1984, 'Rural transportation: another gap in rural America', *Transport Quarterly,* 38: 419–32.

Russwurm, L.H. 1975, 'Urban fringe and urban shadow', pp. 148–64 in Bryfogle, R.C. and Kreuger, R.R. (eds), *Urban problems* (Holt, Rinehart and Winston, Toronto), revised edition.

Russwurm, L.H. 1977, *The surroundings of our cities* (Community Planning Press, Ottawa).

Rutledge, I. 1978, 'Land reform and the Portuguese revolution', *Journal of Peasant Studies,* 5: 79–98.

Rutter, M. and Madge, N. 1976, *Cycles of disadvantage: a review of research* (Heinemann, London).

Ryle, G.B. 1969, *Forest service* (David and Charles, Newton Abbot).

Sale, K. 1975, *Power shift* (Random House, New York).

Sarre, P. 1981, *Second homes: a case study in Brecknock* (Open University, Milton Keynes).

Saunders, P. 1986, *Social theory and the urban question* (Hutchinson, London), 2nd edition.

Saville, J. 1957, *Rural depopulation in England and Wales 1851–1951* (Routledge and Kegan Paul, London).

Savoie, D.J. 1989, 'Rural development in Canada: the case of north-east New Brunswick', *Journal of Rural Studies,* 5: 185–98.

Sayer, A. 1984a, *Method in social science* (Hutchinson, London).

Sayer, A. 1984b, 'Defining the urban', *Geojournal,* 9: 279–85.

Sayer, A. 1985, 'Realism in geography', pp. 159–73 in Johnston, R.J. (ed.), *The future of geography* (Methuen, London and New York).

Scargill, D.I. 1983, 'The ville moyenne', pp. 319–53 in Patten, J.H.C. (ed.), *The expanding city* (Academic Press, London).

Scherer, J. 1972, *Contemporary community: sociological illusion or reality?* (Tavistock Publications, London).

Schools Unit, University of Sussex, 1983, 'The Common Agricultural Policy', *Exploring Europe,* No. 2.

Schmitt, G. 1984, 'Part-time farming in the Federal Republic of Germany', *Land Use Policy,* 1: 25–33.

Schwarzweller, H.K. 1971, 'Tractorization of agriculture: the social history of a German village', *Sociologia Ruralis,* 11: 127–39.

Scott, J.K. 1970, 'Processing of New Zealand's food resources', pp. 51–60 in Bockemuel, H.W. (ed.), *New Zealand's wealth* (Manawatu Branch, New Zealand Geographical Society, Palmerston North).

Scott, A.J. and Roweis, S.T. 1977, 'Urban planning in theory and practice: a reappraisal', *Environment and Planning A,* 9: 1097–119.

Scottish Development Agency (SDA), 1989, *Rural Development in Scotland* (Rural Development Unit, SDA, Edinburgh).

Scottish Office, 1978, *Rural indicators research* (Scottish Office, Edinburgh).

Scottish Tourist Board, 1981, *Scottish leisure survey* (HMSO, Edinburgh).

Searle, G.H. 1981, 'The role of the state in regional development: the example of non-metropolitan New South Wales', *Antipode,* 13: 27–34.

Sedjo, R.A. 1987, 'Forest resources in the World: forests in transition', pp. 7–31 in Kallio, M., Dykstra, D.P. and Binkley, C.S. (eds), *The global forest sector: an analytical perspective* (John Wiley and Sons, Chichester).

Sedjo, R.A. and Radcliffe, S.J. 1981, *Postwar trends in US forest products trade* (Johns Hopkins Press, Baltimore).

Self, P. and Storing, H.J. 1962, *The state and the farmer* (George Allen and Unwin, London).

Sellgren, J.M.A. 1986, 'Areas of Outstanding Natural Beauty: a case for positive management', *Ecos,* 7: 29–33.

Sellgren, J.M.A. 1988, 'Planning policies in Areas of Outstanding Natural Beauty: the role of County and District Councils', pp. 17–23 in Hart, P. and Atherden, M. (eds), *The role of institutions in countryside management* (Rural Geography Study Group, Institute of British Geographers, College of Ripon and York St John, Harrogate).

Sewell, W.K.D. 1988, 'Getting to yes in the wilderness: the British Columbia experience in environmental policy-making', pp. 335–56 in Robinson G.M. (ed.), *A social geography of Canada: essays in honour of J. Wreford Watson* (North British Publishing, Edinburgh).

Shanin, T. 1971, 'Peasantry: a delineation of concept and a field of study', *European Journal of Sociology*, 12: 289–300.

Shanin, T. 1972, *The awkward class: a political sociology of peasantry in a developing society: Russia 1910–25* (Clarendon Press, Oxford).

Shanin, T. 1983, 'Defining peasants: conceptualizations and deconceptualizations', *Sociological Review*, 30: 407–32.

Shanin, T. 1987, *Peasants and peasant society* (Blackwell, Oxford).

Shannon, G.W. and Dever, G.E. 1974, *Health care delivery: spatial perspectives* (McGraw Hill, New York).

Shaw, J.M. 1976, 'Thresholds for village foodshops', in Jones, P. and Oliphant, R. (eds), *Local shops: problems and prospects* (Unit for Planning Retail Information, Reading).

Shaw, J.M. 1979, 'Rural deprivation and social planning: an overview', pp. 175–207 in Shaw, J.M. (ed.), *Rural deprivation and planning* (Geo Books, Norwich).

Shaw, J.M. and Stockford, R. 1979, 'The role of statutory agencies in rural areas: planning and social services', pp. 117–36 in Shaw, J.M. (ed.), *Rural deprivation and planning* (Geo Books, Norwich).

Sheehy, S.J. 1980, 'The impact of EEC membership on Irish agriculture', *Journal of Agricultural Economics*, 31: 297–310.

Sheial, J. 1979, 'The Restrictions of Ribbon Development Act: the character and perception of land-use control in inter-war Britain', *Regional Studies*, 13: 501–12.

Shoard, M. 1976, 'Recreation: the key to the survival of England's countryside?', pp. 58–73 in MacEwen, M. (ed.), *Future Landscapes* (Chatto and Windus, London).

Shoard, M. 1980, *The theft of the countryside* (Maurice Temple Smith, London).

Shoard, M. 1987, *This land is our land* (Maurice Temple Smith, London).

Shortridge, J.R. 1976, 'The collapse of frontier farming in Alaska', *Annals of the Association of American Geographers*, 66: 583–604.

Shucksmith, M. 1981, *No homes for locals* (Gower, Aldershot).

Shucksmith, M. 1983, 'Second homes: a framework for policy', *Town Planning Review*, 54: 174–93.

Shucksmith, M. 1984, *Scotland's rural housing: a forgotten problem* (Rural Forum, Perth).

Shucksmith, M. 1987, 'Rural housing in Scotland: the policy context', pp. 17–27 in MacGregor, B.D., Robertson, D.S. and Shucksmith, M. (eds), *Rural housing in Scotland: recent research and policy* (Aberdeen University Press, Aberdeen).

Shucksmith, M. 1988, 'Current rural land-use issues in Scotland: an overview', *Scottish Geographical Magazine*, 104: 176–80.

Sidaway, R. and Duffield, B.S. 1984, 'A new look at countryside recreation in the urban fringe', *Leisure Studies*, 3: 249–72.

Sillince, J.A.A. 1986, 'Why did Warwickshire's key settlement policy change in 1982? An assessment of the political implications of cuts in rural services', *Geographical Journal*, 152: 176–92.

Simmons, I.G. 1975, *Rural recreation in the industrial world* (Edward Arnold, London).

Simon, J. and Sudman, S. 1982, 'How much farmland is being converted to urban use? An analysis of Soil Conservation Service estimates', *International Regional Science Review*, 7: 257–72.

Sinclair, R.J. 1967, 'Von Thunen and urban sprawl', *Annals of the Association of American Geographers*, 57: 72–87.

Sinclair, P.R. and Westhues, K. 1975, *Village in crisis* (Holt, Rinehart and Winston, Toronto).

Sjoberg, G. 1960, *The pre-industrial city* (Free Press, New York).

Sjoholt, P. 1988, 'Scandinavia' pp. 69–97 in Cloke, P.J. (ed.), *Policies and plans for rural people: an international perspective* (Unwin Hyman, London).

Skinner, B.J. 1976, *Earth resources* (Prentice Hall, Englewood Cliffs, NJ).

Slater, M. 1984, 'Italy: surviving into the 1980s', pp. 61–83 in Williams, A. (ed.), *Southern Europe transformed: political and economic change in Greece, Italy, Portugal and Spain* (Harper and Row, London).

Slee, R.W. 1981, 'Agricultural policy and remote rural areas', *Journal of Agricultural Economics*, 32: 113–22.

Smailes, A.E. 1944, 'The urban hierarchy in England and Wales', *Geography*, 29: 41–51.

Smailes, P.J. 1979, 'The effects of changes in agriculture upon the service sector in South Australian country towns', *Norsk Geografiska Tiddskrift*, 33: 125–42.

Smailes, P.J. and Hugo, G.J. 1985, 'A process view of the population turnaround: an Australian rural case study', *Journal of Rural Studies*, 1: 31–44.

Smart, G. and Wright, S. 1983, *Decision-making for rural areas* (Bartlett School of Architecture and Planning, London).

Smit, B., Rodd, S., Bond, D., Brklacich, M., Conklin, C. and Dyer, A. 1983, 'Implications for food production potential of future urban expansion in Ontario', *Socio-Economic Planning Science*, 17: 109–20.

Smith, N. 1982, 'Gentrification and uneven development', *Economic Geography*, 58: 139–55.

Smith, N. 1984, *Uneven development* (Basil Blackwell, Oxford).

Smith, P.J. 1988, 'Community aspirations, territorial justice and the metropolitan form of Edmonton and Calgary', pp. 179–96 in Robinson G.M. (ed.), *A social geography of Canada: essays in honour of J. Wreford Watson* (North British Publishing, Edinburgh).

Smith, V. and Wilde, P. 1977, 'The multiplier impact of tourism in Tasmania', pp. 165–72 in Mercer, D. (ed.), *Leisure and recreation in Australia* (Sorrett Publishing, Melbourne).

Soper, M.H.R. 1979, *British cereals: wheat, barley and oats* (Association of Agriculture, London).

Soper, M.H.R. 1983, *Dairy farming and milk production* (Association of Agriculture, London).

Soper, M.H.R. 1986, *British agriculture today* (Association of Agriculture, London).

Soper, M.H.R. and Carter, E.S. 1985, *Modern farming and the countryside: the issues in perspective* (Association of Agriculture, London).

Spaven, F.D.N. 1979, 'The work of the HIDB', in HIDB, *The Highlands and Islands: a contemporary account* (HIDB, Inverness).

Spooner, D.J. 1972, 'Industrial movement and the rural periphery: the case of Devon and Cornwall', *Regional Studies*, 6: 197–215.

Stabler, J.C. 1987, 'Non-metropolitan population growth and the evolution of rural service centres in the Canadian Prairie region', *Regional Studies*, 21: 43–53.

Stacey, M. 1960, *Tradition and change: a study of Banbury* (Oxford University Press, Oxford).

Stacey, M. 1969, 'The myth of community studies', *British Journal of Sociology*, 20: 134–7.

Stager, J.K. 1988, 'Co-operatives as instruments of social change for the Inuit of Canada', pp. 295–306 in Robinson G.M. (ed.), *A social geography of Canada: essays in honour of J. Wreford Watson* (North British Publishing, Edinburgh).

Stanley, P.A. and Farrington, J.H. 1981, 'The need for rural public transport: a constraints-based case study', *Tijdschrift Voor Economische en Sociale Geografie*, 72: 62–80.

Stanton, C.R. 1976, *Canadian forestry: the view beyond the trees.*

Sternleib, G. and Hughes, J.W. 1977, *Post-industrial America: metropolitan decline and inter-regional job shifts* (Center for Urban Policy Research, New Brunswick).

Sternleib, G. and Hughes, J.W. and Hughes, C.O. 1982, *Demographic trends and economic reality: planning and markets in the 80s* (Center for Urban Policy Research, New Brunswick).

Stewart, P. 1985, 'British forestry policy: time for a change?', *Land Use Policy*, 2: 16–29.

Stevens, J.R.G. (ed.) 1980, 'Must farmers use less purchased energy?', *Span*, 23: 101–24.

Stiglitz, J.E. 1979, 'Equilibrium in product markets with unperfect information', *American Economic Review*, 69: 339–45.

Stimson, R.J. 1982, *The Australian city: a welfare geography* (Longman Cheshire, Melbourne).

Stirling, P. 1965, *Turkish village* (Weidenfeld and Nicholson, London).

Stockford, D. 1978, 'Social services provision in rural Norfolk', pp. 59–75 in Moseley, M.J. (ed.), *Social issues in rural Norfolk* (University of East Anglia, Norwich).

Stopp Jnr., G.H. 1984, 'The destruction of American agricultural land', *Geography*, 69: 64–6.

Storey, R.J. 1982, 'Community co-operatives: a Highlands and Islands experiment', pp. 71–86 in Sewel, J. and O'Cearbhaill, D. (eds), *Co-operation and community development* (Social Sciences Research Centre, University College, Galway).

Strachan, A.J. 1988, 'Business development in the rural south Midlands', *East Midlands Geographer*, 11: 13–21.

Strachan, A.J. and King, R.L. 1982, 'Emigration and return migration in southern Italy: a multivariate cluster and map analysis', *Occasional Papers, Department of Geography, University of Leicester*, No. 9.

Stupich, D.D. 1975, 'British Columbia passes vital legislation', pp. 328–30 in Bryfogle, R.C. and Kreuger, R.R. (eds), *Urban problems revisited* (Holt, Rinehart and Winston, Toronto).

Sturt, G. 1912, *Change in our village* (G. Duckworth, London).

Sulzberger, J. 1980, *The Richmond years* (Pictorial Publications Ltd., Hastings, NZ).

Summers, G.F. *et al.* 1974, *Industrial invasion of non-metropolitan America: a quarter century of experience* (Center of Applied Sociology, Department of Rural Sociology, University of Wisconsin, Madison).

Summers, G.F. (ed.) 1983, *Technology and social change in rural areas* (Westview Press, Boulder, Col.).

Sylvester, D. 1969, *The rural landscape of the Welsh Borderland* (Macmillan, London).

Symes, D.G. 1982, 'Part-time farming in Norway', *Geojournal*, 6: 351–4.

Symes, D.G. and Marsden, T.K. 1983, 'Complementary roles and asymmetrical lives. Farmers' wives in a large farm environment', *Sociologia Ruralis*, 23: 229–41.

Taafe, E.J., Gauthier, H.L. and Maraffa, T.A. 1980, 'Extended commuting and the intermetropolitan periphery', *Annals of the Association of American Geographers*, 70: 313–29.

Taffin, C. 1985, 'Accession à la propriété et "rurbanisation"', *Econ. Stat.*, 175: 55–67.

Tarditi, S. 1987, 'The Common Agricultural Policy: the implications for Italian agriculture', *Journal of Agricultural Economics*, 38: 407–22.

Tarditi, S. *et al.* (eds) 1989, *Agricultural trade, liberalization and the European Community* (Clarendon Press, Oxford).

Tarrant, J.R. 1980a, 'Agricultural trade within the European Community', *Area*, 12: 37–42.

Tarrant, J.R. 1980b, 'Production and disposal of surplus EEC milk products', *Area*, 12: 247–52.

Taylor, C. and Emerson, D. 1981, *Rural post offices: retaining a vital service* (National Council for Voluntary Organisations, London).

Taylor, J.A. 1978, 'The British Upland Experiment and its management', *Geography*, 63: 338–53.

Taylor, R. 1979, 'Migration and the residual population', *Sociological Review*, 27: 475–89.

Tepicht, J. 1973, *Marxisme et agriculture: le paysan polonais* (A. Colin, Paris).

Thieme, G. 1983, 'Agricultural change and its impact in rural areas', pp. 220–47 in Wild, M.T. (ed.), *Urban and rural change in West Germany* (Croom Helm, London and Canberra).

Thieme, G. and Paul, G. 1980, 'Die landwirtschaft in der Bundesrepublik Deutschland', *Geographische Zeitfragen*, 6.

Thissen. F. 1978, 'Second homes in the Netherlands', *Tijdschrift Voor Economische en Sociale Geografie*, 69: 322–32.

Thomas, D. 1970, *London's Green Belt* (Faber and Faber, London).

Thomas, I.C. 1986, 'Linkages, technology and rural development: the case of Mid-Wales', *Cambria*, 13.

Thomas, I.C., and Drudy, P.J. 1987, 'The impact of factory development in "growth towns" employment in Mid-Wales', *Urban Studies*, 24: 361–78.

Thomas, R. 1966, 'Industry in rural Wales', *Welsh Economic Studies*, 3 (University of Wales Press, Cardiff).

Thompson, F. 1959, *Lark Rise to Candleford* (Oxford University Press, London).

Thorns, D.C. 1976, *The quest for community: social aspects of residential growth* (George Allen and Unwin, London).

Thorns, D.C. 1980, 'Constraints versus choices in the analysis of housing allocation and residential mobility', pp. 50–68 in Ungerson, C. and Karn, V. (eds), *The consumer experience of housing* (Gower, London).

Thraves, B.D. 1988, 'Urban Canada 2001', pp. 197–207 in Robinson, G.M. (ed.), *A social geography of Canada: Essays in honour of J. Wreford Watson* (North British Publishing, Edinburgh).

Thrift, N.J. 1987a, 'Manufacturing rural geography?', *Journal of Rural Studies*, 3: 77–81.

Thrift, N.J. 1987b, book review of Amin, A. and Goddard, J.B., *Technological change, industrial restructuring and regional development*, *Progress in Human Geography*, 11: 307–9.

Till, T.E. 1981, 'Manufacturing industry: trends and impacts', pp. 194–230 in Hawley, A.H. and Mazie, S.M. (eds), *Nonmetropolitan America in transition* (University of North Carolina Press, Chapel Hill).

Tobin, G.A. 1976, 'Suburbanization and the development of motor transportation: transport technology and the suburbanization process', pp. 95–112 in Schwartz, B. (ed.), *The changing face of the suburbs* (University of Chicago Press, Chicago and London).

Todd, D. 1979a, 'On urban spill-overs and rural transportation: a Canadian example', *Regional Studies*, 13; 305–21.

Todd, D. 1979b, 'Regional and structural factors in farm-size variation: a Manitoba elucidation', *Environment and Planning A*, 11: 257–69.

Todd, D. 1980, 'Rural out-migration and economic standing in a prairie setting', *Transactions of the Institute of British Geographers*, new series, 5: 446–65.

Todd, D. 1981, 'Rural outmigration in southern Manitoba: a simple path analysis of "push" factors', *Canadian Geographer*, 25: 252–66.

Todd, D. 1983, 'The small-town viability problem in a prairie context', *Environment and Planning A*, 15: 903–16.

Todd, D. and Brierley, J.S. 1977, 'Farm size and regional structure: a preliminary insight', *Environment and Planning A*, 9: 75–84.

Tomlinson, D.G. and Tannock, P.D. 1982, *Review of assistance for isolated children scheme* (Commonwealth Department of Education, Canberra).

Toujas-Pinede, C. 1974, 'Les repatries d'Afrique du Nord dans l'agriculture du Tarn-et-Garonne', *Revue Géographique Pyrénées Sud-Ouest*, 45: 381–418.

Tourism and Recreation Research Unit (TRRU), 1981, 'The economy of rural communities in the national parks of England and Wales', *TRRU Research Reports*, No. 47.

Townsend, P. 1979, *Poverty in the UK* (Penguin, Harmondsworth).

Townsend, A.R. 1986, 'The location of employment growth after 1978: the surprising significance of dispersed centres', *Environment and Planning A*, 18: 529–45.

Towse, R.J. 1988, 'Industrial location and site provision in an area of planning restraint: part of south west London's metropolitan green belt', *Area*, 20: 323–32.

Toyne, P. 1974, *Organisation, location and behaviour: decision-making in economic geography* (Macmillan, London).

Trewartha, G.T. 1969, *A geography of population: world patterns* (John Wiley and Sons, New York).

Tricker, M.J. 1984, 'Rural education services: the social effects of reorganisation', pp. 111-20 in Clark, G., Groenendijk, J. and Thissen, F. (eds), *The changing countryside* (Geo Books, Norwich).

Tricker, M.J. and Martin, S. 1984, 'The developing role of the Commission', *Regional Studies*, 18: 507–14.

Tricker, M.J. and Martin, S. 1985, 'Rural development programmes: the way forward', pp. 293-306 in Healey, M.J. and Ilbery, B.W. (eds), *The industrialization of the countryside* (Geo Books, Norwich).

Tricker, M.J. and Mills, L. 1987, 'Education services', pp. 37–55 in Cloke, P.J. (ed.), *Rural planning: policy into action?* (Harper and Row, London).

Tryon, R.C. and Bailey, D.E. 1970, *Cluster Analysis* (New York).

Tsoukalis, L. 1981, *The European Community and its Mediterranean enlargement* (Allen and Unwin, London).

Turner, J.R. 1975, 'Applications of landscape evolution: a planner's view', *Transactions of the Institute of British Geographers*, 66: 156–61.

Turner, R.K. 1987, 'Wetlands conservation: economics and ethics', pp. 121–59 in Collard, D., Pearce, D. and Ulph, D. (eds), *Economics, growth and sustainable environments* (Macmillan, London).

Turner, R.K. 1988a, 'Pluralism in environmental economics: a survey of the sustainable economic development debate', *Journal of Agricultural Economics*, 39: 352–60.

Turner, R.K. (ed.) 1988b, *Sustainable environmental management: principles and practice* (Belhaven Press, London).

Turner, R.K. and Brooke, J. 1988, 'Management and valuation of an Environmentally Sensitive Area: A Norfolk Broadland case study', *Environmental Management*, 12: 193–202.

Turner, R.K., Dent, D. and Hey, R.D. 1983, 'Valuation of the environmental impact of wetland flood protection and drainage schemes', *Environment and Planning A*, 15: 871–88.

Tweeten, L. and Brinkman, G.L. 1976, *Micropolitan development theory and practice of greater-rural economic development* (Iowa State University Press, Ames).

Ullman, E.L. 1954, 'Amenities as a factor in regional growth', *Geographical Review*, 44: 119–32.

United Nations, 1955, *Demographic Yearbook 1952* (United Nations, New York).

United States Department of Agriculture (USDA) Forest Service, 1982, *An analysis of the timber situation in the US, 1952–2030* (Forest Resource Report No. 23, USDA, Washington, DC).

Unwin, T. 1985, 'Farmers' perceptions of agrarian change in North-West Portugal', *Journal of Rural Studies*, 1: 339–57.

Unwin, T. 1987, 'Household characteristics and agrarian innovation adoption in north-west Portugal', *Transactions of the Institute of British Geographers*, new series, 12: 131–46.

Unwin, T. 1988, 'The propagation of agrarian change in north-west Portugal', *Journal of Rural Studies*, 4: 223–38.

Urry, J. 1981, 'Localities, regions and social class', *International Journal of Urban and Regional Research*, 5: 455–74.

Urry, J. 1984, 'Capitalist restructuring, recomposition and the regions', pp. 45–64 in Bradley, A. and Lowe, P. (eds), *Locality and rurality: economy and society in rural regions* (Geo Books, Norwich).

Urry, J. 1985, 'Social relations, space and time', in Gregory, D. and Urry, J. (eds), *Social relations and spatial structures* (Macmillan, London).

Urry, J. 1986, 'Locality research: the case of Lancaster', *Regional Studies*, 20: 233–42.

Vallaux, C. 1908, *Géographie sociale: La mer* (F. Alcan, Paris).

Van Der Knapp, G.A. and White, P.E. (eds) 1985, *Contemporary studies of migration* (International Symposia Series, Geo Books, Norwich).

Van Hulten, M.H.M. 1969, 'Plan and reality in the Ijsselmeer polders', *Tijdschrift Voor Economische en Sociale Geografie*, 60: 67–77.

Van Lier, H.N. 1988, 'Land-use planning on its way to environmental planning', pp. 89–107 in Whitby, M. and Ollerenshaw, J. (eds), *Land use and the European environment* (Belhaven Press, London and New York).

Van Lier, H.N. and Steiner, F.R. 1982, 'A review of the Zuider Zee reclamation works: an example of Dutch physical planning', *Landscape Planning*, 9: 35–59.

Vartiainen, P. 1989a, 'Counterurbanisation: a challenge for socio-theoretical geography', *Journal of Rural Studies*, 5: 217–26.

Vartiainen, P. 1989b, 'The end of drastic depopulation in rural Finland: evidence of counterurbanisation?', *Journal of Rural Studies*, 5: 123–36.

Vaudois, J. 1980, 'L'aménagement rural dans la region du Nord-Pas-de-Calais: les plans d'aménagement rural', *Hommes et Terres du Nord*, 34–45.

Veldman, J. 1984, 'Proposal for a theoretical basis for the human geography of rural areas', pp. 17–26 in Clark, G., Groenendijk, J. and Thissen, F. (eds), *The changing countryside* (Geo Books, Norwich).

Vellekoop, C. 1968, 'Migration plans and vocational choices of youth in a small New Zealand town', *New Zealand Journal of Educational Studies*, 3: 20–39.

Vidal de la Blache, P. 1918, *Principles de geographie humaine* (Paris).

Vincent, J.A. 1980, 'The political economy of Alpine development: tourism or agriculture in St Maurice', *Sociologia Ruralis*, 20: 250–71.

Vining, D.R. 1982, 'Migration between the core and the periphery', *Scientific American*, 247: 36–45.

Vining, D.R. and Kontuly, T. 1977, 'Increasing returns to city size in the face of an impending decline in the size of large cities: which is the bogus fact?', *Environment and Planning A*, 9: 59–6.

Vining, D.R., Pallone, P. and Yang, C.H. 1982, 'Population dispersal from core regions: a description and tentative explanation of the patterns in twenty countries', pp. 171–91 in Kawahima, T. and Korcelli, P. (eds), *Human settlement systems: spatial patterns and trends* (International Institute for Applied Systems Analysis, Luxemburg).

Vining, D.R., Plaut, T. and Brien, K. 1977, 'Urban encroachment on prime agricultural land in the United States', *International Regional Science Review*, 2: 143–56.

Vining, D.R. and Strauss, A. 1977, 'A demonstration that the current deconcentration of population in the United States is a clean break with the past', *Environment and Planning A*, 9: 751–8.

Visser, S. 1980, 'Technology change and the spatial structure of agriculture', *Economic Geography*, 56: 311–19.

Vogeler, I. 1981, *The myth of the family farm: agribusiness dominance of US agriculture* (Westview Press, Boulder, Col.).

Volkman, N.J. 1987, 'Vanishing lands in the USA: the use of Agricultural Districts as a method to preserve farm land', *Land Use Policy*, 4: 14–30.

Wade, R. 1980, 'Fast growth and slow development in southern Italy', pp. 197–221 in Seers, D., Schaffer, B. and Kiljunen, M.L. (eds), *Undeveloped Europe: studies in core–periphery relations* (Harvester Press, Hassocks).

Walker, C.A. and McCleery, A. 1987, 'Economic and social change in the Highlands and Islands', *Scottish Economic Bulletin*, 35; 8–20.

Walker, G. 1974, 'Social interactions in the Holland Marsh', *Ontario Geographer*, 8: 52–63.

Walker, G. 1975, 'Social networks in rural space: a comparison of two southern Ontario localities', *East Lakes Geographer*, 10: 68–77.

Walker, G. 1976, 'Social perspectives on the countryside: reflections on territorial form north of Toronto', *Ontario Geographer*, 10: 54–63.

Walker, G. 1977, 'Social networks and territory in a commuter village, Bond Head, Ontario', *Canadian Geographer*, 21: 329–50.

Walker, G. 1979, 'Farmers in southern Simcoe county, Ontario: part-time fulltime comparisons', *Ontario Geographer*, 14: 59–67.

Walker, G. and Beesley, K.B. 1984, 'Urbanites in the rural-urban fringe: the case of northwest Toronto', *Geographical Perspectives*, 53: 44–52.

Walker, R. 1981, 'A theory of suburbanism: capitalism and the construction of urban space in the United States', pp. 383–430 in Dear, M. and Scot, A. (eds), *Urbanization and urban planning in capitalist society* (Methuen, New York).

Wallace, D.B. 1981, 'Rural policy: a review article', *Town Planning Review*, 52: 215–22.

Wallace, I. 1985, 'Towards a geography of agribusiness', *Progress in Human Geography*, 9: 491–514.

Wallach, B. 1981, 'The slighted mountains of upper East Tennessee', *Annals of the Association of American Geographers*, 71: 479–98.

Ward, A.H. 1975, *A command of co-operatives: the development of leadership, marketing and price control in the co-operative dairy industry of New Zealand* (New Zealand Dairy Board, Wellington).

Wardwell, J. 1977, 'Equilibrium and change in non-metropolitan growth', *Rural Sociology*, 42: 156–79. .

Wardwell, J. and Brown, D. 1980, 'Population redistribution in the United States during the 1970s', in Brown, D. and Wardwell, J. (eds), *New directions in urban-rural migration* (Academic Press, New York).

Warner, S.B. 1963, *Streetcar suburbs: the process of growth in Boston, 1870–1900* (Harvard University Press, Cambridge, Mass.).

Warner, W.K. 1974, 'Rural society in a post-industrial age', *Rural Sociology*, 39: 306–18.

Warnes, A.M. 1983, 'Migration in late working age and early retirement', *Socio Economic Planning Sciences*, 17: 291–302.

Warnes, A.M. and Law, C.M. 1975, 'Life begins at sixty: the increase in retirement migrations', *Town and Country Planning*, 43: 531–4.

Warnes, A.M. and Law, C.M. 1976, 'The changing geography of the elderly in England and Wales', *Transactions of the Institute of British Geographers*, new series, 1: 453–71.

Warnes, A.M. and Law, C.M. 1984, 'The elderly population of Great Britain: locational trends and policy implications', *Transactions of the Institute of British Geographers*, new series, 9: 37–59.

Warnes, A.M. and Law, C.M. 1985, 'Elderly population distribution and housing prospects in Britain', *Town Planning Review*, 56: 29–314.

Warren, C.L., Kerr, A. and Turner, A.M. 1989, 'Urbanization of rural land in Canada, 1981–86', *State of the Environment Fact Sheet*, Environment Canada, No. 89-1.

Wathern, P., Young, S.N., Brown, I.W. and Roberts, D.A. 1986, 'The EEC Less Favoured Areas Directive: implementation and impact on upland land use in the UK', *Land Use Policy*, 3: 205–12.

Wathern, P., Young, S.N., Brown, I.W. and Roberts, D.A. 1988, 'Recent upland land use change and agricultural policy in Clwyd, North Wales', *Applied Geography*, 8: 147–63.

Watkins, C. 1984, 'The planting of woodland in Nottinghamshire since 1945', *East Midland Geographer*, 8: 147–58.

Watkins, C. 1986, 'Recent changes in government policy towards broadleaved woodland', *Area*, 18: 117–22.

Watson, W. 1958, *Tribal cohesion in a money economy* (University of Manchester Press, Manchester).

Weber, M. 1922, *General economic history* (Adelphi, London).

Weekley, I. 1988, 'Rural depopulation and counterurbanisation: a review', *Area*, 20: 127–34.

Weinrod, A. 1979, 'Industrial involution in Sardinia', *Sociologia Ruralis*, 19: 246–66.

Weinschenk, G. and Kemper, J. 1981, 'Agricultural policies and their regional impact in Western Europe', *European Review of Agricultural Economics*, 8: 251–81.

Wenger, C. 1980, *Mid-Wales: deprivation or development* (University of Wales Press, Cardiff).

West, P.C., Blahna, D.J. and Fly, J.M. 1987, 'The unemployment impacts of the population turnaround in northern lower Michigan', *Rural Sociology*, 5: 522–31.

Westmacott, R. and Worthington, T. 1984, *Agricultural landscapes: a second look* (Countryside Commission, Cheltenham).

Whatmore, S.J., Munton, R.J.C., Marsden, J.K. and Little, J.K. 1987a, 'Towards a typology of farm businesses in contemporary British agriculture', *Sociologia Ruralis,* 27: 21–37.

Whatmore, S.J., Munton, R.J.C., Marsden, J.K. and Little, J.K. 1987b, 'Interpreting a relational typology of farm businesses in southern England', *Sociologia Ruralis,* 27: 103–22.

Whetham, E.H. 1953, *British farming, 1938–49* (Thomas Nelson, London and New York).

Whitby, M.C., Robins, D.L.J., Tansey, A.W. and Willis, K.G. 1974, *Rural resource development* (Methuen, London).

Whitby, M.C. and Willis, K.G. 1978, *Rural resource development: an economic approach* (Methuen, London).

White, P.E. 1980, 'Migration loss and the residual community: a study in rural France, 1962–1975', pp. 198–222 in White, P.E. and Woods, R.I. (eds), *The geographical impact of migration* (Longman, London).

White, P.E. 1981, 'Rural geography', pp. 296–7 in Johnston, R.J. (ed.), *The dictionary of human geography* (Blackwell, Oxford).

White, P.E. 1985, 'Modelling rural population change in the Cilento region of southern Italy', *Environment and Planning A,* 17: 1401–13.

White, P.E. 1986, 'Rural planning', pp. 414–15 in Johnston, R.J. *et al.,* (eds), *The dictionary of human geography* (Blackwell, Oxford), 2nd edition.

White, P.E. and Woods, R.I. 1980a, 'Spatial patterns of migration flows', pp. 21–41 in White, P.E. and Woods, R.I. (eds), *The geographical impact of migration* (Longman, London).

White, P.E. and Woods, R.I. 1980b, 'The geographical impact of migration', in White, P.E. and Woods, R.I. (eds), *The geographical impact of migration* (Longman, London).

White, P.R. 1978, 'Midland Red's market analysis project', *Omnibus Magazine,* May/June, 61–8.

Whitelegg, J. 1987, 'Rural railways and disinvestment in rural areas', *Regional Studies,* 21: 55–63.

Whitlock, R. 1988, *Lost village: rural life between the wars* (Hale, London).

Whyte, D. 1978, 'Have second homes gone into hibernation?', *New Society,* 45: 286–8.

Wibberley, G.P. 1959, *Agriculture and urban growth: a study of the competition for rural land* (M. Joseph, London).

Wibberley, G.P. 1972, 'Conflicts in the countryside', *Town and Country Planning,* 40: 259–64.

Wibberley, G.P. 1976, 'Rural resource development in Britain and environmental concern' *Journal of Agricultural Economics,* 27: 1–16.

Wibberley, G.P. 1982, 'Countryside planning – a personal evaluation', *Occasional Papers, Department of Environmental Studies and Countryside Planning, Wye College, University of London,* No. 7.

Wild, M.T. 1983a, 'The residential dimension to rural change', pp. 161–99 in Wild, M.T. (ed.), *Urban and rural change in West Germany* (Croom Helm,

London and Canberra).

Wild, M.T. 1983b, 'Social fallow and its impact on the rural landscape', pp. 200–19 in Wild, M.T. (ed.), *Urban and rural change in West Germany* (Croom Helm, London and Canberra).

Wild, M.T. and Jones, P.T. 1988, 'Rural suburbanisation and village expansion in the Rhine Rift Valley: a cross-frontier comparison', *Geografiska Annaler*, 70B: 275–92.

Wilhelm, H.G.H. 1977, 'The growth of the vacation farm as a rural settlement element in the Federal Republic of Germany', *East Lakes Geographer*, 12: 1–10.

Wilkinson, K.P. 1978, 'Rural community change', pp. 115–25 in Ford, T.R. (ed.), *Rural USA: persistence and change* (Iowa State University Press, Ames).

Williams, G. 1986, 'Development agencies and the promotion of rural community development', *Countryside Planning Yearbook*, 5: 62–86.

Williams, J. and Sofranko, A. 1979, 'Motivations for the immigration component of population turnaround in nonmetropolitan areas', *Demography*, 16: 239–56.

Williams, N. and Sewel, J. 1987, 'Council house sales in rural areas', pp. 66–85 in MacGregor, B.D., Robertson, D.S. and Shucksmith, M. (eds), *Rural housing in Scotland: recent research and policy* (Aberdeen University Press, Aberdeen).

Williams, R. 1973, *The country and the city* (Chatto and Windus, London).

Williams, S.W. 1977, 'Internal colonialism, core–periphery contrasts and devolution: an integrative comment', *Area*, 9: 272–8.

Williams, W.M. 1956, *The sociology of an English village: Gosforth* (Routledge and Kegan Paul, London).

Williams, W.M. 1963, *A West Country village: Ashworthy* (Routledge and Kegan Paul, London).

Willis, R.P. 1984, 'Farming in New Zealand and the EEC – the case of the dairy industry', *New Zealand Geographer*, 40: 3–11.

Willis, R. 1988, 'New Zealand', pp. 218–39 in Cloke, P.J. (ed.), *Policies and plans for rural people* (Unwin Hyman, London).

Willits, F.K., Bealer, R.C. and Crider, D.M. 1982, 'Persistence of rural/urban differences', pp. 69–76 in Dillman, D.A. and Hobbs, D.J. (eds), *Rural society in the United States: issues for the 1980s* (Westview Press, Boulder, Col.).

Wilson, J. 1981, 'Some negative aspects of life in an Orkney parish', pp. 24–38 in Moore, R. (ed.), 'Labour migration and oil', *SSRC North Sea Oil Panel, Occasional Papers*, 7.

Wilson, J.W. and Pierce, J.T. 1984, 'The Agricultural Land Commission of British Columbia', pp. 272–91 in Bunce, M.F. and Troughton, M.J. (eds), 'The pressures of change in rural Canada', *Geographical Monographs, Department of Geography, Atkinson College, York University*, No. 14.

Winchester, H.P.M. and Ilbery, B.W. 1988, *Agricultural change: France and the EEC* (John Murray, London).

Winegarten, A. 1978, 'British agriculture and the 1947 Agriculture Act', *Journal of the Royal Agricultural Society*, 139: 74–82.

Wingo, L. 1961, *Transportation and urban land* (Washington, DC).

Winsberg, M.D. 1980, 'Concentration and specialization in United States agriculture, 1939–1978', *Economic Geography,* 56: 183–9.

Wirth, L. 1938, 'Urbanism as a way of life', *American Journal of Sociology,* 44: 1–24.

Wolfenden, Lord 1978, *The future of voluntary organisations* (Croom Helm, London).

Woodruffe, B.J. 1976, *Rural settlement policies and plans* (Oxford University Press, Oxford).

Woods, R.I. 1982, *Theoretical population geography* (Longman, London).

Woods, R.I. 1985, 'Towards a general theory of migration', pp. 1–5 in Van der Knapp, G.A. and White, P.E. (eds), *Contemporary studies of migration* (International Symposia Series, Geo Books, Norwich).

Woollett, S. 1981a, *Alternative rural services* (National Council for Voluntary Organisations, Bedford Square Press, London).

Woollett, S. 1981b, 'Self help: providing your own community services', *The Planner,* 67: 72–3.

Wrathall, J.E. 1978, 'The oilseed rape revolution in England and Wales', *Geography,* 63: 42–5.

Wrathall, J.E. 1986a, 'The CAP and the changing pattern of arable crops in the UK', pp. 28–41 in Bowler, I.R. (ed.), 'Agriculture and the Common Agricultural Policy', *Occasional Papers, Department of Geography, University of Leicester, No. 15.*

Wrathall, J.E. 1986b, 'The Pennine Rural Development Area – a census based atlas of housing and associated socio-economic variables', *Local Information Papers, Department of Geographical Sciences, Huddersfield Polytechnic, No. 8.*

Wrathall, J.E. 1988a, 'Recent changes in arable crop production in England and Wales', *Land Use Policy,* 5: 219–31.

Wrathall, J.E. 1988b, 'Rural Development Areas – the role of the Development Commission in stimulating deprived rural areas', pp. 2–11 in Hart, P. and Atherden, M. (eds), *The role of institutions in countryside management* (Rural Geography Study Group, Institute of British Geographers, College of Ripon and York St John, Harrogate).

Wrathall, J.E. and Moore, R. 1986, 'Oilseed rape in Great Britain – the end of a "revolution"?', *Geography,* 71: 351–5.

Wright, E.O. 1985, *Classes* (Verso, London).

Yeates, M.H. 1985, 'Land in Canada's urban heartland', *Land Use in Canada Series, Lands Directorate, Environment Canada, No. 27.*

Zaring, J. 1977, 'The romantic face of Wales', *Annals of the Association of American Geographers,* 67: 397–418.

Zaslowsky, D. and Wilderness Society, 1986, *These American lands: parks, wilderness and the public lands* (Henry Holt, New York).

Zelinsky, W. 1971, 'The hypothesis of the mobility transition', *Geographical Review,* 61: 219–49.

Ziche, J. 1968, 'Kritik der deutschen bauerntum sideologie', *Sociologia Ruralis,* 8: 105–41.

Zube, E.H. 1974, 'Cross-disciplinary and intermode agreement on the description and evaluation of landscape resources', *Environment and Behavior*, 6: 69–89.

Zuiches, J.J. 1981, 'Residential preferences in the United States', pp. 72–115 in Hawley, A.H. and Mazie, S.M. (eds), *Nonmetropolitan America in transition* (University of North Carolina Press, Chapel Hill).

Index

Subjects

Abercrombie Report on planning Greater London (1944), 279
Accessibility, 86, 102, 344, 349, 351-5
Action for Communities in Rural England (ACRE), 371-2
Active managerial farmers, 166-7
Advance factories, 256-8
Afforestation, 229-33, 236-9, 303
Agrarian Reform Zone (ZIRA), 203
Agrarianism, 13, 165
Agribusiness, 120, 131, 138-49, 166
Agribusinessman, 166-7
Agricultural Districts (USA), 163, 165
Agricultural diversification, 208
Agricultural dualism, 138-9
Agricultural economics, 22-3
Agricultural exports, 5, 178, 181, 210
Agricultural extensification, 334
Agricultural fundamentalism, 165
Agricultural industrialisation, 144
Agricultural Land Reserves (British Columbia), 159
Agricultural mechanisation, 84-6, 130-1, 135-6, 189, 202
Agricultural policy, 145, 173-211
Agricultural quotas, 141, 161
Agricultural restructuring, 137-42
Agricultural specialisation, 134-7
Agricultural subsidies, 130
Agricultural surpluses, 181-4
Agriculture, 78, 87, 129-211, 235, 329-35, 400
Agriculture Act (1947) (UK), 175 et seq, 278-9
Agriculture Act (1977) (Sweden), 160
Agro-towns, 192-3
Alaska Land Bill, 226
Alternative Land Use and Rural Economy (ALURE), 161-2

Amenagement rural, 388
Annual Review of agriculture (UK), 177
Appalachian Regional Development Act (1965), 241
Approved Woodland Schemes, 228
Areas of Outstanding Natural Beauty (AONBs), 318-19, 321-4, 330, 337, 401

Backwash forces, 376-8
Badajoz Plan, 200-1
Barley, 131-2, 176
Barlow Report on the Distribution of the Industrial Population (1940), 279
Beeching Report (1963), 356-7
Binary economy, 78
Birth rates, 102
Boreal forest, 212
'Bottom-up' strategies, 403
Branch plants, 102, 242, 244, 252
British Columbia Natural Resources Corporation, 239-40
Broadleaved Woodland Scheme, 233
Broadleaves, 234, 238
Broads Authority, 334
Broads Grazing Marshes Conservation Scheme, 333
Bus Regulatory Reform Act (1982) (South Dakota), 359
Buses, 353, 358-61

Canada Land Inventory (CLI), 152
Canada Land Use Monitoring Program (CLUMP), 152, 158
Capital Transfer Tax (CTT) (UK), 168
Capitalism, 54-5, 58, 68-70, 73-4, 78, 138-9, 141, 146-7, 151, 170-1, 188-9, 294, 402
Car sharing, 362
Cars, 351-2, 354-5, 358, 361-2
Cassa per il Mezzogiorno, 193-7, 273

Countries